Research Methodology

Research Methodology

Techniques and Trends

Umesh Kumar Bhayyalal Dubey and
D. P. Kothari

CRC Press
Taylor & Francis Group
Boca Raton London New York

CRC Press is an imprint of the
Taylor & Francis Group, an **informa** business
A CHAPMAN & HALL BOOK

First edition published 2022
by CRC Press
6000 Broken Sound Parkway NW, Suite 300, Boca Raton, FL 33487-2742

and by CRC Press
2 Park Square, Milton Park, Abingdon, Oxon, OX14 4RN

CRC Press is an imprint of Taylor & Francis Group, LLC

Library of Congress Cataloging-in-Publication Data
Names: Dubey, Umesh Kumar B., author. | Kothari, D. P. (Dwarkadas
Pralhaddas), 1944- author.
Title: Research methodology : techniques and trends / Umesh Kumar B. Dubey, D.P. Kothari.
Description: First edition. | Boca Raton : Chapman & Hall/CRC Press, 2022.
| Includes bibliographical references and index.
Identifiers: LCCN 2021053390 (print) | LCCN 2021053391 (ebook) |
ISBN 9781138053984 (hardback) | ISBN 9781032233406 (paperback) |
ISBN 9781315167138 (ebook)
Subjects: LCSH: Research—Methodology.
Classification: LCC Q180.55.M4 D83 2022 (print) | LCC Q180.55.M4 (ebook)|
DDC 001.4/2—dc23/eng/20220128
LC record available at https://lccn.loc.gov/2021053390
LC ebook record available at https://lccn.loc.gov/2021053391

ISBN: 978-1-138-05398-4 (hbk)
ISBN: 978-1-032-23340-6 (pbk)
ISBN: 978-1-315-16713-8 (ebk)

DOI: 10.1201/9781315167138

Typeset in Times
by codeMantra

Contents

Statistical Tables

Preface

People talk about research in academic institutions, but little attention is given to research methodology. As a result, a great deal of research tends to be futile.

Entity of the nation is to maximize productivity, which can be improved through several research studies. Every research is a set of activities to study and develop a model or procedures to analyze the results of a realistic problem supported by literature review and analysis of data. So, the objectives are optimized and further make recommendations for its implementation.

The motivation for writing this book has come from the lack of contextually relevant and a comprehensive book on Research Methodology, which are sufficient to provide rather substantial course of study for the students. The book is intended for the students of Management, M.Phil., Engineering, Science & Technology, Medical Science, Pharmacy, Nursing, Commerce, Arts, Social-Science, Operation Research, and Economics.

Most chapters of this book have been developed from the lecture notes used in teaching of UG-/ PG-/Doctorate-level programs. We have used our research and teaching experience and our industrial contacts to create this textbook full of real industry-based examples and research insights.

We have kept the presentation simple and stimulating; nevertheless, the treatment of topics is detailed and up-to-date enough so that experts (researchers, industry professionals, etc.) in the field can use the book as a reference.

The text contains in-depth coverage of concepts and techniques of research methodology. Chapters of the book include examples, figures, and bibliography/references. The text includes the comprehensive coverage of research problems, types of research, and research procedure followed by guidelines of data collection and presentation, which made the platform for further analysis. Well-defined algorithms for the various statistical methods are provided, and the operations of different statistical techniques with relevant examples are demonstrated. In the entire text, a large number of tables and figures are presented and easy-to-understand style has been followed to illustrate the concepts and techniques. The chapter on report writing has been dealt with in a comprehensive manner, which will aid researchers to document their research studies and analyses along with findings.

To enhance the understanding of the subject matter by students belonging to different disciplines, our approach is conceptual. This book provides an understanding of problem-solving methods based on formulation, procedure, and analysis. It provides students the art of using research procedure, methods, and techniques to enable them in developing appropriate methodology for their research. Each chapter contains review questions, self-practice numerical with hints, and answers to help students in self-evaluation. To facilitate ready recall of concepts discussed in each chapter, a summary is provided.

Acknowledgments

In preparing the text of this book, we have benefitted immensely by referring to many books and publications. We express our gratitude to such authors, publications, and publishers, many of whom have been listed in the references. If anybody is left out inadvertently, we seek their pardon.

The authors are highly thankful to Dr. Vinayak Deshpande, Honorable Ex-Pro Vice Chancellor of the Rashtrasant Tukadoji Maharaj Nagpur University, Nagpur, for his constant encouragement and help while developing this text.

My heartfelt thanks to Pro-vice Chancellor Dr. Milind Pande, Associate Dean Dr. Shubhalaxmi Joshi and management of MIT WPU Kothrud Pune, Maharashtra India as the book won't have seen the light of day but for their generous help & cooperation extended to me throughout the preparation of this book.

We express our deep sense of gratitude toward Dr. Prakash L. Neulkar, Associate Professor, Shardchandra Arts & Commerce College, Butibori, Nagpur, who has supported us and has been a source of encouragement.

Finally, we would like to thank CRC Press, Taylor & Francis Group, especially Ms. Astha Sharma and Ms. Isha Singh for facilitating the work.

We are indeed grateful to Mrs. Pinky Umesh Dubey for her timely help in drawing all the graphs and doing formatting of the text, without which this book won't have triggered the process.

We hope that this book will serve readers well and that we will continue to receive their support and suggestions. Suggestions and comments to improve the book in content and in style are always welcome and will be appreciated, acknowledged, and incorporated in the future editions of the book.

Dr. Umesh Kumar B. Dubey
Assistant Professor. Faculty of Science,
School of Mathematics & Statistics,
MIT WPU Pune, Maharashtra (India)

Dr. D. P. Kothari
Chairman BOG Tihri Hydro Institution,
Uttra-Khand Ex-Director in-charge IIT Delhi,
VC VIT Vellore, Principal VNIT Nagpur

Authors

Dr. Umesh Kumar Bhayyalal Dubey earned two Ph.D. degrees; the first in the Faculty of Science, in Statistics, entitled "Some Contribution to Information Theory" awarded on September 29, 2004, and the second in the Faculty of Commerce & Management, in Business Administration & Business Management, entitled "Job Analysis of Industrial workers with special reference to Nagpur District" awarded on January 18, 2020, both from Rashtrasant Tukadoji Maharaj Nagpur University, Nagpur. He had completed his post-graduation, M.Sc. (Statistics), first MBA (Operations Management from IGNOU), and second MBA (HRM & MM from RTM Nagpur University). He has authored the book *Quantitative Techniques in Business, Management, and Finance: A Case-Study Approach*, published by Chapman and Hall/CRC Press Taylor & Francis Group, UK, on December 8, 2016.

He has received a sum of Rs. 4,93,333/- from All India Council for Technical Education (AICTE), New Delhi, as Grant-in-Aid to conduct Conference under the scheme **Grant for Organizing Conference (GOC)** for the year 2018–2019. He has over 21 years of experience in teaching and research in renowned institutions such as ICFAI.

Dr. Dubey is currently an Assistant Professor at the Faculty of Science, School of Mathematics & Statistics, MIT WPU Kothrud, Pune, Maharashtra, India. His research area encompasses Testing of Hypothesis, Research Methodology, Clinical Trials, Stochastic Processes, Sampling, Statistics and Probability, and Operation Research.

Website Link: https://drumeshkumarbdubey.in/

Dr. D. P. Kothari earned his B.E. in Electrical Engineering in 1967, M.E. with specialization in Power Systems in 1969, and Ph.D. from Birla Institute of Technology & Science, Pilani (Rajasthan), India in 1975. From 1969 to 1977, he was involved in teaching and development of several courses at BITS Pilani. Earlier Dr. Kothari served as Vice Chancellor, VIT, Vellore, Director in-charge and Deputy Director (Administration) as well as Head in the Centre of Energy Studies at Indian Institute of Technology, Delhi, and as Principal, VRCE, Nagpur, India. He was visiting professor at the Royal Melbourne Institute of Technology, Melbourne, Australia, during 1982–1983 and 1989. He was also NSF Fellow at Purdue University, USA, in 1992. He also taught at Melbourne University, Australia, for one semester in 1989. Dr. Kothari, who is a recipient of the most Active Researcher Award, has published and presented 830 research papers in various national as well as international journals and conferences, guided 56 Ph.D. scholars and 68 M.Tech. students, and authored 70 books in various allied areas. He has delivered several keynote addresses, 100 plus Webinars and invited lectures at both national and international conferences. He has also delivered 42 video lectures on YouTube with a maximum of 1,00,000 hits!

Dr. Kothari is a Fellow of the National Academy of Engineering (FNAE), Fellow of Indian National Academy of Science (FNASc), Fellow of Institution of Engineers (FIE), Fellow IEEE, Hon. Fellow ISTE, and Fellow IETE. He has received 79 awards to date; his many awards include the National Khosla Award for Lifetime Achievements in Engineering (2005) from IIT, Roorkee. The University Grants Commission (UGC), Government of India, has bestowed the UGC National Swami Pranavandana Saraswati Award (2005) in the field of education for his outstanding scholarly contributions. He is also the recipient of the Lifetime Achievement Award (2009) conferred by the World Management Congress, New Delhi, for his contribution to the areas of educational planning and administration. He received Excellent Academic Award at IIT Guwahati by NPSC-2014.

In 2016, he received six Lifetime Achievement awards from various agencies. On April 20, 2016, he received the *Living Legend Award* in Chennai Conference. Recently he received the *Malviya Award for Excellence in Power System* conferred by IIT (BHU) during the NPSC at IIT Gandhinagar on

December 17, 2020. He is also distinguished Emeritus Professor and adjunct professor at several institutes such as Charutar University, Gujarat and Wainganga College of Engineering, Nagpur. Currently, Dr. Kothari is with the S.B. Jain Institute of Management, Research and Technology, Nagpur, serving as Director of Research and Professor. He is also chairman of the board of governors of THDC Institute of Hydropower Engineering and Technology, Tehri, India.

Wikipedia Link: http://en.wikipedia.org/wiki/D.P.Kothari

1

Introduction to Research

1.1 Introduction

Research is a process through which new knowledge is discovered. Research helps us to organize this new information into a coherent body, a set of related ideas that explain events that have occurred, and predict events that may happen. Conducting research requires to follow a sequence of steps. The exact sequence and steps vary somewhat with the type of research. The steps vary slightly by whether a study involves a quantitative or a qualitative approach and data.

1.2 Meaning of Research

The word research is coined by two syllables: *re* plus *search*. The dictionary defines the former as a prefix meaning again, a new, or over again and the latter as a verb, meaning to examine closely and carefully, to test and try, or to probe. The simplest meaning of research is to search for facts, answers to research questions and solution for the existing problem.

Research is defined as a systematic, controlled, empirical and critical investigation of hypothetical propositions about the presumed relationship about various phenomena. Research is a systematic investigation to find answers to a problem.

We can conclude that Research refers to the systematic method consisting of enunciating the problem, formulating a hypothesis, collecting the fact or data, analyzing the facts, and reaching certain conclusions either in the form of solutions toward the concerned problem or in certain generals for some theoretical formulation.

1.3 Criteria of Good Research

Good research possesses certain qualities that are as follows:

1. Good research is systematic: It implies that research is structured according to a set of rules to follow certain steps in specified sequence. Systematic research also invites creative thinking, and certainly avoids use of guessing and intuition for arriving at the proper findings, conclusion.
2. Good research is empirical: It implies that any conclusion drawn is based on hardcore evidence gathered from information collected from real-life experiences and observations. This provides a basis for external ability to research findings and conclusion.
3. Good research is valid and verifiable: It implies that research involves precise observation and accurate description. The researcher selects reliable and valid instruments to be used for the collection of data and uses some relevant statistical tools for accurate description of the results obtained. Whatever the researcher concludes on the basis of finding is correct and can be verified by himself /herself and others.

DOI: 10.1201/9781315167138-1

4. Good research is logical: It implies that research is guided by the rules of reasoning and logical process of induction general to specific and deduction specific to general that plays an important role in carrying out research. In fact, logical reasoning makes research feasible and more meaningful in the context of quantitative decision making.
5. Good research develops theories and principles, which are very helpful in accurate prediction with regard to the variables under study. On the basis of the sample observed and studied, the researcher makes sound generalizations with regard to the entire populations. Thus, research goes beyond immediate situations, objects, or groups being investigated by formulating a generalization or theory about these factors.
6. Purpose of research should be clearly defined and common concepts that are used should be operationally defined.
7. The research procedure should be precisely planned, focused, and appropriately described in order to enable other researchers to do research for further advancement.
8. Research design should be carefully planned to generate results to maintain objectivity.
9. The research report should be as much frank as possible to gauge effects of the findings.
10. Data analysis in the research report should be adequate to reveal its significance and the method of analysis employed be appropriate.
11. Validity and reliability of data should be examined carefully.
12. Systematic approach: It implies that a planned and organized research saves researcher's time and money. Each step of investigation should be so planned that it leads to the next step. Part of this approach are planning and organization.
13. Objectivity: It implies that true research should attempt to find an unbiased answer to the decision-making problem.
14. Reproducible: It implies that in reproducible research procedure, an equally competent researcher could duplicate, and from it deduce approximately the same results. The information with regard to samples, methods, collection, etc., should be specified.
15. Relevancy has the task of avoiding collection of irrelevant information and saves time and money; it compares the information to be collected with researcher's criteria for action; it enables to see whether the research is proceeding in the right direction.

1.4 Objectives of Research

The main objectives of research are as follows:

1. To gain familiarity with a phenomenon or to achieve new insights into it; studies with this object in view are termed as exploratory research studies.
2. To portray accurately the characteristics of a particular individual, situation or group; studies with this object in view are known as descriptive research studies.
3. To determine the frequency with which something occurs or with which it is associated with something else; studies with this object in view are known as diagnostic research studies.
4. To test a hypothesis of a causal relationship between variables; such studies are known as hypothesis-testing research studies or experimental studies.
5. Exploration: It implies that an understanding of an area of concern in very general terms. For example, we wish to know how to go about doing more effective research on violence on women.
6. Description: It implies that an understanding of what is going on. For example, we wish to know the attitudes of potential clients toward the use of washing machine.

7. Explanation: It implies that an understanding of how things happen. It involves an understanding of cause-and-effect relationships between the events. For example, we wish to know whether a group of people who have gone through a certain program have higher self-esteem than a control group.
8. Prediction: It implies that an understanding of what is likely to happen in the future. If we can explain, we may be able to predict. For Example, if one group had higher self-esteem, is it likely to happen with another group?
9. Intelligent intervention: It implies that an understanding of what or how in order to help more effectively and efficiently.
10. Awareness: It implies that an understanding of the world, often gained by a failure to describe or explain.

Thus, research is the fountain of knowledge for the sake of knowledge and an important source of providing guidelines for solving different business, personal, professional, governmental, and social problems. It is a sort of formal training that enables one to understand the new developments in one's field with ease.

1.5 Types of Research

1.5.1 Exploratory Research

Exploratory research is designed to provide a background, to familiarize and, as the word implies, just "explore" the general subject. An exploratory research is the investigation of relationships among variables without knowing the objective of the study. Typical approaches in exploratory research are the literature survey and the experience survey. The literature survey is an economical and quick way for researchers to develop an excellent understanding of an area of problem in which they have minimum experience. It also familiarizes them with past research results, data sources, and availability of data types.

The experience survey concentrates on an individual who has specific knowledge in that area. Representative samples are undesired. A covering of widely divergent views is always good. Researchers are looking for ideas and not for conclusions.

1.5.2 Conclusive Research

For drawing definite conclusions an exploratory research gives rise to several hypotheses that will have to be tested. These conclusions provide the structure for decision making when tested for validity. To test the hypotheses generated by exploratory research, conclusive research is used. Conclusive research can be classified as either descriptive or experimental.

1.5.2.1 Descriptive Research

Descriptive research is designed to describe something. For example, for a newly launched product, the characteristics of users are described as the degree to which product use varies with Income, Age, Sex, and Other characteristics.

A descriptive study must collect data for a definite purpose for a maximum profit. A specific hypothesis is the guide when descriptive studies vary in degree. Depending on the research problem, it allows both implicit and explicit hypotheses to be tested.

For example, an oil company may find its sales declining. On the basis of market feedback, the company may hypothesize that economically backward family do not consume its oil for the preparation of meal. For testing of such a hypothesis, a descriptive study can then be designed.

1.5.2.2 Experimental Research

Experimentation refers to that process of research, in which one or more variables are manipulated under the conditions, in which the data that show the effects will be collected. Experiments will create artificial situations so that the researcher can obtain specific required data and the data can be measured accurately. Experiments are artificial because the situations are usually created for testing purposes. This artificiality is the essence of the experimental method since it gives researchers more control over the factors under study. If they can control the factors present in a given situation, they can obtain more conclusive evidence of cause-and-effect relationships between them. Thus, the ability to set up a situation for the express purpose of observing and recording accurately the effect on one factor when another is deliberately changed permits researchers to accept or reject hypothesis beyond a reasonable doubt.

1.5.3 Business Research

Business Research is defined as the systematic and objective process of generating information for aid in business decisions. This research information should be Scientific—not intuitive or haphazardly gathered—Objective, and Impersonal.

Business research can be used for any aspect of the enterprise. By providing appropriate information, research should be an aid to managerial judgment although it should not be a substitute for it. Applying the research is a managerial art in itself. All types of organizations that engage in some kind of business activity can use business research.

1.5.3.1 The Scope of Business Research

Business research fulfills the operation manager's need for knowledge of the organization, the market, the economy, or other area of uncertainty. It helps the manager in predicting how individuals, markets, organizational units, or other entities will respond to his business decisions.

The emphasis of business research is to shift decision-makers from risky intuitive decisions based on systematic and objective investigations.

1.5.3.2 Types of Business Research

Several bases can be adopted for the classification such as Nature of data, Branch of knowledge, Extent of coverage, Place of investigation, Method employed, Time frame, and so on.

1.5.4 According to the Branch of Knowledge

Branches of knowledge may broadly be divided into two groups:

1. Life and Physical sciences such as Botany, Zoology, Physics, Mathematics, Statistics, and Chemistry.
2. Social Sciences such as Political Science, Public Administration, Economics, Sociology, Commerce, Management, and Education.

Research in these fields is also broadly referred to as Life and Physical Science Research and Social Science Research. Business education covers both Commerce and Management, which are part of Social Sciences. Business research is a broad term, which covers many areas. The research carried out in these areas is called Management Research, Production Research, Personnel Research, Financial Management Research, Accounting Research, Marketing Research, etc.

1.5.5 Management Research

It includes various functions of management such as planning, organizing, staffing, communicating, coordinating, motivating, and controlling. Various motivational theories are the result of research.

1.5.6 Manufacturing Research

It focuses more on materials and equipment rather than on human aspects. It covers various aspects such as new and better ways of producing goods, inventing new technologies, reducing costs, and improving product quality.

1.5.7 Personnel Management Research

It may range from all types of very simple problems to highly complex problems. It is primarily concerned with the human aspects of the business such as Personnel Policies, Job Analysis, Job Requirements, Job Evaluation, Recruitment, Selection, Placement, Training & Development, Promotion & Transfer, Morale & Attitudes, Wages & Salary Administration and Industrial Relations. Basic research in this field would be valuable as human behavior affects organizational behavior and also productivity.

1.5.8 Management Research

It includes:

1. Financial Institutions, Financing Instruments, e.g., Shares, Debentures,
2. Financial Markets, e.g., Capital Market, Money Market, Primary Market, Secondary Market,
3. Financial Services, e.g., Merchant Banking, Discounting, Factoring,
4. Financial Analysis, e.g., Investment Analysis, Ratio Analysis, Funds Flow, Cash Flow Analysis, etc.

1.5.9 Accounting Research

Accounting information is used as a basis for reports to the management, shareholders, investors, tax authorities, regulatory bodies, and other interested parties. Areas for accounting research include inventory valuation, depreciation accounting, generally accepted accounting principles, accounting standards, corporate reporting, etc.

1.5.10 Marketing Research

It deals with Product Development and Distribution Problems, Marketing Institutions, Marketing Policies and Practices, Consumer Behavior, Advertising and Sales Promotion, Sales Management, After Sales Service, etc. Marketing research includes Market Potentials, Sales Forecasting, Product Testing, Sales Analysis, Market Surveys, Test Marketing, Consumer Behavior Studies, Marketing Information System, etc.

1.5.11 Business Policy Research

It is the research with policy implications. The results of such studies are used as indices for policy formulation and implementation.

1.5.12 Business History Research

It is concerned with the past. For example, to the trade and commerce during the British regime.

1.5.13 According to the Nature of Data

A simple dichotomous classification of research is Quantitative research and Qualitative research or non-quantitative.

1.5.13.1 Quantitative Research

Quantitative research is variables based whereas qualitative research is attributes based. Quantitative research is based on measurement or quantification of the phenomenon under study. That is, it is data based and hence more objective and popular.

1.5.13.2 Qualitative Research

Qualitative research is based on the subjective assessment of attributes, motives, opinions, desires, preferences, behavior, etc. Research in such a situation is a function of researcher's insights and impressions.

1.5.14 According to the Coverage

Macro study is a study of the whole whereas micro study is a study of a part. For example, working capital management in State Road Transport Corporations in India is a macro study whereas Working Capital Management in Maharashtra State Road Transport Corporation is a micro study.

1.5.15 According to Utility or Application

Depending upon the use of research results, i.e., whether it is contributing to the theory building or problem solving, research can be Basic or Applied.

1.5.15.1 Basic Research

Basic Research is also called Pure or Theoretical or Fundamental Research. Basic research includes original investigations for the advancement of knowledge that do not have specific objectives to answer problems of sponsoring agencies.

1.5.15.2 Applied Research

Applied research is also called Action Research and constitutes research activities on problems posed by sponsoring agencies for the purpose of contributing to the solution of these problems.

1.5.16 According to the Place where it is Carried Out

Depending upon the place where the research is carried out and according to the data generating source, research can be classified into Field Studies or Field Experiments, Laboratory Studies or Laboratory Experiments, and Library Studies or Documentary Research.

1.5.17 According to the Use of Research Methods

Depending upon the research method used for the investigation, it can be classified as Survey Research, Observation Research, Case Research, Experimental Research, Historical Research, and Comparative Research.

1.5.18 According to the Time Frame

Depending upon the time period for the study, it can be:

1.5.18.1 One Time or Single Time Period Research

For example, 1 year or a point of time. Most of the Sample Studies, Diagnostic Studies.

1.5.18.2 Longitudinal Research

For example, several years or several time periods, a time-series analysis, industrial development during the 5-year plans in India.

1.5.19 According to the Purpose of the Study

On the basis of purpose or aim or objective of the study. Is it to describe or analyze or evaluate or explore? Accordingly, the studies are known.

1.5.19.1 Descriptive Study

The main purpose of descriptive research is the description of a person, situation, institution, or an event as it exists, e.g., fact-finding studies.

1.5.19.2 Analytical Study

The researcher uses facts or information already available and analyzes them to make a critical examination of the observations, e.g., Ex-Post Facto Studies or Post-Mortem Studies.

1.5.19.3 Evaluation Study

This type of study is generally conducted to examine or evaluate the impact of a particular event, e.g., impact of a particular decision or a project or an investment.

1.5.19.4 Exploratory Study

The little information is known on a particular subject matter. Hence, a study is conducted to know more about it so as to formulate the problem and procedures of the study. Such a study is called exploratory or formulative study.

1.6 Importance of Research

Empirical and theoretical researches are taking place in various fields, such as learning, motivation, perception, concept learning, memory, etc. In the quest of facts, laws and theories, research studies are very helpful in gauging the behavior of human and animal.

Practical gains of research include discoveries such as improved methods of treating disordered people, required designs of vehicles in order to make them easier and safe to use, and new ways of enhancing their performance.

Effective experimental designs developed by the researchers helped to isolate the effect of other variables from independent variables. In researches, rigorous scientific norms and advanced statistical methods are applied in collection, tabulation, organization, analysis, and interpretation of the data.

1.7 Problem or Opportunity Identification

Problem or opportunity identification involves scanning and monitoring the internal and external business environment. Such an analysis helps in identifying opportunities and threats that a company is facing and also in understanding the market trends. The role of research at this stage is to provide information about the problems and the opportunities.

For example, an Indian apparel company that wants to enter the United States in the market, determining the brand awareness about the company among the customers, perception about the company among the potential employees, examining the competitors and their characteristics, and understanding American consumer behavior.

1.8 Problem or Opportunity Prioritization and Selection

In Problem or Opportunity identification step, the organizations would have identified many possible problems and opportunities. However, it is impossible for any organization to address these problems or opportunities in one attempt. So, at this stage the focus would be on prioritizing the problems and the opportunities.

Prioritization of the problems is based on two factors: the influence of problem on the operations and the time factor.

Top priority is given to the problems, which have a major influence on the operations and the problems that need to be addressed in the short term.

Another activity that organizations undertake at this stage is to collect more information about the problems and the opportunities.

For example, if an organization has identified a particular problem, then research would help it to unearth the underlying causes of the problem. If the organization has identified an opportunity then more information about the opportunity is collected. Such an analysis provides greater clarity about the situation. Research at this stage is used to aid the organization to prioritize the problems and identify the right opportunities. Generally, Qualitative and Quantitative Research Studies are undertaken at this stage.

1.9 Problem or Opportunity Resolution

After identifying the problem or opportunity, the next step is to decide on the way to resolve the problem or make use of the opportunity. Two steps are involved in the problem resolution are developing alternatives and evaluating the alternatives.

Based on the problem or opportunity identified in the previous step, several alternate courses of action are considered. These alternatives are evaluated to select the best course of action. The alternatives are evaluated on the basis of certain criteria. The application of research at this stage is mainly to help the organization in evaluating the alternatives available.

For example, a consumer electronics company that wanted to launch a new Washing Machine model faced a dilemma with regard to the advertising strategy: it should adopt, as its marketing staff had suggested, four different advertising programs. To evaluate the advertising program the company undertook a consumer-jury test where target programs were shown to them. They were asked to rate those advertisements on various parameters: likeability, memorability, attentiveness, and believability.

Based on the results of the test, the company finalized the best option among the four advertising programs. Another way in which research aids in evaluating the alternative options is through forecasting.

For example, a company has four different investment options from among which it has to choose the best one. By forecasting the revenue potential of each investment option, the company can select the investment option that has the highest revenue potential.

1.10 Implementing the Course of Action

After deciding upon the best course of action, the organization has to implement it effectively. At this stage, research is mainly used to monitor and control the programs that are being implemented.

Evaluative research studies are undertaken at this stage. One type of evaluative research study used is Performance Research. In this type of research, the performance of a particular activity is measured, so that it can be compared with the objectives set for the activity.

For example, if a company has offered a discount coupon scheme in the market, the coupon redemption rate at the end of the scheme is measured and compared with the objectives that were set for this scheme. This helps in evaluating the performance of the scheme.

Companies also monitor the performance of particular activity continuously so as to identify the opportunities and detect the problems at an early stage. This helps a company in altering the plans or developing new programs.

1.11 Factors Affecting Research

Although research provides many benefits to an organization, it is not a panacea for all the problems that an organization faces. And conducting research also involves cost, time, and effort. Therefore, an organization should decide upon the option of conducting research after considering various factors. These include time constraints, availability of resources, data, nature of information sought expected by the organization, and costs involved.

1.11.1 Time Constraints

Time constraint is a key factor that influences a company's decision with regard to whether to conduct a research study or not. In certain cases, lack of time prompts a company to take decision without making any research study. Sudden changes in competitors' strategies, regulatory changes, change in the market environment, or changes in the company's operations, require immediate action.

For example, a company, namely "A," has drastically cut the prices of its bath soap in India. And a company, namely "B," responded to the price cuts without making any study on the implications of the price cuts on its product sales or image.

1.11.2 Availability of Resources and Data

Another factor that influences the decision to undertake research or not, is the availability of resources. The availability of resources can be in terms of either budgetary allocations or human resources. Lack of financial resources may lead to improper conduct of a research study. The results obtained from such research, in turn, will be inaccurate or biased. Lack of financial resources forces a company to compromises on the way its research project is undertaken, such as taking a smaller sample size where the project demands a larger sample size, using irrelevant methods of data collection, and even comprising on the data analysis process, which is crucial for any research study.

Therefore, before conducting the research study, the company needs to consider the issue of availability of financial resources. A company also needs to consider the availability of human resources while taking decision about the research study. Lack of qualified personnel may affect the data collection and data analysis processes in a research study. Lack of qualified and trained personnel may lead to selection of improper sample, improper use of statistical tools of diagrammatic and graphical representation, and inaccurate analysis and interpretation of data. Therefore, a company needs to look for well-qualified, skilled, and well-trained personnel before conducting a research study.

1.11.3 Nature of Information Sought Expected by the Organization

The information or input that a company wants to obtain from the research study also influences the decision of whether to conduct the research study or not. If the information that a company wants to obtain from the research study can be obtained from the internal records of the company, or from prior studies conducted by the company, then conducting research is a waste of time and effort.

For example, if a company such as Hamdard Laboratories (India) is launching a new Refreshing Twists in India and wants information about the market potential of the product, it can use its knowledge and its prior studies with regard to the beverages market in India, rather than conducting a new market study. In certain cases, the management's experience and intuition are enough to take a particular decision, and there is no need for a new research study.

1.11.4 Costs Involved

The benefits of a research are many. However, a research demands significant efforts and that requires allocation of sufficient budget for the same. Therefore, every company has to make a cost–benefit analysis before taking a decision with regard to the conduct of a research study. Unless the benefits of the research, in the form of the information to be gained that would serve to improve the quality of the decisions to be made, outweigh the expenditure on the research, the research proposal should not be approved.

1.12 Globalization and Research

Globalization of business and the formation of regional trading blocks have had a major impact on all aspects of business and, especially, on research. Companies are increasingly looking out for global markets due to various compelling reasons. As firms overcome the geographic barriers of their operations to cash in on the opportunities in the global market, the need for timely and relevant information from a broader and more diverse range of markets is increasing.

An organization or a market research company conducting global research requires a different set of capabilities and approaches as compared to the ones involved in domestic research. Some of the issues that an organization needs to consider before venturing into global research are as follows:

1. Global research efforts need to be more closely associated with market growth opportunities outsides the industrialized nations.
2. Global researchers need to devise new creative approaches to understand the global markets.
3. Researchers should make use of technological advances in order to undertake global business activities effectively.

Until now, the focus of global research has been confined to industrialized markets including the United States, Europe, and Japan. However, these markets are saturating while emerging markets such as India, and Southeast Asia are showing a high growth potential.

Therefore, multinational firms should concentrate on understanding these markets by devoting greater time and effort in conducting research activities in these markets.

However, conducting research studies successfully in emerging markets requires different approaches than the usual. These markets do not possess well-developed research infrastructure. Moreover, the literacy rate is low. Therefore, researchers while designing the response formats and research instruments for the emerging markets need to keep in mind these aspects. Researchers should also develop innovative tools to understand these markets. Unlike in western markets where quantitative research techniques are used more, qualitative and observation studies are effective in emerging markets.

Researchers can use innovative tools such as videotaping technique to understand consumer behavior in these markets. Researchers can use focus groups to understand views, preferences, and cultures.

Companies can also use projective and elicitation techniques such as collage, picture completion, analogies and metaphors, and psycho-drawing to gain a deeper understanding about these markets.

The use of technology can aid the researchers in effectively implementing the research activities. Researchers can make use of technologies such as CATI (Computer-Assisted Telephone Interviewing), CAPI (Computer-Assisted Personal Interviewing), and the Internet to make the research process faster, efficient, and effective.

1.13 Research and the Internet

The use of the Internet in research studies is increasing. The decline costs of conducting online research activities coupled with the increasing number of Internet users have made the Internet a cost-effective alternative to traditional research methods for research organizations. The following sections discuss the role of the Internet in primary and secondary research.

1.13.1 Primary Research

Primary data are collected directly from respondents using data collection methods such as Survey Interviews, Questionnaires, Measurements Techniques, and Direct Observation or Tabulation.

The use of the Internet for primary data collection is still in its infancy. Although there have been satisfying results of its initial implementation and the future prospects look good it is still used cautiously. There are various advantages in conducting online surveys compared to traditional survey methods. These include:

1. The responses and feedback can be obtained faster.
2. Costs for conducting online surveys are less compared to traditional survey methods.
3. Questionnaires can be delivered to the respondents faster.
4. Confidentiality is maintained as only the receipts read the questionnaire.
5. Respondents can reply to the questionnaire at their convenience.

Apart from online surveys, organizations are also conducting online focus group studies. Although there are several advantages in using the Internet, there are certain drawbacks as well. Online surveys lack face-to-face interaction. Also, lack of accessibility of the Internet among the rural population compared to other media is a major constraint.

1.13.2 Secondary Research

Secondary data is the data that already exists which has been collected by some other personal or researcher or organization for their own use. Secondary data are generally made available to other researchers free or at a concessional rate. Major use of the Internet in research is in the area of secondary research. The research reports and databases maintained by major research companies are also available on the Internet. This makes it faster, economical, and reliable for companies to know about the competitor activities.

The very essence of the Internet as a major source of secondary information probably springs forth from the advantages of its broad scope, covering virtually every topic and the reasonable cost in acquiring them.

Summary

Research can be defined as a systematic and objective process of gathering, recording, and analyzing data to guide decision-making. Qualities of good research are empirical, logical, verifiable, based on theories and principles, and replicable. Research is mainly used to reduce the uncertainty of decisions. Researchers use scientific method to test the ideas developed within the context of discovery and justification.

Researchers must maintain objectivity by keeping complete records, standardize procedures, make operational definitions, minimize biases, and minimize errors. A reliable and valid result can be repeated in similar conditions by independent investigators.

In business, decision-making goes through four key interrelated stages: Problem or opportunity identification, Problem or opportunity selection, Problem or opportunity resolution, and Implementing the course of action.

Research helps the management in each of these stages by providing useful and timely information. Organizations should decide upon the option of conducting research after considering various factors. These include time constraints, availability of resources, availability of data, nature of information that the organization is expecting, and the costs involved.

Globalization of business and the formation of regional trading blocks have had a major impact on all aspects of business and especially, on research. Companies are increasingly looking out for Global markets. As firms overcome the geographic barriers of their operations to cash in on the opportunities in the global market, the need for timely and relevant information from a broader and more diverse range of markets is increasing. The role of the Internet in research studies has also been discussed in the chapter. The declining costs of conducting online research activities coupled with the increasing number of Internet users have made the Internet an attractive option for the research organizations.

Review Questions

1. What is research? Discuss qualities of good research.
2. Explain various criteria of a good research?
3. How can you minimize biases in research?
4. Define the concept of research and analyze its characteristics.
5. Explain the significance of business research.
6. Write an essay on various types of research.
7. Explain the significance of research in various functional areas of business.
8. What is meant by business research? Briefly explain different methods of research.
9. Explain the significance of research in various functional areas of business.

2

Research Process

2.1 Introduction

Research involves a series of steps that systematically investigate a problem or an opportunity facing the organization. The sequence of steps involved in the research process includes problem or opportunity identification and formulation, planning a research design, selecting a research method, selecting the sampling procedure, data collection, evaluating the data, and preparing the research report for presentation. The above steps provide a broad outline applicable to any research project. However, the number and sequence of activities can vary as per the demand of an individual research project.

The research process can be divided into three phases—1. Planning, 2.Execution, and 3.Report preparation. The planning phase is a problem or opportunity identification and leads to selection of the sampling procedure. Data collection and evaluation can be described as the execution phase of the research process, whereas report preparation can be considered as the last phase.

2.2 Steps in the Research Process

2.2.1 Identifying and Defining the Problem or Opportunity

The first and most important step for identifying a problem is asking a question or identifying a need that arises as a result of curiosity and to which it becomes necessary to find an answer.

The research question determines the direction of study and researchers have to struggle a lot in identifying and articulating the same. Essentially two steps are involved in formulating the research problem, i.e., 1. Understanding the problem thoroughly and 2. Rephrasing the same into meaningful terms.

The main function of formulating a research problem is to decide what you want to find out about time and expertise and knowledge available at your disposal. It is equally important to identify any gaps in your knowledge of relevant disciplines, such as statistics required for analysis.

For identifying a good solvable problem, the investigator undertakes the review of the literature. A body of prior work related to a research problem is referred to as literature. Scientific research includes a review of the relevant literature. When a researcher reviews the previous researches in related fields, he becomes familiar with several knowns and unknowns. Therefore, the obvious advantage of review of the literature is that it helps to eliminate duplication of what has already been done and provide guidance and suggestions for further research. The main purpose of the review of the literature is fourfold.

1. It gives an idea about the variables that have been found to be conceptually and practically important and unimportant in the related field. Thus, the review of literature helps in discovering and selecting variables relevant for the given study.
2. It provides an estimate of the previous work and provides an opportunity for the meaningful extension of the previous work.
3. It helps the researcher in systemizing the expanding and growing body of knowledge. This facilitates in drawing useful conclusions with regard to the variables under study and provides a meaningful way of their subsequent applications.

DOI: 10.1201/9781315167138-2

4. It also helps in redefining the variables and determining the meanings and relationships among them so that the researcher can build up a case as well as a context for further investigation that has merit and applicability.

There are different sources of review of the literature such as Journals, Books, Abstracts, Indexes, and Periodicals.

As businesses today operate in a highly volatile environment governed by various macro environmental factors, they need to constantly assess their relative position and identify the various problem areas or opportunities they need to work upon in order to sustain themselves competitively in the market. The managers need to analyze the changing dynamics of business and evolve a strategy to adapt to the areas or opportunities. It is very important for the manager to identify them accurately and at the earliest.

Problem identification precedes the problem definition state. For instance, a company producing new glyceride soap may realize that its new product is not selling, but it may not know the reason for this at the outset. Although it has identified the problem in a broader perspective, it needs to define the problem specifically in terms of what is to be researched.

Once the question has been asked, the next step is to identify the factors that have to be examined to answer the question.

It is important to define the problem in a precise manner. A well-defined problem gives the researcher a proper direction for carrying out investigation. It also helps in utilizing the resources provided for the research effectively. A researcher can focus his or her efforts in collecting relevant information, if the problem is defined properly.

Some research problems such as conducting a survey on the newspaper-reading habits of a given set of the population can be clearly defined. But, if a company wants to define a research problem such as declining sales, it needs to explore the research problem further through exploratory research.

2.2.1.1 Exploratory Research

Exploratory research aims at understanding the topic being researched. Through exploratory research, researcher arrives at a set of questions that are to be answered in order to solve the problem or cash in on an opportunity.

Exploratory research is undertaken in the initial stages of the research process. It is an informal process that helps in defining the identified problem. This process involves evaluating the existing studies on related topics, discussing the problem with an expert, analyzing the situation, and so on.

At the end of this process, the researchers should be clear about what type of information needs to be gathered and how the research process should proceed.

Secondary Data Analysis and Pilot Studies are the most popular tools used in exploratory research. Secondary data is the data that has already been collected previously for some other research purpose. The sources of secondary data are magazines, journals, online articles, company literature, and so on.

Data collected by these secondary sources need to be analyzed so that the researcher has the knowledge to define the problem.

For our problem of low sales, since it is a new product in the market, it may be difficult to obtain information. But a researcher can get some related information, which may help him or her to a certain extent in defining the problem.

In Pilot studies collecting data soruce are the actual respondents in order to gain insight into the topic and help the researcher in conducting a larger study. Here, data is collected informally in order to find out the views of the respondents. The researchers may casually seek the respondent's opinion of the new cellphone wave protectors. Once the research problem is identified and clearly defined, a formal statement containing the research objectives must be developed.

2.2.1.2 Preparing the Statement of Research Objectives

The objectives of the research should be stated in a formal research statement. The statement of objectives should be as precise as possible. Objectives act as guidelines for various steps in the research process, and therefore they have to be developed by analyzing the purpose of the research thoroughly. The objectives of the research must be brief and specific; also, it is preferable to limit the number of objectives.

The research objectives comprise the research questions and the hypothesis. If the objective of the research is to study the perceptions of the customer, a typical research question could be, "Do the customers perceive the adverse effect of newly launched Glycerin Bath Soap to be hazardous to their skin?" Once the objectives and the research questions are identified, a researcher has to develop a hypothesis statement that reflects these research objectives.

2.2.1.3 Formulation of Objectives

Research is not for the sake of research. To achieve something, research is undertaken. Thus, a goal-oriented activity is a research. To give a direction to the research study, we have to identify the specified goal/goals to be achieved. Hence, it is equally important to formulate the research objectives. Once research objectives are stated, the entire research activity will be geared to achieving those objectives.

For example, we intend to examine the working of an Examination Department in a University to know whether it is fulfilling the objectives for which it has been set up.

For this study, we will gather all the relevant information/data such as Setting of question paper, availability of faculties and utilization of amount, facilities provided, expert's opinions, etc.

Similarly, if we are clear about what we want to achieve through the research exercise, then the rest of the things will depend upon the objectives such as identifying sources of data, instruments for the collection of data, and tools for analyzing data.

However, the objectives of the study must be clear, specific, and definite.

2.2.2 Formulation of Hypothesis

When the researcher has identified the problem and reviewed the relevant literature he/she formulates a hypothesis, which is a kind of suggested answer to the problem Hypothesis plays the key role in formulating and guiding any study. The hypotheses are generally derived by the study of earlier research findings, existing theories, and personal observations and experiences.

Hypothesis may be defined as a tentative statement showing a relationship among variables under study. It is stated in the form of a declarative sentence. It is a statement based on some presumptions about the existence of a relationship among variables that can be tested through empirical data.

For instance, the exploratory research for the above problem may have resulted in the hypothesis that consumers perceive that the New Glycerin Bath soap is harmful to the skin. When a researcher is developing a hypothesis, he/she will try to assume an answer for a particular research question and then test it for its validity.

For instance, suppose you are interested to know the effect of hard working for achieving success in life. You have analyzed the past research and found the indication that the variables under study are positively related. You need to convert this idea in terms of a testable statement. At this point, you may develop the following hypothesis. Those who achieved success in their life shall require lesser number hardship than those who are not got successes.

For conducting an unbiased research the researcher must formulate a hypothesis in advance of the data-gathering process. No hypothesis should be formulated after the data are collected.

A hypothesis normally makes the research question clearer to the researcher. For instance, if the research question is—"Why are the sales of refrigerators going up in winter?" In this case, the hypothesis could be—"The sales of refrigerators are going up during winter due to off-season discounts." This makes the research question much clearer. The formulation of a hypothesis allows the researcher to make a presumption or "guess" and can thus ensure that all the relevant aspects of the research are included in the research design.

For instance, the above example gives the researcher scope to include a question on off-season discounts in the questionnaire during the research design phase.

If a research study is to be conducted about the consumption patterns of tea and coffee in India, the hypothesis could be: "Consumption of tea is higher in North India and coffee in South India because of the varying lifestyles and Geographical area of these regions." This hypothesis adds factors of geographic location and lifestyle to the research problem. For any research question, several hypotheses can be developed, but there are limits to the number of hypotheses that can be validated. Researchers should avoid including any hypothesis that has already been validated by other similar studies.

However, a hypothesis cannot be developed for every research question. Moreover, a vague hypothesis may be of no use at all.

For example, if a company wants to know whether its sales will increase in the current year, then a hypothesis, "The sales will increase in current year," versus "The sales will not increase in current year," will add little value to the research question as they are almost the same as the research question itself.

Before proceeding to the next stage it is essential to consider the following points:

1. To assess the value of information that is being sought. In this stage it is important to conduct a cost–benefit analysis, wherein the costs incurred in obtaining the required information are compared with the benefits accruing to the organization. If the costs are more than the benefits, then it is better to halt the research, while the subsequent phases of the research process can be carried on if the benefit is greater than the cost.
2. To ensure that the required information does not already exist as it would make the research effort futile.

2.2.3 Identifying, Manipulating, and Controlling Variables

While talking about the hypothesis you will encounter this word, i.e., variable.

Variables are defined as those characteristics that are manipulated, controlled, and observed by the experimenter. The types of variables that must be recognized are dependent variable, independent variable, and extraneous variable.

2.2.3.1 Dependent Variable

The variable about which the prediction is made on the basis of the experiment is called as Dependent Variable. In other words, the dependent variable is the characteristic or condition that changes as the experimenter changes the independent variables.

2.2.3.2 Independent Variable

The independent variable is the condition or characteristic that is manipulated or selected by the experimenter in order to find out its relationship to some observed phenomena.

2.2.3.3 Extraneous Variable or Relevant Variable

The extraneous variable is an uncontrolled variable that may affect the dependent variable. The experimenter is not interested in the changes produced due to the extraneous variable, and hence, he or she tries to control it as far as practicable. The extraneous variable is also known as the relevant variable.

2.2.4 Formulation of a Research Design

Once the problem or opportunity identification and definition stage is complete, the process of research design begins. Research design is a crucial step in the research design process. A research design is the actual framework of a research that provides specific details with regard to the process to be followed in conducting the research.

Formulation of a research design may be regarded as the blueprint of those procedures that are adapted by the researcher for testing the relationship among the dependent and the independent variable.

The research design is based on the objectives formulated during the initial phases of the research. The research design includes all the details with regard to the research such as the place where the information should be obtained, the time and budget allotted for conducting the research, the appropriate measurement techniques, and the sampling process.

Factors determining the selection of an appropriate research design are the research objective, the importance of the decision, the cost involved in conducting the research, and the availability of data sources.

There are several kinds of experimental designs and its selection is based on the purpose of the research, types of variables to be controlled and manipulated, and the conditions under which the experiment is to be conducted.

The main purpose of experimental design is to help the researcher in manipulating the independent variables freely and to provide maximum control of the extraneous variables so that it may be said with all certainty that the experimental change is due to only the manipulation of the experimental variable. The main function of a research design is to explain how you will find answers to your research questions. The research design sets out the logic of your inquiry. A research design should include the logistical arrangements that have to made according to the proposed research design, the measurement procedures, the sampling strategy, the frame of analysis, and the time frame.

For any investigation, the selection of an appropriate research design is crucial in enabling you to arrive at valid findings, comparisons, and conclusions. A faulty design may derive misleading findings. Empirical investigation is primarily evaluated in light of the research design adopted. While selecting a research design, it is important to ensure it is valid, workable, and manageable.

2.2.5 Constructing Device for Observation and Measurement

When the research design has been formulated, the next step is to construct or choose appropriate tools of research for scientific observation and measurement. Questionnaire and Interview schedule are the most common tools that have been developed for the research. If readymade tools are not available, then the researcher may have to develop appropriate tools before undertaking the study. All these tools of research are means through which data are collected by asking for information required through a person rather than observing them.

2.2.6 Selecting the Research Method

After developing an appropriate formulation of research design, it is important for the researcher to select a proper research method. The basic methods of conducting a research study are Surveys, Experiments and observation, Secondary data studies, and Observation techniques.

The research design method is chosen based on the objectives of the study, the costs involved in conducting the study, the availability of the data, and the importance and urgency of the decision.

2.2.6.1 Surveys

A survey is a research technique that is used to collect information through a sample of respondents by employing a questionnaire. Surveys are normally carried out to obtain primary data.

Primary data are the data gathered first-hand to answer the research question being investigated. Surveys are conventionally conducted by meeting the respondents in person or contacting them through the telephone. In the past, the Internet has been widely used for conducting surveys through email. A researcher can personally meet the respondents to survey their preferences of television channels. Another researcher may use a telephone/mobile to ask the consumer about his or her satisfaction levels related to a newly launched purchased product. Yet, another researcher may send an email to a respondent to check whether he/she is interested in a newly launched product. These methods have their own advantages and disadvantages. Researchers adopt any of these methods depending on their requirement.

2.2.6.2 Experiments

In research, experiments can be conducted for studying cause-and-effect relationships. Analyzing the changes in a variable, by manipulating another variable, helps in identify cause-and-effect relationships through experiments.

For instance, analyzing the sales targets achieved by an individual salesperson by manipulating their monetary rewards is a typical example of experimentation. Test marketing conducted by companies to test the viability of their newly launched product in the market is a form of experimentation.

2.2.6.3 Secondary Data Studies

A secondary data study is concerned with the analysis of already existing data that are related to the research topic in question. In secondary data studies, secondary data are studied in order to analyze the future sales of a newly launched product.

For instance, for the newly launched Glycerin Bath Soap research, secondary data with the market setup, market network, potential customers of the product may be essential for assessing the future sales trends of the Bath Soap. Secondary data studies help in projecting future sales trends using some mathematical models.

2.2.6.4 Observation Techniques

Observation technique is a process where the respondents are merely observed without any interruption by the observers.

For instance, the shopping patterns of customers in D-mart or Big-bazaar assessed by the observers or by counting the number of vehicles passing through a junction can qualify as observation research.

The advantage of this method is that the observers do not depend on the respondents for their responses as they are only observed and are not asked to participate in the research process. Although the observation technique is useful, it cannot be used for studying several other factors such as motivations, attitudes, and so on.

2.2.6.5 Analyzing Research Designs

Although several research designs are available for a researcher to choose, it is very difficult to say that a particular research design best suits a particular research problem. Therefore, researchers should be cautious while selecting a research design. The best method to select a research design is to work backward; that is, a research design should be selected based on the end result that needs to be obtained.

For example, to study the Cigarette-Smoking/Tobacco Chewing habit of people in public places, an observation technique would be a better method than a survey research as it would save on research costs and would not require the researchers to rely on the responses of the respondents.

Once the researcher selects a research method that is most appropriate for his/her research, he/she need not develop a sampling procedure. Sampling is the most important activity pertaining to the planning phase of the research process.

2.2.6.6 Selecting the Sampling Procedure

After deciding the tools for the study the researcher also decides about the participants of the study. Usually a small sample is drawn, which represents the population. The participants could be children, adolescents, college students, teachers, managers, clinical patients, or any group of the individual in whom/where the phenomenon under investigation is prevalent.

Sampling is generally a section of the research design but is considered separately in the research process. Sampling is a process that uses a small number of items or a small portion of a population to draw conclusions, with regard to the whole population. Alternately, a sample can be considered as a subset of a larger set called the population. A well-defined sample has the same characteristics as the population as a whole and, therefore, when a research is conducted on such a sample the results obtained will represent the characteristics of the whole population.

But, if errors are made in selecting the sample, then the research results will be wrong, since a wrongly selected sample does not represent the characteristics of the population as a whole.

For instance, to study the petrol and diesel consumption patterns of people, if a sample is selected as a list of vehicle owners, it may not represent the whole population, since there are several others who use petrol or diesel for running generators or for purposes other than traveling.

It is therefore very important to define the population before selecting the sample; otherwise, the research results may not be helpful for the manager in taking effective decisions.

For example, a computer or laptop manufacturing company wanting to assess its future sales potential may select a sample as a population of household who have no computer or laptop. But there may be several computer or laptop users who may want to buy a another computer or laptop or replace the existing, and if they are not included in the population, then the research results may not be accurate.

Another important aspect of sampling is to decide on the size of the sample. What should be the size of the sample for research study? The larger the sample size, the greater will be its precision. But for practical reasons it is not feasible to select large samples. Therefore, a sample that is selected using probability-sampling technique will be sufficient for obtaining effective results.

A sample can be selected among a population through probability sampling or through non-probability sampling.

When the subsets of a population are chosen in such a way that it ensures a representative by giving every element in the population a known and equal chance of being selection, it is called probability sampling.

When subsets of a population in which little or no attempt is made to ensure a representative element are chosen, it is called nonprobability sampling.

All the steps in a research process till selecting the sampling procedure constitute the planning phase. The execution phase of the research process begins with data collection, which is the next logical step following the sampling procedure. Once a researcher decides on a sample, he/she needs to obtain data by this sample.

2.2.7 Data Collection

After preparing a suitable sample, the researcher collects the data among the units in this sample.

As there are several research techniques, there are a number of data collection methods as well. Depending on the nature of research problem a researcher may choose a particular method, for example, observation, experiment, case study, and survey, for data collection. The researcher also decides on how the tools need to be administered for collecting data, which might be individual or group. In data collection phase, the researcher must consider the recruitment of staff and assignment, way of increasing response rate, cost of training of staff, etc. The effect of each of these must be evaluated in terms of cost, accuracy, reliability, and validity.

For example, in the survey method, the data are collected by asking the respondents to fill out a questionnaire administered to them, whereas in the observation technique, the respondents are just observed without their direct participation in the research.

Whatever the method used to collect the data, it is very important that the data are collected without any errors.

Errors may creep in during the data-collection process in several forms. Potential data-collection errors may arise if the interviewee does not understand the question or if the interviewer records the answers inaccurately.

The stages of Data Collection are Pretesting and Main study.

Pretesting involves collecting data among a small sub-sample to test whether the data-collection plan for the main study is appropriate. This helps the researchers to minimize any potential errors that may crop up during the main study. The pretest results may also be used to decide on a way of tabulating the collected data. If the results of a pretest are not appropriate for decision-making, then the researcher may consider altering the research design.

2.2.8 Evaluation of the Data

The most important aspect of data evaluation is to convert the data collected into a format that will facilitate the manager in effective decision making. The reason for analyzing the data is to obtain research results and to prepare the research report. Several mathematical and statistical models are used to evaluate the data. Evaluation of data normally starts with editing and coding of the data. Editing is undertaken to verify the data and check for any potential errors or for any inconsistencies and so on.

Another task of editing is to remove any errors that may have cropped up during the interview such as recording the answers under the wrong columns of a questionnaire and so on.

Coding is a process of assigning different symbols to different sets of responses. The coding process is done so that the data can be fed in and interpreted easily using computers. Presently, technological advances have made it possible for data to be collected and directly fed into computers, removing the possibility of human error.

For instance, an interviewer may question respondents through telephone and record the answers directly into a computer, where the data are processed almost immediately, thus eliminating the scope for errors, which may arise if conventional methods of data collection are used.

2.2.9 Data Analysis and Interpretation

The interpretation of the data that has been collected by using different analytical techniques according to the requirements of the management/organization/client is called Analysis.

After making observation, the data collected are analyzed with the help of various Quantitative Methods, Statistical Methods, Statistical Tools and Qualitative techniques, in order to make the analysis suitable for effective decision making. Statistical Analysis of the data may range as Simple Frequency Distribution Tables to Complex Multivariate Analysis. Careful scrutiny of the data is a critical aspect of scientific method. The purpose of the analysis is to make sense of the data and see what light they throw on the problem and the hypotheses of the study and draw conclusion accordingly. Data analysis can be done by using univariate analysis in which research deals with a single characteristic of interest, bivariate analysis in which researcher deals with the characteristics of interest, and by using multivariate analysis in which more characteristics are involved.

Depending on the nature of data and purpose of the experiment, either a parametric statistic or a non-parametric statistic is chosen for statistical analysis. In general, the purpose of carrying out the statistical analysis is to reject the null hypothesis so that the alternative hypothesis may be accepted.

2.2.10 Drawing Conclusion

The investigator, after analyzing the results, draws some conclusions. In fact the investigator wants to make some statement about the research problem that he/she could not make without conducting his/her research. Whatever conclusion is drawn, researcher generalizes it to the whole population.

During this phase, hypotheses are accepted or rejected. At the same time, the conclusions of the study are related to the theory or research findings among which the hypotheses originally came. Depending on the new findings the original theory may have to be modified.

2.2.11 Preparing and Presenting the Research Report and Publication

After the evaluation of the data, this is the last and the major phase that comes into picture is the preparation of a research report. The research reports can be presented in either oral or written format.

The researcher documents all the steps of his or her research in clear terms: this report informs what you have done, what you have discovered, and what conclusion you have drawn among the findings.

The research report should contain a brief description of the objectives of the research, a summary of the research design adopted, a summary of the major findings, and conclude with the limitations and recommendations.

The purpose of conducting any research is to obtain information that can aid in efficient decision-making. Therefore, it is very important to carefully analyze the information obtained and present it according to the requirements of the management of the company. At this stage, the research report should be developed most efficiently and should portray the research findings most effectively. Often researchers fill the research reports with all the technical details. This should be avoided to the maximum possible extent, as the management/organization/client is more interested in the actual research results and they have to be presented lucidly in a concise format. The amount of information provided in the research report should be based on the requirements of the manager/client. A research report also acts as a historical document, in the sense that the manager may refer to this document in the future, if a research on the same lines is being conducted sometime in the future.

If you are clear about the whole process you will also be clear about the way you want to write in your report. This helps the reader to understand the study and use it for various purposes. It allows reader to replicate the study. The publication of study in scientific journals or books and in public domain makes the work available for wider dissemination.

Summary

The research process can be considered as the framework of the entire topic of research. It involves a series of steps starting with the identification of the problem or opportunity to the stage of preparing the research report. These stages are identification and definition of the problem or opportunity, formulation of hypothesis, identification manipulation and controlling of the variable, planning the research design, constructing devices for observation, selecting a research method, selecting a sampling procedure, data collection, evaluating the data, and finally preparing and presenting the research report and publications. Any research is primarily conducted for taking decisions with regard to various problems or opportunities identified by the organization. Whenever a company identifies a potential problem or opportunity, it recognizes the need for conducting a research study. Once the problem is clearly identified, the manager can check whether the required information is already present; if such information is easily accessible the manager need not spend a lot of resources in obtaining the same information again. After clearly identifying the problem it needs to be defined accordingly, and subsequently the objectives of the research are determined. Development of the hypothesis plays a crucial roles in the research process. Once this is done, the research boundaries are defined followed by estimating the value of information to be obtained against the costs incurred on conducting the research. At this stage the most important section of the research begins; this is planning the research design and involves the selection of the sample and the measurement technique. After the data are collected and evaluated, they are later presented in the form of a report to the company's management for decision making.

Review Questions

1. List the steps involved in research process.
2. Explain the importance of research questions in research.
3. What is the role of review of literature in research process?
4. Why formulation of a hypothesis is necessary?
5. How the steps in the research process help a person to get knowledge?

3

Research Design

3.1 Introduction to Research Design

Conceptualizing a research design is the important steps in planning a research study. The main function of a research design is to explain how we will find answers to the research question. For any investigation, the selection of an appropriate research design is crucial in enabling us to arrive at valid findings and conclusion. Research can be explained as Re+ Search = Again + Explore. Different names known for Research design are Research outline, Plan, and Set of proposal.

This plan or design is generally vague and tentative in the beginning. As the study progresses and insights into it deepen it undergoes many modifications and changes. A series of decisions in working out of a plan include what, why, where, when, who, and how of the research.

3.2 Meaning of Research Design

Research design is a logical structure of an enquiry. The essence of the research design constitutes the given research question or theory, type of evidence required to answer the question or to test the theory in a convincing manner.

Let us use an analogy to understand the term "research design." While constructing a building, the first decision to be arrived at is whether we need a high-rise office building, a factory, a school, or a residential apartment, etc.

Until this decision is made, we cannot sketch a plan layout and order construction material or set critical dates for completion of the project.

Similarly, a researcher needs to have clarity on the research questions and then the research design will flow to the research questions. To ensure that the evidence obtained enables us to answer the initial research questions as unambiguously as possible is the main function of a research design. We need to answer the research question, to test a theory, to evaluate a program, or to accurately describe some phenomenon for obtaining relevant evidence entails specifying the type of evidence.

The evidence that need to be collected to answer the research question constitute Sampling issues, Data collection method, e.g., Questionnaire, Observation, Document analysis, Questionnaire design, etc.

Thus, the research design "deals with a logical problem and not a logistical problem". Research design aims to test and eliminate alternative explanation of results, apart specifying the logical structure of the data.

Research design is the plan, structure, and strategy of investigation conceived so as to obtain answers to research questions and to control variance. The plan and the structure of enquiry are formulated in order to obtain answers to research questions. The plan is the overall scheme or the program of the research. It includes an outline of what the investigator will perform, starting with writing the hypothesis and their operational implications to the final analysis of data. The structure of the research is the outline of the research design, and the scheme is the paradigm of operation of the variable.

The planning process includes the framework of the entire research process, starting with developing the hypothesis to the final evaluation of collected data. Strategy includes the methods to be used to gather and analyze the data.

Research design can be understood as that which gives the blueprint for collection, measurement, and analysis of data. The design helps researchers to utilize available resources efficiently to achieve research

DOI: 10.1201/9781315167138-3

objectives. The research design outlines the actual research problem on hand and details the process for solving it.

3.3 Need for Research Design

Research design is essential because it facilitates the smooth flow of various research processes. A good design means that good research results can be obtained with minimum utilization of time, money, and effort. Therefore, it can be said that design is highly essential for planning research activities. An ideal research design can be developed by considering the available resources such as time, manpower, and money before beginning the design. The validity of research results is based on the initial research design. If the initial research design is not properly prepared, it will jeopardize the entire research process and will fail to meet the objectives. Hence, a research design has to be developed with the utmost care, as it forms the foundation for the entire research work process that follows.

3.4 Characteristics of a Good Research Design

Some important characteristics of a good research design are Flexibility, Adaptability, Efficiency, Being economical, etc. A good research design should minimize bias and maximize accuracy of the requirement, and it should provide adequate information so that the research problem can be analyzed on a wide perspective.

A research design is considered to be good if it provides specific answers to the research question or questions, adequately tests the hypothesis, presents the appropriate research question or research problem, adequately controls the extraneous independent variable, generalizes the results of a study to other subjects, and provides internal and external validity.

An ideal design should identify the exact research problem to be studied, the objective of the research, the process of obtaining information, the availability of adequate and skilled manpower, and the availability of adequate financial resources for carrying out the research.

A good research design will clearly describe the techniques to be used for selecting samples, collecting data, and managing costs and other aspects that are essential for conducting research.

Example

In the case of exploratory research, which is usually carried out for discovering ideas for further research, the research design should be flexible enough to consider various aspects of the problem situation. In cases where accuracy in research results is of paramount importance, i.e., cases where investment is huge at stake, a research design, which minimizes bias and maximizes reliability of data, will be appropriate.

For research with regard to testing of hypothesis for measuring causal relationships among the variables, the research design should allow for inferences about causality along with minimizing bias and maximizing reliability. These are some important characteristics of good research design. However, practically it is a very crucial task to clearly specify a specific form of research study for a particular research problem. Some research topics involve issues that can be resolved only by employing more number of research method.

3.5 Function of a Research Design

The purpose of a research design is to obtain dependable and valid answers to research questions. Research problems are stated in the form of null and alternate hypothesis. The research design guides the researcher on how to collect data for testing and formulation of the hypothesis.

Another function of research design is to control variance. Research design provides the researcher with a set of proposal for studying research questions. It dictates boundaries of research activity and enables the investigator to channel his or her energies in a specific direction. It enables the investigator to anticipate potential problems in the implementation of the study and to assist the investigator in providing answers to various kinds of research questions.

3.6 Research Design Concepts

It is essential to acquire knowledge about certain important concepts relating to research design, in order to ensure better understanding of various research designs.

3.6.1 Dependent and Independent Variables

If a variable is dependent on the result of some other variable, it is then called a dependent variable. An independent variable is not dependent on any other variable with reference to that particular study.

For example, height and weight are dependent on age, but age is not dependent on height and weight. Therefore, age is an independent variable, whereas weight and height are dependent variables.

3.6.2 Extraneous Variable

Extraneous variables are independent variables that are not directly linked with the study but may influence the dependent variable.

For example, assume that a hypothesis was framed with a stated relationship among the progress in academic performance of children and their self-study. Here, academic performance is the dependent variable and self-study is the independent variable. Apart self-study concept, grasping power may also affect academic performance, but, grasping power is not related to the study's purpose or objective. Therefore, grasping power can be referred to as an extraneous variable. If any direct or indirect effect occurs on the dependent variable because of the extraneous variable, it is called an "experimental error."

3.6.3 Control

Control is essentially devised to minimize the effects of extraneous variables. This is an important characteristic of a good research design.

3.6.4 Confounded Relationship

When a dependent variable is affected by the influence of an extraneous variable, then the relation among the dependent and independent variables is confused or confounded by an extraneous variable.

3.6.5 Research Hypothesis

If a hypothesized relationship or prediction or an assumption has to be tested using scientific methods, it is called research hypothesis. A research hypothesis links an independent variable to a dependent variable. It should generally contain a dependent and an independent variable.

3.6.6 Experimental and Nonexperimental Hypothesis

If the primary objective of conducting research is to test a hypothesis, it is termed research hypothesis testing. This can be done for both experimental and nonexperimental research. When an independent variable is manipulated during research, it is called as an experimental hypothesis testing research. Non-experimental research hypothesis testing pertains to non-manipulation of an independent variable in research.

For example, assume that a researcher wants to study whether the daily intake of balanced diet by the students influences their skills in sports. For this study, if the researcher selects a random sample of 50 students, it is called non-experimental research hypothesis testing. This is because the independent variable, sporting skills, is not manipulated. If he or she selects a sample of 50 students and divides them into equal groups P and Q (say), where group Q is provided with a training in National Sports Academy and Group P is a control group with no external manipulations, it is called experimental research hypothesis testing as the independent variable, sporting skills, is being manipulated.

3.6.7 Experimental and Control Groups

While conducting experimental research hypothesis testing, if the group is studied under usual conditions then it is called a Control Group. When the group is studied under special conditions, then it is called an experimental group.

For example, if one considers the above illustration, Group P is the control group as there is no external manipulation and normal conditions prevail. Group Q is the experimental group as there is an external manipulation in the form of a National Sports Academy Coaching providing training for the students. Research studies can be designed consisting of only experimental groups or involving both experimental and control groups.

3.6.8 Treatments

Treatments refer to the conditions to which the experimental and control groups are subjected. For example, if one considers the above example, treatments were available. First treatment is the training program without a coach and the other treatment is a training program with a coach.

3.6.9 Experiment

The process involving checking the validity of a hypothesis statement of a research problem is called an experiment.

For example, if one wants to study the impact of a Sports Academy Coaching Trainer on the performance of various sports such as wrestling, football, cricket, etc., one can conduct an experiment. The experiment can be an absolute experiment or a comparative experiment. The study of the impact of a Sports Academy Coaching Trainer on the various sports performances is called an absolute experiment. The study of the impact of Sports Academy Coaching Trainer on the sport performance compared to the impact of another sports trainer on performance is called a comparative experiment.

3.6.10 Experimental Units

Prespecified plots or blocks, where various treatments are used, are called experimental units. Experimental units need to be defined very carefully.

3.7 Classification of Research Designs

Several research design approaches are available, which can be classified as follows:

1. Exploratory studies, which include techniques such as Secondary Data Analysis, Experience Surveys, Focus Groups, and 2-Stage Design;
2. Descriptive studies;
3. Casual studies under which causal relationships are studied, such as symmetrical reciprocal and asymmetrical relationships.

3.7.1 Exploratory Studies

To make problem suited to more precise investigation or to frame a working hypothesis an operational perspective, exploratory research is carried out. Exploratory studies are helpful in understanding and assessing the critical issues of problems. These studies are not suitable for cases in which a definite result is desired. However, the study results are used for subsequent research to attain conclusive results for a particular problem situation. In short, to obtain necessary information and to develop a proper foundation for conducting detailed research later, exploratory research can be used.

The following are the main reasons for which exploratory studies are conducted: analyze a problem situation, evaluate alternatives, and discover new ideas.

For example, the top management of an organization has ordered the research department to evaluate the Organization or Factory production pattern. For researchers, this is not a clearly defined problem situation. Therefore, they will have to first conduct exploratory studies to understand every aspect in connection with the Organization or Factory production process, beginning it with purchasing raw materials, inventory management, processing them into finished goods, and stocking them.

Using both qualitative and quantitative techniques, exploratory research can be conducted. Qualitative techniques are mostly used for conducting exploratory research.

For example, Indepth interviews, Projective techniques, Elite interviewing, and Document analysis need to be done. A combination of these techniques gives rise to important exploratory techniques, such as Secondary data analysis, Experience surveys, and Focus groups.

3.7.1.1 Secondary Data Analysis

To look for availability of secondary data is the initial process of exploratory research. Secondary data are those that are already available as a result of research carried out by others for their own use. The rationale behind searching for secondary data is that it does not make sense to collect primary data on a subject, when secondary data are already available. The search for such data can start with the database of the researcher's organization. Research reports of previous studies also provide adequate information about a particular problem on hand.

Previous studies further help researchers to identify methods that have proved successful or unsuccessful. A significant amount of secondary data can be obtained derving out of journals, magazines, and other periodicals, of which there are many in number.

Online sources and search engines provide endless information on almost any topic. After collecting adequate secondary data, the researcher can use that data as a strong basis for further processing of the research.

3.7.1.2 Experience Surveys

To gain additional knowledge on a particular subject area among experts in that field, experience surveys are usually conducted. Since a lot of vital information about a research area will not be available freely, experience surveys are conducted.

When adequate secondary data are difficult to obtain, experience surveys in such situation come in handy. While interviewing experts for eliciting crucial information with regard to a particular area of study in experience surveys, it is essential to obtain their views and perceptions about the important aspects and issues of the research study. The format of the questionnaire used in experience surveys should be flexible enough to include several dimensions of the subject that may arise during the interaction process.

Several questions cropping up as a result of experience survey may completely alter the existing perceptions about a particular issue or may change the way people look at the problem. These surveys are generally conducted by obtaining the necessary information belonging to experts in similar fields.

For example, assume a manufacturer of steel is trying to assess the performance of machinery in a plant. The people who can provide the necessary information are the existing administrative manager, production managers, the shop floor workers, people who deal with subunits of the machinery, the top-level managers, the manufactures of such machines, and finally all those who are associated with the machinery in different ways.

3.7.1.3 Focus Groups

Focus groups originate at Sociology studies. They have been extensively used in Marketing Research. Recently, they have been used to study research problems of a wide range of subjects. Generally, to evaluate the potential of a newly launched product idea or concept, focus group studies are conducted. A focus group comprises several persons, who are led by a trained moderator. The moderator's task is to lead the team in generating and exchanging ideas on a particular issue. The process starts by issuing a topic for discussion among participants by the moderator. In such discussions, the moderator's role will be to watch the proceedings silently and ensure that the discussion proceeds as expected.

However, the moderator needs to intervene to ensure that every individual in the group participates. Once the focus group's observations and recommendations are obtained, the information is evaluated by the moderator. This forms the basis for further research.

Unfortunately, in recent times, focus groups are being criticized for being ineffective.

Usually, face-to-face interaction takes places in focus groups. Due to constraints of space, time, and geographical problems, these studies are conducted through Teleconferences, Videoconferences, and online technology such as the Internet and e-mail.

3.7.1.4 2-Stage Design

For designing research, a 2-stage design is a beneficial approach. Here, the exploration is conducted in stages: clearly defining the research problem and developing the research design.

When the problem is vaguely defined and the researcher is not clear about the particular topic that has to be studied, a 2-stage design is beneficial. In these circumstances, the first stage will clearly define the problem for study and the another stage will develop the research design.

3.7.2 Descriptive Studies

Unlike exploratory studies, descriptive studies come under formal research, where the objectives are clearly established. A researcher gathers details about every aspect of a problem situation in descriptive studies. It is necessary to design the research efficiently, irrespective of the problem's complexity. Although descriptive studies look easier than experimental studies, they are equally important. Descriptive studies form the basis for analytical, experimental, and quasi-experimental studies. They help in developing hypothesis.

3.7.3 Causal Studies

To identify the cause-and-effect relationship uniting variables is the basic aim of causal studies.

For example, studying the effect of purchasing price, expenses on advertisement, and expenses on marketing on sales comprise causal studies. Hence, a thorough knowledge of the subject area of research is essential for researchers. The basic premise of the causal relationship is that when we do a particular act or thing or cause, it gives rise to another act or thing or effect. Scientifically, it is highly impossible to prove a causal relationship. Researchers develop evidence to understand causal relationships.

For example, if a researcher wants to establish a relationship that proper consumption of vitamins, i.e., cause, leads to perfect body development, i.e., effect, among children, the researcher should then be able to prove that proper consumption of vitamins precedes perfect body development.

3.7.3.1 Causal Relationships

The casual analysis is the process of determining how specific variable influences the corresponding changes in another variable. The cause-and-effect relationship is less explicit in research. The types of possible relationships that can arise linking variables are Symmetrical, Reciprocal, and Asymmetrical.

3.7.3.2 Symmetrical

When pair of variables fluctuate together, it is called as symmetrical variable. However, it is assumed that the changes in either variable are not due to changes in the other. When the binary variables become alternate indicators of another cause of independent variable, then usually in such cases symmetrical conditions occur.

For example, the irregular or less attendance of students in an MSc Statistics class for a particular subject but active participation in parties and events other than education is the result of dependence on factors such as study environment and family background.

3.7.3.3 Reciprocal

There is an existence of a reciprocal relationship when binary variables influence or reinforce each other.

For example, a reciprocal relationship exists when a person reads a particular newspaper advertisement, which leads him to buy that newly launched brand product. Later, after using the product, it consequently sensitizes the person to notice and read the successive newspaper advertisement of that particular brand or company.

3.7.3.4 Asymmetrical

When changes in single variable, i.e., independent variable, are responsible for changes in another variable, i.e., dependent variable, then asymmetrical relationship exists. The following are the types of asymmetrical relationships.

3.7.3.4.1 Stimulus–Response Relationship

It represents an event that results in response arising out of some object. For example, an increase in product price may lead to decline in quantity of sales.

3.7.3.4.2 Property–Disposition Relationship

A property is the enduring nature of a subject, which does not depend on circumstance for its activation. Under certain circumstances, a disposition is an inclination to respond in a certain manner.

For example, family status, age, gender, religion, caste, etc., can be considered personal properties. Attitudes, thoughts, opinions, values, ethics, etc., are portion of disposition. For property disposition, examples include the effect of age on attitude with regard to savings, gender, and its impact on attitude toward social issues, etc.

3.7.3.4.3 Disposition–Behavior Relationship

Consumption patterns, work performance, interpersonal skill, working techniques, etc., are division of behavior responses. For example, include a person's perception about a newly launched product and its purchase, job satisfaction, productivity, etc.

3.7.3.4.4 Property–Behavior Relationship

Examples of property–behavior relationship are family life cycle and purchase of goods; and social class and family-saving pattern, etc.

3.8 Selection of Specific Research Design

It is important for the researcher to select a specific research design for conducting the research while initiating a research process. A researcher is confronted with various research designs. The following are the different categories of research designs.

3.8.1 Desired Degree of Formality

Based on the research structure and on the immediate objectives to be achieved, a research study can be exploratory or formal in nature. Exploratory studies are conducted for developing guidelines for future research and are based on a loose structure. The immediate objective of an exploratory study is to develop a hypothesis for future research. On the other hand, a formal study is initiated with a proper hypothesis and is based on a formal procedure for conducting the research.

3.8.2 Objective of Study

The objective of the study also determines the selection of a research design.

For example, if the study is being conducted, to find out what, when, where, who, or how much of something, then best suited design is Descriptive Study. A causal study is suitable when the study aims at finding out "Why."

For example, how many employees are leaving an organization or company, when they are leaving, and such other issues, are under the purview of descriptive studies. To analyze why the turnover of employees is very high requires a causal study.

3.8.3 Data Collection Method

Out of the subject of study, data can be collected in different ways. Sole method is to observe subjects on certain parameters, which is called observation studies. In such studies, the subjects are not questioned.

Another method of data collection is to elicit responses deriving out of subjects, i.e., respondents, by asking them questions through a questionnaire. Here, the researcher can adopt either method based on the study that needs to be conducted.

For example, if research has to be done on the traffic flow at a particular junction, then the observation method is best. On the other hand, a questionnaire for obtaining consumer responses is the best method when consumer perceptions about a newly launched product are to be estimated.

3.8.4 Variable Control

In an experiment, a researcher can control variables by manipulating them. To analyze whether certain variables have the capacity to influence other variables, experimental designs are conducted customarily. Researchers cannot control or manipulate variables in other forms of research such as the ex-post-factor design. In this type of research manipulation amounts to bias.

3.8.5 Time Dimension

Time dimension consists of cross-sectional studies and longitudinal studies. Whereas longitudinal studies are repeated over a period of time, cross-sectional studies are conducted once. They keep track of changes taking place during that period.

3.8.6 Scope of the Study

By the research design adopted the scope of the study is limited. Therefore, it is very important to select an appropriate research design. In several instances, a statistical study is most appropriate, where an attempt is being made to study the sample that is representative of the entire population, and inferences are made on their general characteristics based on the research.

3.8.7 Environment Conditions of Research

According to conditions of research—environmental, field, or laboratory—the selection of research design varies. Under field conditions or actual conditions, the research is carried out as it is, whereas in laboratory conditions, real-life situations are enacted and an artificial environment is created for the research.

3.8.8 Subject's Perceptions

According to the perceptions of the subjects under study lies the effectiveness of a research study. For instance, if the respondent is/are aware that a research is being carried out, then he or she may respond differently, reducing the effectiveness of the survey. To ensure the study's effectiveness, it is the researcher's duty to be watchful about such changes.

3.9 Benefits of Research Designs

There are several benefits of a good research design. A well-formulated research design acts as a bridge or conduit in the middle of the final objective of the research and the method of study to achieve the objectives. The success rate probability of an organization will improve by developing a good research design. Good research design anticipates the client's requirements in terms of results and helps in proper data analysis and interpretation, so that it can be represented as a useful finding for the client. The researcher cannot have a clarity on what he or she is required to do unless a good research design is developed. A research design that is put on paper will act as a thorough guideline.

The research design is particularly helpful to a researcher in identifying the type of data that need to be obtained for conducting a research. The purpose of the research will be completely undetermined if irrelevant data are gathered. In tabulating, analyzing, and interpreting data, a proper research design is helpful. As and when the situation arises, a good research design will provide the added advantage of being flexible. Finally, we can conclude that a good research design will act as a guideline for effective research and will be immensely beneficial.

Summary

Research design can be defined as the arrangement of conductions for collection and analysis of data in a manner that aims to be relevant to the research purpose with economy in procedure. Research design is primarily essential because it facilitates the smooth flow of various research processes. Some important characteristics of good research design are flexibility, adaptability, efficiency, economy, and so on. A good research design should minimize bias and maximize accuracy of data and contain the least number of errors.

Some important research design concepts are dependent and independent variables, extraneous variables, control, confounded relationship, research hypothesis, experimental and non-experimental hypothesis testing research, experimental and control groups, treatments, experiments, and experimental units.

While initiating a research process, it is important for the researcher to select a specific research design. Before the start of research activity, it is crucial to select an appropriate and particular research design that should be used. These approaches can be classified as (a) Exploratory studies under which we have secondary data analysis, experience surveys, focus groups, and 2-stage design; (b) Descriptive studies; and (c) Casual studies under which causal relationships such as symmetrical, reciprocal, and asymmetrical relationships exist.

Review Questions

1. Define research design.
2. Explain basic objectives of research design.
3. How can we check the criteria of a good research design?
4. Define research design and indicate its purposes.

5. What are the functions of a research design?
6. Define treatment in research design.
7. Which constituent of research study does guide the research design?

4

Measurement Concepts in Research

4.1 Measurement

A proper measurement system has to be developed before actually venturing into the field to collected data. For measuring the characteristics that are relevant to the research study, at this stage a researcher has to address some fundamental issues relating to the variables that need to be measured and the different measurement scales that have to be used.

Measurement is the process of assigning numbers to various attributes of people, objects, or concepts. Technically, it is the process of mapping aspects of a domain to other aspects of a range according to some rule of correspondence.

Measurement is the process of assigning numbers or labels to different objects under study to represent them quantitatively or qualitatively. Measurement thus can be understood as a means to denote the amount of a particular attribute that a particular object possesses.

4.2 Identifying and Deciding on the Variables to Be Measured

The primary step in the measurement process is to identify the area or the concept that is of interest for the study is Concept and Construct. The Concepts is a general idea derived or an abstract inferred or derived from specific instances.

For example, once a concept has been identified, studying the motivational levels of employees in an organization, the researcher can focus on developing a construct.

Certain types of concepts which exist at different levels of thought that are developed to simplify complex situations concerning the area of study can also be considered as constructs. For theoretical usage as well as for explaining the concepts themselves, constructs are developed.

A constitutive definition of the concept will define the central theme of the study and specify the research boundaries. Concept will help the researchers in framing and addressing the research question in an appropriate manner.

For example, if we want to study the education system in Rajasthan State of India, this will not help at all in developing a research question, since it has to be clearly defined as to what education system needs to be studied, is it primary education or secondary or higher education, or is it related to an Adult Continuing Education & Extension Program, and so on.

4.3 Research Measurement Issues

Various concerns arise in the process of measurement, which need to be addressed by the researcher. Some of the important issues include the following:

1. The underlying characteristics of the concept allowing ordinal level or nominal level.
2. The features of the concept are discrete or continuous. Because it allows the use of more powerful statistical techniques for analysis, efforts should be made that measurement represents the highest scale of measurement for a concept.

DOI: 10.1201/9781315167138-4

3. The number of indicators for measurement of a concept. Some simple concepts can be measured by one indicator whereas abstract concepts are measured with more than one indicator. The number of indicators appropriate to measure a concept is to be decided by the researcher.
4. A valid and reliable source of measure.
5. The measurement should be free from errors such as invalidity error and unreliability.
6. The measurement validity is distinct from internal validity and external validity because these are separate research design issues.
7. The proper use of available data. For that, a researcher should be well aware of the types of data available with the various data compilation agencies and available data sources.

4.4 Need Development of Measurement Scales

Scale is defined as a set of numbers or symbols developed in a manner so as to facilitate the assigning of these numbers or symbols to the units under research following certain rules. Generally, it is very easy to measure certain parameters such as sales of a particular product, the profitability of a firm, the productivity and development of the employees in an organization, and so on. These are relatively easier, as they can be measured quantitatively by applying different scales of measurement. However, it is relatively difficult to measure some aspects such as the employee's motivational level in an organization, the customer's attitude toward a newly launched product, or the acceptance levels of the customer of a newly designed product, and so on.

Measurement of concept is very difficult, as the respondents may not be able to put their feelings across exactly in onwards, and sometimes the scales may not be capable of drawing the right response from the respondent.

At times, the respondents might not be willing to reveal their opinions to the researcher. To overcome these difficulties, a researcher's primary objective is to seek the cooperation of the respondent and create an environment of trust and mutual understanding. The interviewer should try to reduce all the negative feelings of the respondent and develop a situation wherein the respondent feels free to share all his feelings relevant to the research with the interviewer. It is also important for the researcher to clearly specify what information he needs and why. If the research design permits, organizations generally develop scaling techniques to measure certain crucial aspects of business such as measuring customer retention.

4.5 Measurement Scales

The postulates of measurement are equalities or identities, rank order, and additivity.

The design of a measurement scale depends on the following:

1. Objective of the research study,
2. The mathematical or statistical calculations that a researcher expects to perform on the data collected using the scales.
3. The objective of the research study may be as simple as classifying the population into various categories, or as complex as ranking the units under study and comparing them to predict some trends.

The process of observing and recording the observation that are collected is called as Measurement. The four levels for measuring variable are Ratio scale, Interval scale, Ordinal scale, and Nominal scale.

4.5.1 Nominal Scale

Under the nominal scale, the data are recorded into categories, without any order or structure. If against any question or statement, response is recorded as simply yes or no, then the scale will be nominal.

There is no distance between yes and no and also has no order. The statistical techniques can be used, with nominal scales data are Mode, Cross tabulation—with Chi-square (χ^2), Logistic Linear Regression Model, Principal component analysis, Factor analysis, etc.

Nominal scale is a qualitative scale without order. In nominal scale numbers are used to name, identify, or classify persons, objects, groups, gender, and an industry or an organization type. The numbers do not really mean anything, even if we assign unique numbers to each value. This scale neither has any specific order nor does it have any value.

In the case of nominal measurement, statistical analysis is attempted in terms of Counting or Frequency, Percentage Proportion, Mode, Coefficient of Contingency, Chi-square (χ^2), etc.

Under nominal measurement scale, addition, subtraction, multiplication, and division are not possible. The scale helps to segregate data, which are mutually exclusive and exhaustive into categories. This scale assigns numbers to each of these categories and these numbers do not stand for any quantitative value, and hence they cannot be added subtracted or divided.

For example, a nominal scale designed to measure the nature of occupation, employment status, may be given as below.

4.5.1.1 Occupation

(1) Public Sector, (2) Private Sector, (3) Entrepreneurs, (4) Jobless or Non-employee or Idle, (5) Others.

In the above example, the numbers 1, 2, 3, 4, and 5 only serve as labels to the various categories of employment status, and hence, to perform any type of mathematical or statistical operations on those numbers, those numbers cannot be used by the researcher.

Any relationship between the variables is not given by the nominal scale, and the frequency of items appearing under each category is the only quantitative measure, i.e., the number of people in public sector jobs, etc. Using nominal scale one can only calculate the mode for the collected data.

4.5.2 Ordinal Scale

Ordinal scale comes next to nominal scale. An ordinal scale is a qualitative scale with order. An ordinal scale is used to arrange objects according to some particular order.

Thus, the variables in the ordinal scale can be ranked. The type of scale that gives ranks is called an ordinal measurement scale. In ordinal scale, numbers denote the rank order of the objects or the individuals. Numbers are arranged from highest to lowest or lowest to highest.

The statistical operations that can be applied in ordinal measurement are Median, Mode, Range, Percentile, Rank correlation coefficient, and Nonparametric correlation and modeling techniques. Ordinal scales do not provide information about the relative strength of ranking. This scale does not convey that the distance between the different rank values is equal. Ordinal scales are unequal interval measurements. Ordinal scale does not incorporate absolute zero point. Ordinal variables can only give us the information with regard to relative position of the participants in the observation, but they do not give any information with regard to the absolute magnitude of the difference between the first and the second position, or second and third position, and so on.

For example, in terms of their academic performance achievement, students may be ranked first, second, third and so on.

If someone says that a person came second in the Skating Competition, then we can understand that there was another person who came first and some others were there who were ranked after him. A respondent may rank these players depending on his or her experience or perception of them.

On ordinal measurement a significant amount of consumer-oriented research relies. Here, to rank items, numbers, letters, or any other symbols are used. Ordinal scale tells us whether an object or event has more or less of a characteristic than some other object or event. Ordinal scale does not indicate how much more and how much less we have of the characteristics the objects or events pose.

For example, assume that 150 consumers are divided according to their income and the classification is as given in Table 4.1:

TABLE 4.1

Classification of Consumer According to Income

Monthly income (Rs)	<5000	5001–10,000	10,001–15,000	15,001–20,000	20,001 and above	Total
Number of consumers (Rs)	25	40	35	30	20	150

4.5.3 Interval Scale

In a situation when we not only talk about differences in order but also differences in the degree of order, it is referred to as interval scale.

For example, if we are asked to rate our satisfaction with a preference of bath soap on a seven-point scale, from dissatisfied to satisfied, we are using an interval scale. Interval scale includes all the characteristics of the nominal and ordinal scale of measurement. In interval scale numerically equal distance on the scale indicate equal distances in the attributes of the object being measured. The data obtained from an interval scale are known as interval data. The appropriate measures used with interval scale data are: Arithmetic Mean, Standard Deviation, Rank order variance, Range, Karl Pearson's coefficient of correlation, Regression, Analysis of Variance, Factor analysis techniques, Discriminant analysis, and tests such as t-test, z-test, F-test. We cannot apply coefficient of variation in the interval scale. Interval scales are similar to ordinal scales to the extent that they also arrange objects in a particular order. However, in an interval scale, the intervals between the points on the scales are equal. This is the scale where there is equal distance between the two points on the scale.

Examples of interval scales are given below:

1. A scale represents marks of students using the attributes range 0–10, 10–20, 20–30, 30–40, 40–50, 50–60, and so forth. The midpoints of each range, i.e., 5, 15, 25, 35, 45, 55, etc., are equidistance from each other.
2. Fahrenheit and Celsius scales are used to measure temperature. In these scales the difference between the intervals is the same, i.e., the difference between 30° and 50° is the same as the difference between 15° and 35°. But at the base point, boiling point of water is represented by 212°F and 100°C and freezing of water is represented by 32°F and 0°C. Thus, there is no natural zero base for these scales.
3. Similarly, we can design an interval scale with points placed at an interval of 1 point.

[12]-[11]-[10]-[9]-[8]-[7]-[6]-[5]-[4]-[3]-[2]-[1]

Ask the respondents to place the mobile telephone service providers on this scale of 12–1.

4.5.4 Ratio Scale

A ratio scale can be defined as a scale that measures in terms of equal intervals and an absolute zero point of origin exists. A ratio scale also possesses a conceptually meaningful zero point in which there is a total absence of the characteristic being measured. This zero is common to a distance scale using yards, meters, etc. Since there exists an absolute zero on the ratio scale the data collected can be subjected to any type of mathematical operation, say, addition, subtraction, multiplication, and division. Ratio scale has all the characteristics of nominal, ordinal, and interval scale. Ratio scales are common among physical sciences rather than among social sciences. The statistical techniques used in interval scale can easily be used in ratio scale also.

A ratio scale is the top level of measurement and satisfies the following properties:

1. It is possible to work out the ratio of two observations when the measurement of each observation of a variable is in numerals or quantitative terms. For a variable X taking two values X_1 and X_2 the ratio will be X_1/X_2.

2. Measurement of distance between two observations X_1 and X_2, i.e., $(X_2 - X_1)$.
3. Ascending or descending of the elements of a variable is an indication of natural ordering. Therefore, comparison such as $X_2 > X_1$ or $X_1 > X_2$ is meaningful.

All measures of central tendencies that can be used in this scale include geometric and harmonic means. Ratio scales have a fixed zero point and also have equal intervals. Unlike the ordinal scale, the ratio scale allows for the comparison of two variables measured on the scale. This is possible because the numbers or units on the scale are equal at all levels of the scale.

The examples of ratio scales are the measures of height, weight, money scales, distance, and so on. A very good example of ratio scale is distance; for instance, not only can we say that the difference between 3 and 6 miles is the same as the difference between 6 and 9 miles but we can also say that 9 miles is thrice as long as 3 miles.

Note: Interval and Ratio scale data are sometimes referred to as parametric. Nominal and Ordinal data are referred to as non-parametric.

4.6 Criteria for Good Measurement

The three major criteria are Reliability, Validity, and Practicality.

4.7 Reliability

4.7.1 Meaning of Reliability

Reliability refers to the degree to which the measurement or scale is consistent or dependable. If we use same construct again and again for measurement, it would lead to same conclusion. Reliability is consistency in drawing conclusion. It is considered that, when the outcome of a measuring process is reproducible, then the measuring instrument is reliable. Reliable measuring scales provide stable measures at different times under different conditions.

For example, if a Tea or Coffee vending machine gives the same quantity of Tea or Coffee every time, then it can be concluded that the measurement of the Tea or Coffee vending machine is reliable.

Thus, reliability can be defined as the degree to which the measurements of a particular instrument are free from errors and consistent results is produced. However, in certain situations, poor data collection methods give rise to low reliability. If the respondents do not understand the questions properly and given irrelevant answers to them, then the quality of the data collected can become poor. Any significant results must be repeatable is the idea behind reliability. Other researchers must be able to perform exactly the same experiment, under same conditions and generate the same results. This will indicate the findings and ensure that all researchers will accept the hypothesis. Experiment and research have not fulfilled all of the requirements of testability, without this replication of statistically significant results. This prerequisite is essential to a hypothesis establishing itself as an accepted scientific truth.

Examples of Reliability are as follows:

1. If you are "performing a critical time experiment," you will be using some type of Alarm Clock or Digital Watch. Generally, it is reasonable to assume that the instruments are reliable and will keep true and exact time. However, scientists take measurements a number of times, in order to prevent the chances of malfunction and to maintain the validity and reliability.
2. If a test is constructed to measure a particular trait; say, Stress, then each time it is administered, it should yield same results. A test is considered reliable if we obtain the same result frequently.

For determining the overall validity of a scientific experiment and enhancing the strength of the results, reliability is a necessary ingredient. Reliability is the consistency of your measurement, or the degree to which an instrument measures the same way each time it is used under the same condition with the same matter. In short, reliability is the repeatability of measurement. A measure is considered reliable if a person's score on the same test given twice is similar. It is important to remember that reliability is not measured, it is estimated.

4.7.2 Methods of Estimating Reliability

4.7.2.1 External Consistency Procedures

External consistency procedures compare findings from two independent process of data collection with each other, as a means of verifying the reliability of the measure. Following are the two methods.

4.7.2.1.1 Test–Retest Reliability

If the result of a research is the same, even when it is conducted for the second or third time, it confirms the repeatability aspect.

For example, if percent of a sample say that they do not watch movie, and when the research is repeated after sometime and the result is same or almost the same again, then the measurement process is said to be reliable.

However, there are certain problems with regard to the test–retest method of testing reliability, the first and foremost issue is that it is very difficult to obtain the cooperation and locate all the respondents for a second and subsequent round of research. Apart from this, the responses of these people may have difference on the second occasion, and sometimes environment factors may also influence the responses. The most frequently used method to find the reliability of a test is by repeating the same test on the same sample, on two different time periods. The reliability coefficient in this case would be the correlation between the score obtained by the same individual on two administrations of the test. When the same test is administered on the same individual sample, test–retest reliability is estimated. Therefore, it refers to the consistency of a test among on two different time periods different administrations. The assumption behind this approach is that there will be no substantial changes in the measurement of the construct in question, upon administration on different occasions. The time gap that is given between measures is of critical value, the smaller the time gap, higher the correlation value and vice versa. If the test is reliable, the scores that are attained on first administration should be more or less equal to those obtained on second time also. The relationship between the two administrations should be highly positive.

Limitations of Test–Retest Reliability

1. **Memory effect or carry-over effect**

   One of the common problems with test retest reliability, is memory effect. When two administrations takes place within little span of time, this argument particularly holds true.

   For example, when a memory-related experiment including nonsense syllables is conducted, whereby the subjects are asked to remember a list in a serial-wise order, and the next experiment is conducted within 15 minutes, most of the times, subject is bound to remember his or her responses, as a result of which there can be prevalence of artificial reliability coefficient since subjects give response from memory instead of the test. The condition is the same, when pretest and post-test for a particular experiment is being conducted,

2. **Practice effect**

   When repeated tests are being taken for the improvement of test scores, as is typically seen in the case of Quiz Competition, where there is improvement in the scores this happens as we repeat these tests.

3. **Absence**

   Individuals remain absent for retests.

4.7.2.1.2 Parallel Forms Reliability

Various names of Parallel-Forms Reliability are Alternate Forms Reliability, Equivalent Form Reliability, and Comparable Form Reliability.

Two equivalent forms of a test that measure the same attribute are compared by Parallel forms reliability. The two forms use different items. However, the same rules used to select items of a particular difficulty level. When two forms of the test are available, one can compare performance on one form versus the other. The two forms are administered to the same group of people, sometimes on the same day. The Pearson Product Moment Correlation Coefficient is used, as an estimate of the reliability. When both forms of the test are given on the same day, the only sources of variation are random error and the difference between the forms of the test. Sometimes the two forms of the test are given at different times. In these cases, error associated with time sampling is also included in the estimate of reliability. The method of Parallel Forms provides one of the most rigorous assessments of reliability commonly used.

Unfortunately, the use of parallel forms occurs in practice less often than is desirable. To retest the same group of individuals, often test developers find it burdensome to develop two forms of the same test, and practical constraints make it difficult.

Many test developers instead prefer to base their estimate or reliability on a single form of a test. In many ways you can assess the different sources of variation within a single test. To evaluate the internal consistency of the test by dividing it into subcomponents, is one of the method.

Some of the shortcomings of test–retest reliability can be overcome using parallel form reliability. In parallel form reliability, two measurement scales of a similar nature are to be developed.

For example, if the researcher is interested in finding out the perceptions of consumers on newly launched technologically advanced products, then he or she can develop from reliability questionnaires. Each questionnaire contains a different question to measure their perceptions. The two questionnaires can be administered with a time gap of about 10–15 days.

The reliability in this method is tested by measuring the correlation of the scores generated by the two instruments. The major problem with parallel form reliability is to frame two totally equivalent questionnaires, which is almost impossible.

4.7.2.2 Internal Consistency Procedures

Internal consistency of data can be established, when the data give the same results even after some manipulation.

For example, after a research result is obtained for a particular study, the result can be split into two parts. Then, the result of one part can be tested against the result of the other; if they are consistent, then the measure is said to be reliable.

The reliability of this method is completely dependent on the way the data are divided up or manipulated, which is the problem with internal consistency. Sometimes, it so happens that different splits give different results. Many researchers adopt a technique called as Cronbach Alpha to overcome such problems with split halves, which needs the scale items to be at equal intervals.

Kuder Richardson Formula 20 (KR-20)

An alternate method called Kuder Richardson Formula 20 (KR-20) is used to calculate how consistent subject responses are among the questions on an instrument, in case of difficulty in obtaining the data at equal interval of time. Items on the instrument must be dichotomously scored, i.e., 0 for False and 1 for True. Rather than comparing half of the items with the other half of the items, all items are compared with each other. Kuder–Richardson reliability coefficient is actually the mean of all split-half coefficients, and can be shown mathematically. The idea behind internal consistency procedures is that the items measuring same phenomena should produce similar results. The following internal consistency procedures are used for estimating reliability.

4.7.2.2.1 Split Half Reliability

We randomly divide all items that intends to measure same construct into two sets, in Split Half Reliability. The split half reliability is then, simply the correlation between these two scores, when the complete instrument is administered on sample of people and total scores are calculated for each randomly divided half.

4.7.2.2.1.1 Problem in Split Half Reliability The problem with this approach is that, when the tests are shorter, they run the risk of losing reliability, and in case of long tests only, it can most safely be used. Hence, it is more useful, in case of long tests as compared to shorter ones. However, to rectify the defects of shortness, enabling correlation as if each part were full length, Spearman–Brown's formula can be employed.

$$r = \frac{2rhh}{1 + rhh}$$

where *rhh* is the correlation between two halves.

4.7.2.2.2 Kudar–Richardson Estimate of Reliability

The coefficient of internal consistency could also be obtained, with the help of Kudar–Richardson Formula-20. Item difficulty index is one of the techniques for item analysis. Item difficulty is the proportion or percentage of those answering correctly to an item.

With the help of Kuder–Richardson formula number--20, the following formula is used to compute reliability.

$$\text{KR} - 20 = \frac{N}{N-1}\left[1 - \frac{\sum pq}{\sigma^2}\right]$$

Where,

$N \rightarrow$ The number of items on the test,
$\sigma^2 \rightarrow$ The variance of scores on the total test,
$p \rightarrow$ The proportion of examinees, getting each item correct,
$q \rightarrow$ The proportion of examinees, getting each item wrong.

Kuder–Richardson formula--20 is an index of reliability that is relevant to the special case where each test item is scored 0 or 1 (e.g., True or False).

4.7.2.2.3 Cronbach's Alpha (α)

Coefficient alpha may be thought of as the mean of all possible split-half coefficients, corrected by the Spearman–Brown formula. The formula for coefficient alpha is

$$r_\alpha = \frac{N}{N-1}\left[1 - \frac{\sum \sigma_j^2}{\sigma^2}\right]$$

where

$r_\alpha \rightarrow$ Coefficient alpha,
$N \rightarrow$ The number of items,
$\sigma_j^2 \rightarrow$ The variance of one item,
$\sum \sigma_j^2 \rightarrow$ The sum of variances of all items, and
$\sigma^2 \rightarrow$ The variance of the total test scores.

Coefficient alpha can varies between 0.00 and 1.00, as with all reliability estimates. Type of tests with items that are not scored as 0 or 1, Coefficient alpha extends the Kuder–Richardson method.

For example, coefficient alpha could be used with an attitude scale in which examinees indicate on each item whether they strongly agree, Agree, Disagree, Strongly Disagree.

4.8 Validity

Validity refers to the degree to which a test measures and what it claims to measure. Validity refers to the extent an instrument or scale tests or measures what it intends to measure. This means validity is the extent to which differences found with a measuring instrument reflect true differences among those being tested.

For its proper administration and interpretation, it is very necessary for a test to be valid.

Validity coefficients is the correlation coefficients computed between the test and an ideal criterion. Some measure of the trait or group of the traits or outside the test, which the test itself claims to measure referred as an Independent Criteria. Validity of the measurement is the ability of a scale or a measuring instrument to measure, what it is intended to measure.

Examples of Validity are as follows:

1. Students may complain about the validity of a syllabus of an exam, stating that, it did not measure their understanding of the subject, but only their memorizing ability.
2. A researcher who tries to measure the morale of employees based on their absenteeism alone; in this case too, the validity of the research may be questioned, as absenteeism cannot be purely attributed to decrease morale, but also to other conditions such as prolonged health issues, family internal conflicts, and so on.

Classification of Validity

There are six types of validity, i.e.,

1. Content validity
2. Criterion-related validity
   - 2.1 Concurrent validity
   - 2.2 Predictive validity
3. Construct validity
   - 3.1 Convergent validity
   - 3.2 Discriminate validity, and
4. Face validity
5. Internal validity
6. External validity

4.8.1 Content Validity

Content validity indicates the extent to which it provides coverage of the issues under study. Content validity refers to the adequacy in the selection of relevant variables for measurement. Selected scale should have the required number of variables for measurement.

For example, if the Education Department wants to measure whether all the schools in the city have proper required facilities, and for measuring this, it develops a scale to measure the attributes such as the attractiveness of school name hoardings for easy reach, the frequency of Alumni Students meets and the varieties of beverages, snacks, and eatables that are prepared in the schools canteen, and so on.

Here, it is clear that these variables will not serve the purpose of the research because these variables considered for measurement do not possess any content validity. Instead, the scale should be developed to measure the aspects such as the number of classrooms; laboratories, drinking water facility; washroom facility; the number of qualified and experienced teachers on roll, the area and capacity of the playground and so on.

For any research process, it is often difficult to identify and include all the relevant variables that need to be studied.

4.8.2 Criterion-Related Validity

How well a given measure relates to one or more external criterion based on empirical observations is examined by Criterion Validity. The criterion related validity refers to the degree to which a measurement instrument can analyze a variable that is said to have a criterion. If a new measure is developed, one has to ensure that it correlated with other measures of the same construct. The idea that a valid test should relate closely to other measure of the same theoretical concept is criterion-related validity. A valid test of intelligence should correlate highly with other intelligence test. If a test demonstrates effective predicting criterion or indicators of the construct then we can say that the test has Criterion-related validity.

For example, with the help of Vernier caliper, Metric rulers, Scales, Measuring tapes, and Odometers, length of an object can be measured and if a new technique of measure is developed then one has to ensure that this new measure correlated with other measures of length.

Different types of criterion validity is given below.

4.8.2.1 Concurrent Validity

The relationship between the predictor variable and the criterion variable is related with concurrent validity. At the same point in time both the predictor variable and the criterion variable are evaluated. When criterion measures are achieved at the same time as the test scores, the occurrence of concurrent validity is found. It reflects the degree to which the test scores estimate the individual's present status with regard to criterion.

For example, it would be said to have concurrent validity, if a test measures nervousness, if it rightly reflects the current level of nervousness experienced by an individual.

4.8.2.2 Predictive Validity

An extent to which a future level of a criterion variable can be predicted by a current measurement on a scale is the Predictive validity. It occurs when criterion measures are obtained at a time after the test.

For example, rather than focusing on all the areas that need repair, a builder may give preference to only those repairs that may attract new tenants in the future.

4.8.3 Construct Validity

Construct validity is closely related with factor analysis. It refers to the degree to which a measurement instrument represents and logically connects through the underlying theory. Construct validity assesses the underlying aspects relating to behavior. It measures why a person behaved in a certain way rather than how he has behaved.

For example, whether a particularly newly launched product in the market was purchased by a consumer is not the consideration, but why he or she has not purchased the product is taken into account to judge construct validity. This helps to remove any extraneous factors that may lead to incorrect research conclusions.

Several ways to determine whether test-generated data that have construct validity is given below:

1. The test should actually measure, whatever theoretical construct it supposedly tests, and not something else. For example, a test of teaching ability should not actually test extraversion.
2. A test that has construct validity should measure what it intends to measure but not measure theoretically unrelated constructs. For example, a test of singing aptitude should not require too much reading ability.
3. In predicting results related to the theoretical concepts it is measuring, a test should prove useful. For example, a test of singing ability should predict who will benefit from taking singing lessons, should differentiate groups who have chosen singing as a career from those who haven't should relate to other tests of singing ability, and so on.

The types of construct validity are as follows:

1. "Convergent Validity" and
2. "Divergent Validity" or Discriminant Validity.

4.8.3.1 Convergent Validity

An extent of correlation among different measures that are intended to measure the same concept is the convergent validity. It means the extent to which a measure, which is theoretically predicted to correlate with, is correlated with another measure.

4.8.3.2 Discriminant Validity

The lack of or low correlation among the constructs that are supposed to be different denotes the discriminant validity.

Consider a multi-item scale that is being developed to measure the tendency to stay in low-cost hostels. This tendency has personality variables such as high level of self-confidence, low need for status, low need for distinctiveness, and high level of adaptability.

This tendency to stay in low-cost hostels is not related to brand loyalty or high-level aggressiveness. The scale can be said to have construct if it correlates highly with other measures of tendency to stay in low-cost hostels such as reported hostels patronized and social class. This means that convergent validity has a low correlation with the unrelated constructs of brand loyalty and a high level of aggressiveness, i.e., discriminant validity. This explains the extent to which the operationalization is not correlated with other operationalization that theoretically it should not be correlated.

4.8.4 Face Validity

The collective agreement of the experts and researchers on the validity of the measurement scale is referred as face validity. Face validity is considered the weakest from of validity. Here, experts determine whether the scale is measuring what it is expected to measure or not. Face validity refers to what appears to be measured superficially. It depends on the judgment of the researcher. Until the researcher is satisfied that it is an accurate measure of the desired construct, each question is scrutinized and modified. On the subjective opinion of the researcher, the determination of face validity is based.

4.8.5 Internal Validity

Since internal validity is concerned with the logic of the relationships between the independent variable and dependent variable, it is the most fundamental type of validity. Internal validity is an estimate of the degree to which inferences about causal relationship can be drawn, based on the measures employed and research design. A higher degree of internal validity makes possible for properly suited experimental techniques, where the effect of an independent variable upon the dependent one is observed under highly controlled conditions.

4.8.5.1 Threats to Internal Validity

These include Confounding, Bias in selection, History, Maturation, Repeated testing, Instrument change, Regression toward the mean, Mortality, Diffusion, Compensatory rivalry, and Experimenter bias.

4.8.5.1.1 Confounding

The problem of confounding particularly occurs in research where the experimenter cannot control the independent variable.

4.8.5.1.2 Bias in Selection

Selection bias may interact with the independent variable and thus influence the observed outcome and creates bias. For example, Gender, Personality, Mental Capabilities, and Physical Abilities, Motivation Level, and Willingness to Participate. If at the time of selection an uneven number of subjects to be tested have similar subject-related variables, there could be a threat to the internal validity.

4.8.5.1.3 History

Participants' responses, attitudes, and behavior may be influenced during process of experiment, events outside the experiment, or between repeated measures of dependent variables. For example, Natural Disasters, Political Changes, etc. In this condition, it becomes impossible to determine whether change in dependent variable is caused by independent variable or historical event.

4.8.5.1.4 Maturation

Usually, during the course of an experiment or between measurements, subject change happens. For example, young students might grow up as a result of their experience, abilities, or attitudes, in longitudinal studies, which are intended to be measured. Permanent changes such as physical growth and temporary changes like fatigue and illness may alter the way a subject would react to the independent variable. Thus, a researcher may have trouble in ascertaining whether the difference is caused by time or other variables.

4.8.5.1.5 Repeated Testing

Participants may be driven to bias, owing to repeated testing. Participants may remember correct answers or may be conditioned as a result of incessant administration of the test. Moreover, it also causes possibility of threat to internal validity.

4.8.5.1.6 Change of Instrument

If any instrument is replaced/changed during the process of experiment, then an alternative explanation is easily available, but it may affect the internal validity.

4.8.5.1.7 Regression toward the Mean

During the experiment, if subjects are selected on the basis of extreme scores, then there are chances of occurrence of such an error. For example, when subjects with minimum mathematical abilities are chosen, at the end of the study. If there is any improvement, chances are that it would be due to regression toward the mean and not due to effectiveness of the course.

4.8.5.1.8 Mortality

It should be kept in mind that there may be some participants who may have dropped out of the study before its completion. If dropping out of participants leads to relevant bias between groups, which account for the observed differences then alternative explanation is possible.

4.8.5.1.9 Diffusion

It might be observed that if treatment affects spread from treatment groups to control groups, then there will be a lack of difference between experimental and control groups. However, this does not mean that independent variable will have no effect or that there would not be a no-relationship between dependent and independent variable.

4.8.5.1.10 Compensatory Rivalry or Resentful Demoralization

If the control groups alter as a result of the study, then there will be a change in the behavior of the subject. For example, control group participants may work extra hard to see that expected superiority of the experimental group is not demonstrated. However, this does not imply that the independent variable created no effect or that there would be no relationship between dependent and independent variable. Vice versa, due to a demoralized control group, changes in the dependent variable may only be effected, working less hard, or demotivated.

4.8.5.1.11 Experimenter Bias

Experimenter bias happens that in turn affects the results of the experiment, while experimenters without any intention or reluctance behave differently to the participants of control and experimental groups. Experimental bias can happen by keeping the experimenter from knowing the condition in the experiment or its purpose and by standardizing the procedure as much as possible.

4.8.6 External Validity

External validity concerns whether results of the research can be generalized to another situation, different subjects, settings, and times, and so on. In fact, experiments using human participants often employ small samples collected from a particular geographic location or with idiosyncratic features, e.g., volunteers, lacks external validity. Because of this, the conclusions drawn about cause-and-effect relationships are actually applicable to the people in other geographic locations or in the absence of these features, which cannot be made sure.

4.9 Practicality

From a practical view point, a measure should be economical, convenient, and interpretable; the measure can be done by highly specialized persons and it should not be lengthy and difficult.

4.10 Sensitivity

Sensitivity is an instrument's ability to accurately measure variability in stimuli or responses. Sensitivity is not high for instruments involving "Agree" or "Disagree" types of response. The instrument is appropriately altered when there is a need to be more sensitive to subtle changes.

For example, the categories whose inclusion increases the scale's sensitivity are strongly agree, mildly agree, agree, mildly disagree, strongly disagree, and none of the above.

4.11 Generalizability

Generalizability is referred to as the amount of flexibility in interpreting the data in different research designs. Ability to collect data from a wide variety of respondents and with a reasonable flexibility to interpret such data, the generalizability of a multiple item scale can be analyzed.

4.12 Relevance

Relevance is refers to as the appropriateness of using a particular scale for measuring a variable. It can be represented as Relevance = Relability × Validity.

If correlation coefficient is used, then the scale can have relevance from 0 to 1, where 0 is the low or no relevance level to 1, which is the high relevance level for analyzing both reliability and validity. If either reliability or validity is low, the scale will have little relevance.

4.13 Errors in Measurement

The types of errors in any measurement are Measurement invalidity and Unreliability. Measurement invalidity refers to the degree to which the measure incorrectly captures the concept. Unreliability refers to inconsistency, in what the measure produces under repeated uses. A measure is said to be unreliable if on an average it gives some score for a case on variable, when it is used again and gives another score.

4.13.1 Respondent-Associated Errors

A majority of research studies rely on eliciting information from respondents. If the researchers are able to obtain the cooperation of respondents and elicit truthful responses from them, then the survey can easily achieve its targets. Two respondent-associated errors arise when researchers do not obtain the information as stated above. These respondent errors are nonresponse error and response bias.

4.13.2 Nonresponse Errors

Nonresponse errors arise when the survey does not include one or more pieces of information from a unit that has to be part of the study. The research results will have some bias to the extent that those not responding are different from those who respond. Failure to respond completely or even failure to respond to one or more questions of the surveyor includes in nonresponse errors.

Nonresponse occurs when a person selectively responds to only certain questions of the survey and will not respond to one or more questions of the survey. The reasons for not responding to some questions may be the lack of knowledge or it may be that the respondent doesn't want to answer.

Nonrespondent error may become an important source of bias in the result of the survey.

4.13.3 Response Bias

When the respondents consciously or unconsciously misrepresent the truth, then it amounts to response bias, and gives rise to Response bias. Sometimes, respondents deliberately mislead researchers by giving false answers, so as not to reveal their ignorance or to avoid embarrassment and so on.

4.13.4 Errors Associated with Instrument

Due to defective measuring instrument, errors occur. These error occur due to poor questionnaire design, improper selection of samples, etc. Even a simple issue like lack of adequate space in the questionnaire for registering the answers of the respondent can result in errors. Another type of instrument error occurs as this can result in a lot of confusion for the respondent, if the questionnaire is complex or ambiguous. Due to the misinterpretation of such questions by the respondent, they will inadvertently lead to errors, if the questions in the questionnaire use complicated words and sentences.

4.13.5 Situational Errors

Situational error occur when the interviewer and the respondent are not in good rapport. Plenty of errors arise due to situational factors. In the interview process, the respondent may not provide proper responses or lead to inappropriate responses, if without any invitation, a third person is present during the interview or sometimes the third person might himself or herself participate. Other factors such as the location of the interview also play a crucial part.

For example, the respondents may not respond as properly as they would, if they were interviewed in their homes or if the researcher is conducting intercept interviews in public places. When the researcher does not assure the respondent that the data provided will be kept confidential, in such a situation the respondent may not part with certain information that may be crucial for the research.

4.13.6 Measurer as Error Source

Measurer as error source happens due to Wrong coding, Faulty tabulations and statistical calculations, Interviewer behavior or attitude, the common mistakes committed by interviewers, after the collection of the data the interviewer might reword or rephrase the responses, failing to record the full response of the respondent, inappropriate tabulations, and inappropriate application of irrelevant statistical tools for measurement.

During the process of the interview the interviewer might encourage or discourage the respondent through body language and gestures-smiles to encourage certain responses, while giving responses to certain questions, frowning to discourage certain responses and so on.

Summary

Various concerns arise in the process of measurement. Important among them are: the scale of measurement, i.e., validity, reliability and practicality, number of indicators for measurement, etc.

Measurement scales are normally categorized into four types, namely, nominal scale, ordinal scale, interval scale, and ratio scale. In nominal scale measurement, numbers are assigned to particular attributes. The ordinal scale is used when the attributes can be ranked. The interval scale is used to measure certain attributes such as temperature, where the interval between the variables is the same, although the base is not zero. The ratio scale is used for comparison between two variables measured on a scale.

For analyzing the goodness of measurement there are five major criteria: reliability, validity, sensitivity, generalizability, and relevance. The attribute of consistency of measurement is referred to as Reliability. There are various types of reliability. To gauge the consistency of test scores, the Pearson product-moment correlation coefficient can be used. This form of reliability is referred to as test–retest reliability. The validity of a test is the degree to which it measures what it claims to measure is. A test is valid to the extent that inferences made from it are appropriate, meaningful, and useful.

Content validity is determined by the test designed to sample, the degree to which the question, task, or items on a test are representative of the universe of behavior. A test has face validity if it looks valid to test users, examiners, and especially the examinees. When a test is effective in predicting performance on an appropriate outcome measure.

Criterion-related validity is demonstrated. External validity concerns whether the results of the research can be generalized to another situation, different subjects, settings, times, and so forth. An instrument's ability to accurately measure variability in stimuli of responses refers to as Sensitivity. The amount of flexibility in interpreting the data in different research designs refers to Generalizability. The appropriateness of using a particular scale for measuring a variable refers to Relevance.

By keeping the experiment from knowing the conditions in the experiment or its purpose and by standardizing procedure as much as possible, experimenter bias can be reduced.

Review Questions

1. Define reliability. Discuss any two methods of estimating reliability of test scores.
2. What is meant by internal consistency reliability? Discuss any two methods of assessing internal consistency reliability.
3. What are some problems associated with reliability assessed via the test–retest.
4. State the strengths and drawbacks of parallel forms reliability.
5. Write short notes on:
    i. K-R formula-20
    ii. Spearman Brown formula
    iii. Cronbach's Alpha (α)
6. Define validity.
7. Explain construct validity. How does it differ from content validity?
8. What is internal validity? Discuss various threats of internal validity.
9. What is external validity?

10. What do you mean by measurement?
11. Give examples of interval scale and ratio scale.
12. How will you examine the validity of a measurement?
13. Identify three measure concerns that need to be addressed in the process of measurement.
14. State the various measurement scales of data.
15. Write short notes on:
    a. Convergent and divergent validity,
    b. Concurrent and predictive validity.

5

Measurement of Attitude

5.1 Introduction to Measurement of Attitude

Attitudes project a positive or negative behavior consistently toward various objects of the world. Attributes are the characteristics of the object under consideration. Attitudes are generally develop as a combination of several interrelated beliefs and they formed on a permanent basis. Toward different aspects of the world people in society have different attitudes. In a person's good or bad behavior attitudes play a major role, based on the standards set by society. Judgments made by a user with regard to the object possessing certain attributes or not is referred to as belief. The predisposition or mental state of individuals or users toward a product or idea or attribute of an object refers to the term attitude. It also implies the mental readiness to act in a particular manner and influences the individual's behavior toward the object or group or organization or person under consideration.

To understand and measure the attitudes of its customers toward its products and services is very important for the organization. If the customers have an unfavorable attitude or a poor image about the company, then the company will not be able to sustain for a long time. It is not only essential for the company to ensure that consumers have a favorable attitude toward its products and services, but also to anticipate their future preferences.

Measuring of attitude is a very difficult task because we cannot measure product or customers but we can measure their opinion.

5.2 Components of Attitude

Attitude is the degree of positive or negative affect associated with some psychological object. To evaluate some object or symbol or aspect of his world in a favorable or unfavorable manner, is a predisposition of an individual. Following are the components of attitude:

1. **A Cognitive Component**: A person having belief or information, about the object
2. **An Affective Component**: A person feeling like "good" or "bad" "like" or "dislike," "strong" or "weak," about the object.
3. **A Behavioral Component**: Readiness of a person to respond behaviorally to the object. If a person says that he loves Cadbury Chocolates because they are delicious, sweeter, and tastier and will always eat them, the statement comprises all these three components of an attitude. To improve marketing communications to attract customers and to develop a competitive advantage, firms usually study components of attitudes in consumers.

5.2.1 Cognitive Component

Cognitive component involves the knowledge and perceptions acquired by a combination of direct experience with the attitude object and related information from various sources based on cognition of an individual. A belief that a particular type of behavior leads to a particular outcome commonly leads to knowledge in a person. The cognitive component of attitude held by a person with regard to an object or

DOI: 10.1201/9781315167138-5

an issue consists of beliefs, opinions, knowledge, and information as knowledge comprises awareness about the existence of the object and apart from the relative importance the person gives to each characteristics belief about its different characteristics and features of the product. To attract the customer and overtime try to nurture positive beliefs about their products and services in the minds of customers. Marketers use various marketing mix variables.

5.2.2 Affective Component

An affective component comprises a person's emotions or feelings toward an object. Such feelings of an individual's researchers treat as their favorable or unfavorable assessment of an object. Such feelings may transform themselves into emotionally charged states, such as anger, happiness, shame, distress, guilt, and so on, and are called as the affective components of attitude. These types of experiences will influence one's perception of an object and that person's later behavior.

For example, a housewife might say that she loves shopping in Big Bazaar and that Shoppers' Stop does not have as wide a range of apparel as Big Bazaar does. The housewife's overall emotional feelings form an affective component.

5.2.3 Behavioral Component

A person's future actions and intentions comprise the behavioral component. It is concerned with the likelihood or tendency that an individual with regard to an attitude object will behave in a particular fashion.

For example, if Jayant wants to fly Indian Airlines in the future or the housewife wants to shop from Big Bazaar in her next shopping excursion, these are the behavioral components of attitude. These intentions, however, have limited timeframes.

5.3 Relationship between Attitude and Behavior

Since it is assumed that there is a relationship between attitude and behavior, the study and measurement of attitude is important. The research, however, indicates that such a relationship holds more at aggregate level than at the individual level. One of the factors influencing behavior may be an attitude. Besides attitude, there could be other factors that may be more powerful in influencing behavior.

For example, due to economic considerations, an individual having a favorable attitude toward a product may not buy it. The attitude–behavior relationship relates to measuring of cognitive and affective components and being able to predict future behavior for the purpose of marketing decision.

To analyze the relationship between attitude and behavior is difficult. It is relatively easier to analyzing the future behavior of a group of people than analysis for a single individual. Following are the certain critical aspects governing the attitude and behavior of consumers discovered by the researcher.

1. If the person develops a positive attitude toward a product, then the product or service usage will be maximum; the converse is also true.
2. Attitudes of consumers toward products that they have never tried will be neutral.
3. Based on actual trial and experience of a product, when attitudes are developed, then these attitudes predict behavior effectively. On the other hand, consistency in attitude and behavior is considerably reduced, when attitude is based on advertising.

5.4 Changing Attitude

The most important activity of businesses across the world today is changing customer attitude and changing them positively toward a company and its products. It becomes imperative for marketers to identify ways and means to overcome the downturn, whenever sales of a product fall or market share declines. In company's development efforts, changing attitudes of stakeholders becomes the top priority. The ways by which companies can attempt to change the attitudes of customers toward a product are about a product altering existing beliefs, changing attitudes by changing the importance of beliefs, and adding new beliefs.

5.4.1 Altering Existing Beliefs about a Product

To convert the neutral or negative belief that a customer holds about the product into a positive belief is the fundamental responsibility of a marketer. For this, the marketer may attempt the following:

1. Several tactics can be used to change consumer perceptions about the product or service. However, marketers need to understand that customer beliefs cannot be changed by advertising alone. If there is no tangible quality in the product to support advertising claims, then any change achieved cannot be sustained.
2. Marketers trying to change consumer beliefs should ensure that the change is incremental rather than drastic.

For example, an aggressive advertising campaign may meet with customer resistance, aimed at changing traditional beliefs of a community. Therefore, the change process should be slow and preferably take the customer through all the stages in the learning process.

5.4.2 Changing Attitudes by Changing the Importance of Beliefs

Another strategy is to change customer attitude by changing the importance of beliefs that a customer holds about a particular product feature.

For example, when a foreign company product entered India, it faced a lot of problems initially to sell its products. This resulted in improving its sales significantly.

5.4.3 Adding New Beliefs

To develop new beliefs in customers about products, an altogether different strategy is adopted by marketers for changing customer attitudes. For marketers adding new beliefs is an important job. Once such new belief is clearly communicated to customers, since customers who previously did not bother to buy a product may now choose to buy it; there is a likelihood of higher sales.

For example, traditionally salt has been promoted on the taste attribute. Tata promoted salt with iodine content as essential for health, thus completely changing common beliefs about salt, as Iodine helps the growth, development, and functioning of the thyroid gland.

5.5 Association between Measurement of Beliefs and Situation

Owing to a number of reasons, the match between what the researcher finds and what actually happens is low. A respondent should first have a felt need for the product for projecting a favorable attitude toward a product.

For example, a respondent should need a laptop, only when might he or she displays a positive or negative attitude toward a particular brand of laptop. A person may have a favorable attitude toward a laptop, but this is not sufficient and has to be backed with the ability to purchase the product. Often, while measuring attitudes, certain parameters of the purchase process are neglected.

For example, a person might have decided to buy a product of a particular brand, but on the day of purchase, by a competitor's better promotional campaign, the person may be lured away Or the person may decide to buy a brand that is cheaper and use the money saved for some other purpose. Sometimes, while making a purchase, the person's attitude may change or be influenced by other family members. These are some issues that will reduce the intensity of association between measurement of beliefs and the actual situation.

5.6 Attitude Scales

To gain an advantage in the market, marketers try to understand attitudes and influence them. Unlike measurement scales in the physical sciences such as measuring height, weight, etc., measurement scales for attitude are less precise, measuring attitude is a highly difficult process.

5.6.1 Definition of Attitude Scale

A set of items, questions, or statements, which probe a single aspect of human behavior, attitudes, or feelings, is known as an attitude scale.

5.6.2 Definition of Scaling

The process of measuring quantitative aspects of subjective or abstract is known as Scaling. It is a method to assign numbers or symbols to some attributes of an object. Scaling involves developing a continuum, based on which measured objects are located.

For example, we might develop a 6-point scale, if we want to measure the satisfaction levels of customers using a product, where respondents can choose the 2, if they are least satisfied; 4, if they are moderately satisfied; and 6, if they are highly satisfied. Scales are unidimensional or multidimensional. The former, as the name suggests, is used to measure one particular attribute of an object.

For example, to measure several attributes of an object, a multidimensional scale is used. Consider a scale locating customers of a Big Bazaar according to the characteristic "agreement to the satisfactory quality of products provided by the Big Bazaar." Each customer interviewed may respond with a semantic as "strongly agree," or "somewhat agree," or "somewhat disagree," or "Strongly disagree." We may even assign each of the responses a number. We may assign strongly agree as "1," agree as "2" disagree as "3," and strongly disagree as "4." Therefore, each of the respondents may assign 1, 2, 3, or 4.

5.7 Types of Attitude Scales

The types of scaling techniques used in research are Comparative scales and Noncomparative scales.

5.7.1 Comparative Scales

The respondent is asked to compare one object with another, in comparative scaling.

For example, the researcher can ask the respondents whether they prefer brand A or brand B of a Bath Soap.

Comparative scales involve the comparison of objects directly with one another. In comparative scales, small differences between stimulus objects can be detected.

Various comparative scaling techniques are Paired Comparison Scale, Rank Order Scale, Constant Sum Scale, and Q-sort Scale.

5.7.1.1 Paired Comparison Scale

The paired comparison is most commonly used scaling technique. This method is simply a binary choice as it is used when the study requires to distinguish between the two objects. In this method, respondents choose the stimulus or items in each pair that have the greater magnitude on the choice dimension they are instructed to use. The data obtained in this method is ordinal in nature.

For example, in a study of consumer preferences for the two brands of milk product, i.e., Amul and Mother Dairy, a consumer is asked to indicate which one of the two brands he or she would prefer for personal use:

1. On the basis of taste, which milk product of the following do you prefer? Please tick mark (√)

Amul	Mother Dairy

2. On the basis of packaging, which milk product will you prefer? Please tick mark (√).

Amul	Mother Dairy

3. On the basis of price, which product will you prefer? Please tick mark (√).

Amul	Mother Dairy

When the researcher wants to compare two or more than two objects this technique is useful. If there are more than two objects (e.g., n objects) to compare, the total comparison will be;

$$\text{Number of comparison} = \frac{n(n-1)}{2}, \; n = \text{number of objects.}$$

This is a comparative scaling technique in which, according to some criterion, a respondent is presented with two objects at a time and asked to select one object (rate between two objects at a time). The data obtained are ordinal in nature.

Disadvantages of paired comparison scale

1. The order in which the objects are presented may bias the results.
2. The number of items or brands for comparison should not be too many. Geometrically, as the number of items increases, the number of comparisons increases. If the number of comparisons is too large, the respondents may become fatigued and no longer be able to carefully discriminate among them.
3. This scale has little resemblance to the market situation, which involves selection from multiple alternatives.
4. Respondents may prefer one item over certain others, but they may not like it in an absolute sense.

5.7.1.2 Rank Order Scale

In this method, respondents are provided various objects and asked to rank the objects in the list. Rank order method is less time-consuming. In this method, if there are n objects, only $(n-1)$ decisions need to be made. Respondent can easily understand the instructions for ranking.

For example, rank the various brands of Refrigerator in order of preference. The most preferred can be ranked 1, the next as 2, and so on. The least preferred will have the last rank. No two brands should receive the same rank number (Table 5.1).

TABLE 5.1

Brands of Refrigerator

Sr. No.	1	2	3	4	5
Brand	LG	Godrej	Whirlpool	Samsung	Panasonic
Rank					

Like paired comparison, the rank order scale is also comparative in nature. The resultant data in rank order are ordinal data. This method is more realistic in obtaining the responses and yields better results when direct comparison is required between the given objects. Only ordinal data generated are the disadvantage of Rank Order Scale.

5.7.1.3 Constant Sum Scale

In constant sum scaling, on the basis of predefined criterion, respondents are asked to allocate a constant sum of units to a specific set of objects. If an object is not important, the respondent can allocate zero point and if an object is most important the respondent may allocate maximum points out of the fixed points. The total fixed points are 100. The total may be taken as some other value depending on the study.

For example, allocate preferences with regard to a song based on various predefined criteria. Respondents were asked to rate each criteria in such a way that the total becomes 100 (Table 5.2).

The data obtained in this method are considered an ordinal scale.

Advantage of Constant Sum Scale is it is time saving, because it provides fine discrimination among objects without consuming too much time.

Disadvantages of Constant Sum Scale are the respondents may allocate more or fewer points than those specified and allocation over a large number of objects may create confusion for the respondent. This method cannot be used as a response strategy with illiterate people and children.

5.7.1.4 Q-Sort Scale

This is a comparative scale that uses a rank order procedure with respect to some criterion to sort objects based on similarity. It is more important to make comparisons among different responses of a respondent than the responses between different respondents is the important characteristic of this methodology. Therefore, it is a comparative method of scaling rather than an absolute rating scale. In this method, for describing the characteristics of a product or a large number of brands of a product, the respondent is given statements in a large number.

For example, you may wish to determine the preference from among a large number of Bath Soaps. The following format summarized in Table 5.3 may be given to a respondent to obtain the preferences. The bag given to you contains pictures of 50 Bath Soaps. Please choose 10 Bath Soaps you "prefer most," 11 Bath Soaps you "like," 10 Bath Soaps to which you are "neutral (neither like nor dislike)," 10 Bath Soaps you "dislike," and 9 Bath Soaps you "prefer least." Please list the sorted magazine names in the respective columns of the form provided to you.

Note that the number of responses to be sorted should not be <60 or not more than 140. A reasonable range is 60–90 responses that result in a normal or quasi-normal distribution. As compared to paired comparison, this method is faster and less tedious. It also forces the subject to conform to quotas at each point of scale to yield a quasi-normal distribution.

TABLE 5.2

Preference Allocation Regarding a Song

Criteria	Respondent Preference
Build energy	15
Chord progressions proceed for fragile to strong	10
Show a steady harmonic rhythm	15
Strong relationship between melodic shape, lyrics and chords	20
Tonic note	10
Swara, Taal, selection of raga, composition, and overall impression	20

TABLE 5.3

Preference of Bath Soap Using Q-Sort Scale Procedure

Prefer Most	Like	Neutral	Dislike	Prefer Least
----------	----------	----------	----------	----------
----------	----------	----------	----------	----------
----------	----------	----------	----------	----------
----------	----------	----------	----------	----------
----------	----------	----------	----------	----------
----------	----------	----------	----------	----------
----------	----------	----------	----------	----------
----------	----------	----------	----------	----------
----------	----------	----------	----------	----------
----------	----------	----------	----------	
	----------			
(10)	(11)	(10)	(10)	(9)

5.7.2 Noncomparative Scale

Respondents need only evaluate a single object in noncomparative scaling. Their evaluation is independent of the other object that the researcher is studying. Respondents employ whatever rating standard seems appropriate to them, using a noncomparative scale. Noncomparative techniques consist of continuous and itemized rating scales. Noncomparative scaling techniques involve scaling of objects independent to some specific standard. Respondent evaluates only one object at a time.

For example, in a study of consumer preferences for different Mobile Service Provider, a consumer may be asked to rate a list of factors that he or she would consider while choosing a particular Mobile Service Provider. Rate 1 to least important and 5 to most important (rating could be on any scale; in this example, we have used a 5-point scale). In this scaling technique, data are usually in an interval scale. It can be continuous, metric, or numeric also. The commonly used noncomparative scaling techniques are Continuous Rating Scale, Itemized Rating Scale, Category Scale, and Cumulative Scale. The noncomparative scaling techniques can be further divided into Continuous Rating Scale and Itemized Rating Scale.

5.7.2.1 Continuous Rating Scale

Continuous Rating Scale is also known as graphic rating scale. In continuous rating scale, by marking at an appropriate position on a line respondents indicate their rating. From one extreme criterion to the other, the line is labeled at both ends. The line may contain points 0, 10, 20, 100.

For example, how would you rate a Novel with regard to its quality? (Table 5.4)

If the respondents are literate to understand and accurately differentiate the objects, then very large number of ratings are possible. The data generated from continuous rating scale can be treated as numeric and interval data. Based on the categories under which the ratings fall, the researcher can divide the line into as many categories as desired and assign scores. It is very simple and highly useful. In continuous rating scale, the respondents rate the objects by placing a mark at the appropriate position on a continuous line that runs from one extreme of the criterion variable to the other.

TABLE 5.4

Rating a Novel with Regard to Its Quality

Quality Indicators	Scale Measurement
Content Coverage	Most ------------------------------X-----------------------------------Least
	Above 80 60 40 20 0
Language	Most -------------X--Least
Presentation Style	Most ------------------------------X-----------------------------------Least

5.7.2.2 Itemized Rating Scale

In this scale, respondents are provided with a scale having numbers or descriptions associated with each category. The respondents are asked to select the best fitting category with the object. The commonly used itemized rating are Likert-Type Scale (Summated Scale), Semantic Differential Scale, and Stapel Scale.

Scale having numbers or brief descriptions associated with each category is an Itemized rating scale. In terms of scale position, the categories are ordered and the respondents are required to select one of the limited number of categories that best describes the product, brand, company, or product attribute being rated. Itemized rating scales are widely used in marketing research. The form of an itemized rating scales are Graphic, Verbal, or Numeric (Figure 5.1).

Some rating scales may have only two response categories such as agree and disagree. Inclusion of more response categories provides the respondent more flexibility in the rating task.

As an example consider the following questions:

1. How often do you visit the Big-Bazaar located in your area of residence?
   - Never, • Rarely, • Sometimes, • Often, • Very often
2. In your case how important is the price of brand X Bath Soap when you buy them?
   - Very important, • Fairly important, • Neutral, • Not so important

Since they provide more information, each of the above category scales is a more sensitive measure than a scale with only two responses.

5.7.2.3 Stapel Scale

Stapel scales are named after John Stapel who developed these scales. Stapel scale consists of a single criterion in the center with ten categories numbered from −5 to +5 without a neutral point (zero). Usually the scale is vertically presented. The respondent inaccurately describes the object indicates Negative rating and the respondent describes the object accurately indicates Positive rating. In Stapel scale, data generated is interval data. In this method, data can be collected through telephonic interview.

Itemized Graphic Scale	**Itemized Verbal Scale**	**Itemized Numeric Scale**
 Favorable	Completely Satisfied	−5
		−?
	Somewhat Satisfied	−?
		−?
Indifferent 	Neither Satisfied nor Dissatisfied	−?
		?
	Somewhat Dissatisfied	??
		??
Unfavorable 	Completely Dissatisfied	??
		??
		?5

FIGURE 5.1 Itemized rating scales.

+?	+ ?
+?	+ ?
+?	+ ?
+?	+ ?
+?	+ ?
Friendly Personnel	Competitive Loan Rates
??	? ?
??	? ?
??	? ?
??	? ?
??	? ?

FIGURE 5.2 Format of stapel scale.

The data obtained from Stapel scale can be analyzed in the same way as semantic differential scale. This scale was originally developed to measure the direction and intensity of an attitude simultaneously. When it is difficult to create pairs of bipolar adjectives, modern versions of the Stapel scale place a single adjective as a substitute for the Semantic differential.

The modified Stapel scale places a single adjective in the center of an even number of numerical values (say, +3, +2, +1, 0, −1, −2, −3). How close to or how distant from the adjective a given stimulus is perceived to be is measured by this scale.

For example, select a plus number for words that you think describe Sales & Service of Automobile promptly. The more prompt and perfect you think the word describes the Automobile, the larger the plus number you should choose. Select a minus number for words you think do not describe the Automobile promptly. The less accurately you think the word describes the Automobile, the larger the minus number you should choose (Figure 5.2).

5.7.2.4 Category Scale

Objects are grouped into a predetermined number of categories on the basis of their perceived strength along certain dimension, in Category scaling method. Category Scale is a dichotomous scale. Category Scale is useful for sociodemographic questions. Typically we get "Yes" or "No" type of response in this category. Data in this category are either nominal or ordinal. Under Category Scale, instructions and response tasks are quick and simple, and many options can be included.

For example, in Single Category Scale a single item measures the opinions of the respondents through just one rating scale. Do you own a Flat?

Yes	No

In Multiple Category, please indicate your Annual income by marking (√) in the income group you fall (Table 5.5)

TABLE 5.5

Annual Income Group

15,000–20,000	20,000–25,000	25,000–30,000	30,000–35,000	35,000–40,000

5.7.2.5 Cumulative Scale or Guttman Scale

Cumulative Scale consists of a series of statements to which a respondent expresses his/her agreement or disagreement. The statements are in a form of cumulative series, i.e., in a way, an individual who replies favorably to say item no. 3 also replies favorably to item no. 2 and 1 and so on. By counting the number of points concerning the number of statements he/she answers favorably, the individual's score is worked out. We can estimate as to how a respondent has answered individual statements constituting cumulative scales, knowing the total score. For example,

1. Permitting immigrants to live in your country.
2. Permitting immigrants to live in your community.
3. Permitting immigrants to live in your neighborhood.
4. Permitting immigrants to live next door to you.
5. Permitting your son or daughter to marry to an immigrant.

Advantages of Cumulative scale or scalogram analysis is that it assures that only a single dimension of attitude is measured. Since the scale is determined by the replies of respondents, researcher's subjective judgment is not allowed to creep in the development of scale.

Disadvantages of Cumulative scale or scalogram analysis are as follows:

1. Perfect cumulative or one-dimensional scale is very rarely found.
2. Approximation is used in practice and in comparison to other scaling method, its developmental procedure is cumbersome. The construction of the scale requires a lot of time and effort.
3. There may be very few items existing that may fit the model.
4. They can make only rather gross distinctions among respondents since such scales seldom have more than eight items.

5.7.3 Multi-Item Scales

In the multiple item scale, pertaining to attitude toward an object, a number of statements are used. Each statement contains a rating scale attached to it. The disadvantage of single item rating scale is that it is a crude measure of the feelings or opinions of the respondent. With multiple item scales, this is not the case. Widely used multiple item scale are Thurston Equal Appearing Interval Scale, Likert Scale, Semantic Differential Scale (Bi-polar Scale), and Fishbein's Scale.

The object is measured against each characteristic, one at a time, which has been discussed till now in the attitude measurement scales. The measurement process tells little about the relative importance of different characteristics or how the characteristics relate to each other. One takes recourse to multidimensional scaling, when these aspects become important to describe a group of analytical techniques used to study attitudes, especially those relating to perceptions and preferences. To identify the object attributes that are important to the respondents and to measure their relative importance, these techniques are useful. The major application of multidimensional scaling in managerial research comes in marketing research.

Multidimensional scaling applied here in advertising to answer questions, such as Selection of media for getting the desired reach and Choice of magazines, newspapers for advertisement, if written media, is chosen.

5.7.3.1 Thurstone Equal Appearing Interval Scale

This scale of attitude measurement was first described by Thurstone and Chave. We are interested in scaling respondents and not on statements in this scale.

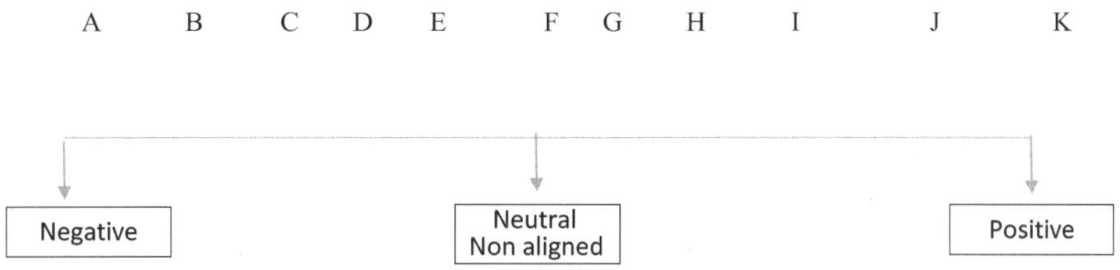

FIGURE 5.3 Thurstone scale.

Steps to construct the scale are as follows:

1. The first step in the scale construction is to scale the attitude statements along the attitude continuum. This is done by asking some "judges" to evaluate the items along some continuum. Pertaining to the subject of enquiry, a large number of statements are collected. This is done by using existing literature on the subject, discussion with knowledgeable persons, personal experience, and through focus-group interviews through exploratory research.
2. From one extreme of favorable attitude to other extreme of unfavorable attitude is the range of statement.
3. The statements should be large in number.
4. Each statement should be written on a separate card and subjects are asked to sort these statements into a number of intervals.
5. The number of intervals could be 11 put on cards labeled A–K where A corresponds to most negative attitude, K corresponds to most positive attitude, and F represents neutral or non-aligned attitude (Figure 5.3).
6. Indicating the respondent has rated each statement on an 11-point rating scale, the piles bearing letter from A to K could be regarded as having numbers from 1 to 11. Therefore, a 11-point rating scale becomes the psychological continuum on which statements have been judged.

   The statements are printed on some cards and the judges are asked to sort the statements into 11 groups. The extreme piles represent the most favorable and the most unfavorable statements. The judges are expected to make the equal intervals between the groups. The mean rating is taken as the scale point for each item by judges. Items are dropped, which are found to be ambiguous or irrelevant.
7. In respect of each statement, an average value is calculated. This value is treated as the scale value or median value for the example to follow of the statement.

The items selected for the final scale are such that

1. Each item has a small standard deviation of ratings over judges and
2. The mean ratings spread evenly from one end of the rating continuum to the other. To form the final scale, the selected items are listed in a random order.

The latter to mark only the items with which he or she agrees needs the administration of the scale for the measurement of the attitude of a respondent. Then, the score for the respondent is taken as the scale value of the median item endorsed or the average scale value of the items endorsed.

For example, suppose a respondent agrees with items that have scale values as 9, 10, and 11. Assuming that score of 11 implies the most positive attitude, this would imply that he or she has a favorable attitude to the object. The Thurstone scales are prepared with an odd number of positions, the usual number being 11.

Disadvantages of Thurstone Equal Appearing Interval Scale:

1. The time requirement being fairly high, the influencing of scale positions by the attitudes of the judges, and
2. No information on the degree or intensity of agreement with the different items.

3. Thurstone scale uses a two-stage procedure and therefore it is both time-consuming and expensive to construct.
4. As there is no explicit response to each item, it does not have much diagnostic value.
5. The scale has also been criticized on account of a method of scoring. Respondents are merely asked to select those statements with which they agree. Therefore, there is a possibility of two or more respondents having the same attitude score. For example, if respondent A agrees with statements having a score value of 4, 6, 9 and another respondent B agrees with statements having scale values of 7, 7, and 5, both will have the same attitude toward the object, which may in fact not be true.

5.7.3.2 Likert Scale

Likert Scale is a simple and straightforward method for scaling attitudes. This scale is also known as summated rating scale. Summated scales consist of a number of statements. Each statement express either a favorable or unfavorable attitude toward the given object. The respondent is asked in terms of several degrees of agreement or disagreement, to respond to each of the statements. By the questionnaire, depending on the wording of an individual item, an extreme answer of strongly agree or strongly disagree will indicate the most favorable response on the underlying attitude measured.

A Likert scale may include a number of items or statements. Each statement is assumed to represent an aspect of an attitudinal domain.

For example, Table 5.6 summarizes the items in a Likert Scale to measure opinions on food products.

Each respondent is asked to circle his or her opinion on a score against each statement. The sum of their ratings for all the items is the final score for the respondent on the scale. The very purpose of Likert Scale is to ensure the final items evoke a wide response and discriminate among those with positive and negative attitudes.

Because items lack clarity or elicit mixed response patterns are detected from the final statement list, they are poor. This will ensure us to discriminate between high positive scores and high negative scores. However, many business researchers do not follow this procedure and you may not be in a position to distinguish between high positive scores and high negative scores because all scores look alike. Many patterns of response to the various statements can produce the same total score (Table 5.7).

Advantages of Likert Scale:

1. This scale is relatively easy and quick to compute and the data in this scale is of interval scale.
2. This scale is relatively easy to construct as it can be performed without a panel of judges.

TABLE 5.6

Items in a Likert Scale to Measure Opinions on Food Products

	Strongly Agree	Agree	Neither Agree nor Disagree	Disagree	Strongly Disagree
Firms too should reduce the price of the food products, if the price of raw materials fall	1	2	3	4	5
For food products, there should be uniform price throughout the country	1	2	3	4	5
While manufacturing food products, the food companies should concentrate more on keeping hygiene	1	2	3	4	5
Before food products are delivered to consumers in the market, the expiry dates should be printed on the food products	1	2	3	4	5
In keeping acceptable quality and on the prices, there should be government regulations on the firms	1	2	3	4	5
Now-a-days most food companies are concerned only with profit making rather than taking care of quality	1	2	3	4	5

TABLE 5.7

Difference between Likert Scale and Thurstone Scale

Point of Distinction	Thurstone Scale	Likert Scale
Respondents expectation	A respondent is expected to endorse those statements which best reflect his or her feelings toward the attitude object	A respondent is to answer every item or statement and a high total score indicates a high favorable attitude toward the object
Time and labor	More time-consuming and more laborious	Less time-consuming and less laborious
Reliability	The reliability coefficient is less	The reliability coefficient is more

3. This scale is considered more reliable because respondents answer each statement included in the instrument.
4. This scale can easily be used in respondent centered and stimulus centered studies.
5. This scale takes less time to construct.

Disadvantages of Likert Scale:

1. It is difficult to know what a single summated score means and it takes longer time to complete than other itemized rating scales because respondents have to read each statement.
2. We can simply examine with this scale, whether respondents are more or less favorable to a topic, but it is difficult to tell how much more or less they are.
3. This does not rise more than to an ordinal scale structure, since a given total score can be secured by a variety of answer patterns, the total score of an individual respondent has little clear meaning.

5.7.3.3 Semantic Differential Scale (Bipolar Scale)

Semantic Differential Scale includes a seven-point scale in comparison to Likert 5-point scale. Semantic Differential Scale is also called bipolar scale. The scale may be used in cases such as comparison of brands, comparison of company's images, and determine the effectiveness of advertising on attitude change, etc. The scale is similar to Likert scales as it consists of a series of items to be rated by respondents. This scale is based on the proposition that an object can have several implied or suggestive meanings to an expressed opinion. This scale can be scored on either –3 to +3 or 1–7. This scale can also provide comparison between products organizations. By the factor analysis, the results of this scale are further analyzed. Semantic differential scale provides a very convenient and quick way of gathering impressions on one or more than one concept. From this scale, the data generated can be considered as numeric in some cases, and to arrive total scores adjectives can be summed. It must define a single dimension and each pair must be bipolar opposites labeling the extremes.

For example, Table 5.8 summarizes examples of Semantic Differential Scale.

TABLE 5.8

Examples of Semantic Differential Scale

Modern	---- ---- ---- ---- ---- ---- ----	Outmoded
Reliable	---- ---- ---- ---- ---- ---- ----	Disgusting
Fresh	---- ---- ---- ---- ---- ---- ----	Dusty
Necessary	---- ---- ---- ---- ---- ---- ----	Irrelevant
Costly	---- ---- ---- ---- ---- ---- ----	Affordable
Convenient	---- ---- ---- ---- ---- ---- ----	Worthless
Forceful	---- ---- ---- ---- ---- ---- ----	Fragile
Instant	---- ---- ---- ---- ---- ---- ----	Stagnant

Respondent's Positive Responses	In my experience, the use of body lotion of Brands A, B, and C was (Stapel Scale)							Respondent's Negative Responses
	+[illegible]	+[illegible]	+[illegible]	[illegible]	[illegible]	[illegible]	[illegible]	
Helpful	----	----	--A--	----	-- B--	-- C --	----	Unusable
Engaging	----	--A--	--B--	----	----	--C--	----	Unfashionable
Submissive	----	----	--AB--	----	--C--	----	----	Agile
Favorable	----	----	--B--	--A--	--C--	----	----	Adverse
Captivating	----	--B--	--A--	----	--C--	----	----	Dull
Boring	----	--A--	--B--	----	----	--C--	----	Acute
Delicious	----	----	--A--	--B--	--C--	----	----	Irritating
Cool	----	----	--A--	--B--	--C--	----	----	Burning
Useful	----	--A--	--B--	--C--	----	----	----	Disgusting
Nice	----	--A--	----	--B--	--C--	----	----	Dirty
	Brand-A		Brand-B		Brand-C			

FIGURE 5.4 The experiences of 100 consumers on three brands of body lotion.

Only extremes have names in the semantic differential scale. The extreme points representing the neutral position denote the bipolar adjectives with the central category. The in-between categories have blank spaces.

In the semantic differential scale in the form of positive and negative phrases used to describe the object form a basis for attitude formation. Sometimes on the left side of the scale put a negative phrase and sometimes on the right side. This is done to prevent a respondent with a positive attitude, without reading the description of the words, from simply checking the left side and a respondent with a negative attitude checking on the right side. Depending on the attitude the respondents are asked to check the individual cells. Then, for comparisons of different objects, one could arrive at the average scores. Figure 5.4 shows the experiences of 100 consumers on three brands of body lotion.

In the above example, first, for each dimension the individual respondent scores are obtained and then the average scores of all 100 respondents for each dimension and for each brand are plotted graphically. The maximum score possible for each brand is +30 and the minimum score possible for each brand is –30. Brand-A has score +13. Brand-B has score +6, and Brand-C has score –12. We can identify from the scale which phrase needs improvement for each brand.

Advantages of Semantic Differential Scale

1. It is an efficient and easy way to secure attitude from a large sample and the total set of responses at both directions gives a comprehensive picture of the meaning of an object as well as a measure of the subject of the rating.
2. It is a standardized technique that can be easily repeated. However, this technique escapes many of the problems of response distortion found with more direct methods. This is a 7-point rating scale that has semantic meaning with end points associated with bipolar labels such as good and bad, complex and simple.

TABLE 5.9

Difference between Likert Scale and Semantic Differential Scale

Point of Distinction	Semantic Differential Scale	Likert Scale
Composition	Comprises of a series of bipolar adjectives or phrases that pertain to the attitude toward the object	Had complete statements
No. of category	Each pair of opposite adjectives is separated by a seven category scale	At times with five category or nine category scale
Scale Description	Some of the individual scales have favorable descriptors on the right-hand side while the others have it on the left-hand side. The rational for having this sort of an arrangement is similar to that of Likert scale	No numerical labels or verbal labels other than anchor labels are used for representing various categories
Respondents to answer	Respondents are asked to put a cross on one of the seven category that best describes his or her views about the attitude object along the continuum implied by bipolar adjectives	A respondent is to answer every item or statement and a high total score indicates a high favorable attitude toward the object

3. For a variety of purposes, the semantic differential scale is used. It can be used to find whether a respondent has a positive or negative attitude toward an object. In comparing brands, products, and company images, it has been widely used. To develop advertising and promotion strategies and in a new product development study, it has also been used (Table 5.9).

5.8 Profile Analysis

Profile analysis is a process, where two or more objects are rated by respondents on a scale. It can be considered as an application of the semantic differential scale. Based on different attributes, in this approach visually comparing different objects is possible. It is very difficult to interpret the profiles as the number of objects increases; this is a major disadvantage.

5.9 Considerations in Selecting Attitude Measurement Scale

Scaling techniques have some advantages and disadvantages. Virtually to measure the attitudes, any technique can be used. But all techniques are not suitable for all purposes, at the same time. A scaling technique that will yield the highest level of information feasible in a given situation, as a general rule, should be used. Also, the use of a variety of statistical analysis is permitted by the technique. The choice of scaling technique is decided by a number of issues. Some significant issues are:

5.9.1 Problem Definition and Statistical Analysis

The problem definition and the type of statistical analysis likely to be performed determine the choice between ranking, sorting, or rating techniques.

For example, ranking provides only ordinal data that limit the use of statistical techniques.

5.9.2 The Choice between Comparative and Noncomparative Scales

Rather than a noncomparative scale, sometimes it is better to use a comparative scale. For example, how satisfied you are with the brand-X washing powder that you are presently using ? (Table 5.10)

Since it deals with a single concept, this is a noncomparative scale; the brand of a washing powder. On the other hand, a comparative scale asks a respondent to rate a concept.

TABLE 5.10

Noncomparative Scale

Completely satisfied	Somewhat satisfied	Neither satisfied nor dissatisfied	Somewhat dissatisfied	Completely dissatisfied

For example, you may ask: Which one of the following brands of washing powder do you prefer?

Brand-X	Brand-Y

In this example, you are comparing one brand of washing powder with another brand. Therefore, in many situations, comparative scaling presents "the ideal situation" as a reference for comparison with actual situation.

5.9.3 Type of Category Labels

In constructing measurement scales, we have discussed different types of category labels used such as verbal categories and numeric categories. Many researchers use verbal categories because they believe that these categories are well understood by the respondents. This decision influences the maturity and the education level of the respondents.

5.9.4 Number of Categories

Based on the research concept, the number of categories that have to be included in the scale should be decided. If a scale is developed with very few categories say, 2, good or bad, yes or no, right or wrong, pass or fail, etc., then it does not reveal the respondents' true attitudes.

At the same time, if a scale contains ten or more categories, the respondent might get confused and will not be able to accurately assign items to the different categories. Therefore, developing a scale that contains between five and nine categories is always better.

Traditional guidelines suggest that there should be between five and nine categories while there is no single, optimal number of categories. Also, for at least some of the respondents, if a neutral or indifferent scale response is possible, an odd number of categories should be used. However, for a specific problem, the researcher must determine the number of meaningful positions that are best suited.

5.9.5 Odd or Even Number of Scale Categories

If a scale does not have a neutral point, it means that, it has an even number of categories. This restricts and forces the respondents to choose a negative or a positive aspect of a scale. So, respondents who are actually neutral cannot express this feeling. Adding a neutral point in the scale helps respondents. Some respondents feel they can take the easy way out by saying that they are neutral and need not concentrate on their inner and real feelings. Deciding whether to have on the scale, an odd number or an even number of categories, is dependent on the nature of research to be conducted.

For example, if a company has recently changed the product package design and is attempting to study whether the customers have liked it, it cannot expect the respondents to be highly emotional toward the product package design. Therefore, it needs to have an odd number or neutral category.

While, if a company only wishes to find out how strongly the consumers like or dislike a product, then adding a neutral category will not serve the purpose.

5.9.6 Balanced Versus Unbalanced Scale

To obtain objective data in general, the scale should be balanced. A scale, which has the same number of positive and negative categories, is said to be balanced, while an unbalanced scale is weighted toward one or the other end. In situations where a broad range of responses are expected, a balanced scale is used. In situations where the results of preliminary research lean more toward one side of the scale than the other, an unbalanced scale is used.

For example, to measure the performance of a new solar cooker, if preliminary research conducted indicates that it is generally liked by the people, then a scale with categories such as (a) Excellent, (b) Very Good, (c) Good, (d) Fair, and (e) Poor is developed.

5.9.7 Forced Versus Unforced Choice

If respondents are given adequate choice for selecting a response, then it becomes an unforced choice. If respondents are not given any choice for selecting a response, then it becomes a forced choice. An unforced decision, which a respondent chooses can either be in the form of "Neutral," or if he or she is not inclined toward either object or "Don't know," which a respondent can choose if he or she lacks the knowledge to answer the question. It becomes an unforced choice when these two categories are included in the scale, as the respondents do not have to select a positive or negative opinion when they don't have any opinion.

It obviously becomes a forced choice when a neutral or don't know category is not included in the scale. Although essential in some research studies, it should be avoided as a rule, restricting the choice of respondents. Where the respondents are expected to have no opinion, in such situations the accuracy of data may be improved by a nonforced scale that provides a "no opinion" category.

For the measurement of attitudes, a number of different techniques are available. Each has some strengths and some weaknesses. For the measurement of any component of attitudes, almost every technique can be used. But, for all purposes, all techniques are not suitable. The selection of the scale depends on the stage and the size of the research project.

The necessary things for the measurement of attitudes are the costs of developing and implementing the instrument, reliability and validity of the instrument, and the statistical analysis.

For preliminary investigation generally, Thurstone's scale, Q-sort, and Semantic Differential Scale are preferred. For item analysis, the Likert Scale is used. For specific attributes, the Semantic Differential Scale is very used. Overall the Semantic Differential Scale is widely used as it is simple in concept and the results obtained are comparable with more complex, one-dimensional methods.

5.9.8 Limitations of Attitude Measurement Scales

The main limitation of attitude measurement scales is the emphasis on describing attitudes rather than predicting behavior. This is primarily because of a lack of models that describe the role of attitudes in behavior.

Summary

An attitude is a mental state involving beliefs, feelings, values, and dispositions to act in a certain way. Attitude comprises three components, i.e., a cognitive component, an affective component, and a behavioral component. It is very important, especially for marketers, to measure the attitudes of consumers so that they act according to customer interest and develop products that suit them. Scaling can be defined as the process of measuring the quantitative aspects on subjective or abstract concepts. Researchers normally tend to use scales that are easy to administer and develop. A number of issues, such as problem definition and statistical analysis, choice between comparative and noncomparative scales, type of category labels, number of categories, etc., discussed in this chapter, should be considered before you arrive at a particular scaling technique. The measurement scales, commonly used in marketing research, can be divided into two types: comparative and noncomparative scales. This is followed by three important tools or scales of attitude measurement, i.e., Thurstone's Equal-Appearing Interval, Semantic Differential, and Likert's scale technique. There are four types of scales used in marketing research to infer attitude toward a particular product or service: Nominal, Ordinal, Interval, and Ratio. A brief discussion of multidimensional scaling follows. Finally, the issues of the selection of an appropriate attitude measurement scale and the limitations of these research tools are discussed.

Review Questions

1. What do you understand by the terms attitude and attitude measurement? Explain.
2. Which type of managerial research and decisions utilize attitude measurement? Explain with examples.
3. Review briefly the different types of issues in attitude measurement.
4. Compare and contrast the various attitude measurement techniques. When will you use each of them? Discuss briefly.
5. In which type of study will you use multidimensional scaling? Discuss.
6. What are the four different levels of measurement? Discuss the mathematical operations that may or may not be used under each level of measurement.
7. Explain the three criteria of measuring the usefulness of an attitude scale.
8. Discuss briefly different issues you consider for selecting an appropriate scaling technique for measuring attitudes.
9. How do you select an appropriate scaling technique for a research study? Explain the issues involved in it.
10. Differentiate between ranking scales and rating scales. Which one of these scales is better for measuring attitudes?
11. In what type of situation is the Q-sort technique more appropriate?
12. Name any four situations in commerce where you can use the Likert scale.
13. Construct a Rank Order Scale to measure toothpaste preferences. Discuss its advantages and disadvantages.
14. Construct a semantic differential scale to measure the experiences of respondents in using Brand-A of Face Wash; assume that all the respondents use that brand.
15. What is the distinction between comparative measuring scale and noncomparative measuring scale?
16. What do you mean by Likert Scale? Give two examples.

6

Sampling Design

6.1 Introduction to Sampling

Where it is not possible to study the entire population, in such situations researchers use the concept of sampling. For a variety of reasons, researchers usually cannot make direct observations of every unit of the population they are studying. Instead, they collect data from a subset of population called as a sample and use these observations drawn to make inferences about the entire population.

Ideally, the characteristics of a sample should correspond to the characteristics of a population from which the sample was drawn. In that case, the conclusions drawn from a sample are probably applicable to the entire population.

Sampling is the backbone of marketing research. In this chapter, you will be introduced to various sampling concepts. A brief mention of sampling and nonsampling errors will be made. The various probability and nonprobability sampling designs as applicable to marketing research will be introduced. Since the choice of sample size involves various elements such as time, money accuracy, etc., an important decision while taking a sample is to know how large a sample should be taken. Therefore, the determination of sample size would also be discussed.

6.2 Basic Definitions and Concepts

Researchers usually cannot make direct observation of every individual in the population under study. Instead, they collect data from a subset of individuals called as a sample and to make inferences about the entire population using those observations.

6.2.1 Element

The unit about which information is collected is called as an element. According to a well-defined procedure this provides the basis for analysis. Elements should be well defined and the possibility of identifying them physically is important. For example, in a retail stores survey, a shop may be considered as a unit, whereas in a family budget enquiry a household may be treated as a unit.

6.2.2 Population or Universe

Population is the entire aggregation of items from which samples can be drawn. A population is a group of individual persons, objects, items, or any other units from which samples are taken for measurement. Pertaining to a given characteristic, population is a well-defined setup of all elements. It refers to the whole that includes all observations or measurements of a given characteristic. Population is also called universe. It may be defined as any identifiable and well-specified group of individuals.

For example, all primary school teachers, number of all college teachers, and all university students. A population may be finite or infinite. A finite population is one where all the members can be easily counted. An infinite population is one whose size is unlimited, and cannot be counted easily. For example, the population of college teachers, school teachers, family members, etc., are examples of finite population and the number of stars in the sky, the number of fishes in the sea, etc., are examples of infinite population.

DOI: 10.1201/9781315167138-6

6.2.3 Sample and Sampling

A part of the population is a sample. It can be an individual element or a group of elements selected from the population. Although sample is a subset, it is representative of the population. Sample is suitable for research in terms of cost, convenience, and time. Based on a probability or a nonprobability approach the sample group can be selected. A sample usually consists of various units of the population. The size of the sample is represented by "n." A sample is any number of persons selected to represent the population according to some rule of plan. Thus, a sample is a smaller representation of the population. A measure based on a sample is known as a statistic.

6.2.4 Sample Size and Sampling Design or Strategy

Sample size is the number of selected individuals from whom you obtain the required information; it is usually denoted by the letter (n). For example, number of students, number of family members, etc.

6.2.4.1 Sampling Design or Strategy

The sampling design or strategy is the way researcher selects the sample or students or families, etc. It refers to the techniques or procedures the researcher would adopt in selecting some sampling units from which inferences about the population are drawn.

6.2.5 Sampling Units

At some stage of the sampling process, a sampling unit is that element or elements considered available for selection. Sampling units and the elements are same in single-stage sampling. For example, the relevant population for conducting a socioeconomic survey in a region. In this case a sample of households may be selected in three stages. First, a sample of Tehsils is selected. Then, from each selected Tehsil a sample of villages is selected after making a list of all the villages in it. Finally, from each selected village, a sample of households is selected after listing all the households in it. In this example, Tehsils are taken as first stage unit, villages as second stage unit, and households as the third or the final stage unit. Each individual or case that becomes the basis for selecting a sample is called sampling unit or sampling elements.

6.2.6 Sampling Frame

Sampling frame, at a stage of sampling process, is a list of all sampling units belonging to the population to be studied with their proper identification and available for selection. In fact, the actual sample is drawn from the sampling frame. Therefore, the sampling frame contains all the sampling units of the population under consideration. It should exclude units of any other population. The sampling frame should be up to date and free from errors of omission and duplication of sampling units. In fact, a lot of time and effort is spent on preparing a suitable sampling frame, in marketing research studies. For example, a list of registered voters, a map, an organization's employee list, etc.

6.2.7 Study Population

The aggregation of elements from which the sample is actually drawn is called as study population. Population was defined as prior to the selection of sample, the aggregate of elements possessing certain characteristics. The actual sample is selected from somewhat different population from the one defined prior to the selection of sample due to certain unavoidable problems. This is because it is very seldom that every element actually has a chance of being selected, which satisfies our definition of a population. As some elements are likely to be omitted from a list of population, our list may be incomplete because of certain reasons such as some people may have unlisted phone numbers, a map may not include a new street, a list of registered voters may be incomplete.

Therefore, the aggregation of elements from which the sample is actually drawn and it is with reference to this study population that the inferences are drawn is the study population.

6.2.8 Bias

How far the average statistic lies from the parameter, the estimation of which is termed as Bias. This means the error that arises when estimating a quantity. In the long run, errors from chance will cancel each other but those from bias will not. Bias can take different forms.

6.2.9 Precision

A measure of how close an estimate is expected to be, to the true value of a parameter, is termed as Precision. Precision is a measure of similarity. Precision is usually related to the standard error of the estimate and expressed in terms of imprecision. Less precision is reflected by a larger standard error.

6.3 Sampling Designs

Researcher could get most accurate data from surveying the entire population of interest, if money, time, trained manpower, and other resources were not a concern.

The researcher is forced to go for sampling when the resources are scarce. To know the characteristics of the population is the real purpose of the survey. With what level of confidence the researcher will be able to say that the characteristics of a sample represent the entire population. The researcher can collect data that actually represent the characteristics of the entire population from which the sample was taken, by using a combination of tasks of hypotheses and unbiased sampling methods.

It is necessary that the sample is unbiased and sufficiently large to ensure a high level of confidence that the sample represents the population. If we increase the sample size, we shall be much closer to the characteristics of the population that was scientifically proved. Ultimately, if we cover each and every unit of the population, then the characteristics of the sample will be equal to the characteristics of the population. That is why in a census there is no sampling error. Thus, "generally speaking, the larger the sample size, the less sampling errors we have."

The statistical meaning of bias is error. To make it unbiased, the sample must be error free. In practice, even while using unbiased sampling methods, it is impossible to achieve an error-free sample. However, by employing appropriate sampling methods, we can minimize the error. The sampling designs describe the procedure by which sample is selected. There exist two classes of methods by which samples can be selected. The various sampling methods classified into probability sampling methods or random sampling methods and nonprobability sampling methods or nonrandom sampling methods.

6.3.1 Probability Sampling Methods or Random Sampling Methods

The random sampling method is also often called probability sampling. A sampling in which every member of the population has a calculable and nonzero probability of being included in the sample is known as probability sampling. In random sampling all units or items in the population have a chance of being chosen in the sample. In other words, a random sample is a sample in which each element of the population has a known and nonzero chance of being selected. Random sampling always produces the smallest possible sampling error. In the real sense, the size of the sampling error in a random sample is affected only by a random chance. Because a random sample contains the least amount of sampling error, we may say that it is an unbiased sample. It is to be noted that we are not saying that a random sample contains no error, but rather the minimum possible amount of error.

Advantages of Probability Sampling Methods

1. It is possible to quantify the magnitude of the likely error in the inference made and this will help in building confidence in drawing inferences.
2. Every element of the population has a known chance of being selected. It should be noted that the term known chance does not mean equal chance. Equal chance probability sampling is a special case of probability sampling, called simple random sampling.

3. In probability sampling methods, there is no chance of arbitrary or biased selection and therefore the laws of probability holds true. Therefore, it permits us to measure the sampling error, which is the difference between the population value and the sample value.

Under probability sampling methods there are a number of different sampling procedures. Some of these methods are Simple Random Sampling, Systematic Sampling, Stratified Sampling, Cluster Sampling, Area Sampling, and Multistage Sampling.

Probability sampling method helps in estimating sampling errors and evaluating sample results in terms of their precision, accuracy, efficiency, etc.

6.3.1.1 Simple Random Sampling

Simple random sampling is a sampling process in which each element in the target population has an equal chance or probability of inclusion in the sample. Under this sampling design each member of the population has known and equal probability of being included in the sample.

A simple random sample is one in which each item in the total population has an equal chance of being included in the sample. In addition, the selection of one item for inclusion in the sample should in no way influence the selection of another item. Simple random sampling should be used with a homogeneous population. Homogeneous population comprises items that possess the same attributes that the researcher is interested in. The characteristics of homogeneity may include such as age, sex, income, social or religious or political affiliation, geographical region, etc. The best way to choose a simple random sample is to use a random number table. A random sampling method should meet the criteria such as every member of the population must have an equal chance of inclusion in the sample and the selection of one member is not affected by the selection of previous members.

Random Numbers

The random numbers are a collection of digits generated through a probabilistic mechanism.

The following are the properties of random numbers:

1. The probability that each digit 0, 1, 2, 3, 4, 5, 6, 7, 8 or 9 will appear at any place is the same. That is 1/10.
2. The occurrence of any two digits in any two places is independent of each other. A unique number is assigned to each member of a population. The members of the population chosen for the sample will be those whose numbers are identical to the ones extracted from the random number table in succession until the desired sample size is reached.

Example of a random number is provided in Table 6.1.

TABLE 6.1

Table of Random Numbers

	1	2	3	4	5	6	7	8	9	10
1	96268	11860	83699	38631	90045	69696	48572	05917	51905	10052
2	03550	59144	59468	37984	77892	89766	86489	46619	50236	91136
3	22188	81205	99699	84260	19693	36701	43233	62719	53117	71153
4	63759	61429	14043	44095	84746	22018	19014	76781	61086	90216
5	55006	17765	15013	77707	54317	48862	53823	52905	70754	68212
6	81972	45644	12600	01951	72166	52682	37598	11955	73018	23528
7	06344	50136	33122	31794	86723	58037	36065	32190	31367	96007
8	92363	99784	94169	03652	80824	33407	40837	97749	18361	72666
9	96083	16943	89916	55159	62184	86206	09764	20244	88388	98675
10	92993	10747	08985	44999	35785	65036	05933	77378	92339	96151

Following are the steps to select a random sample using a simple random sampling method.

1. Determine the population size (N).
2. Determine the sample size (n).
3. Number each member in serial order of the population under investigation. Suppose there are 100 members number them from 00 to 99.
4. By randomly picking up a page from random number tables determine the starting point of selecting sample and dropping your finger on the page blindly.
5. Choose the direction in which you want to read the numbers, i.e., from left to right, or right to left, or down or up.
6. Select the first "n" numbers whose X digits are between 0 and N. If $N=100$ then X would be 2; if N is a four-digit number then X would be 3 and so on.
7. Do not use the number again, once it is chosen.
8. Before obtaining "n" numbers if you reach the end point of the table, pick another starting point and read in a different direction. Then use the first X digit instead of the last X digits. Then continue until the desired sample is selected.

For example, suppose you have a list of 60 students, and by using a simple random sampling method, you want to select a sample of ten students, then use the following steps:

1. First, assign each student a number from 00 to 59.
2. To draw a sample of ten students using a random number table, you need to find ten two-digit numbers in the range 00–59.
3. You can begin anywhere and go in any direction.
   Start from the first row and first column of the random number Table 6.1.
4. Read the last two digits of the numbers.
5. If the number is within the range (00–59), include the number in the sample. Otherwise skip the number and read the next number in some identified direction. If a number is already selected omit it.

In the example, the following numbers are considered to select ten numbers for sample, starting from first row and first column and moving from left to right direction (Table 6.2).

The selected numbers for the sample are indicated by the bold-faced digits in the one's and ten's place value. Therefore, the following are the ten numbers chosen as sample.

31, 45, 17, 05, 52, 50, 44, 19, 36, 01

In research simple random sampling is not widely used because of the following reasons.

1. We usually select individuals, households, shops, or areas as the sampling units, in consumer research studies. As it is very difficult to get lists of households, individuals, and shops, although areas may be completely represented through maps, it may not be easy to prepare a sampling frame.

TABLE 6.2

Numbers Considered to Select Ten Numbers for Sample (The selected numbers for the sample are indicated by the bold-faced digits in the one's and ten's place value)

96268	11860	83699	386**31**	900**45**	69696	48572	059**17**	519**05**	100**52**
035**50**	591**44**	59468	37984	77892	89766	86489	466**19**	502**36**	91136
22188	81205	99699	84260	19693	367**01**	43233	62719	53117	71153

2. We know that an industry comprises various firms of different sizes. One might like to choose sampling design, where there is a higher probability of a larger firm being selected to study some aspects of an industry. If that is the case, in such situations, the very concept of simple random sampling becomes inapplicable.
3. The simple random sampling has some applications in Industrial Marketing where generally purchasing agents or companies or areas, which are usually not very big in number are the sampling units.

Therefore, it becomes easy to prepare a sampling frame thus facilitating the use of simple random sampling.

A simple random sample is a probability sample. A simple random sample requires a complete listing of all the elements, an equal chance for each element to be selected, and a selection process whereby the selection of one element has no effect on the chance of selecting another element.

Advantages of Simple Random Sampling

1. Simple random sampling is advantageous as it is free of classification error and requires minimum advance knowledge of the population.
2. As any other of being selected in the sample each person has an equal chance.
3. Simple random sampling serves as a foundation against which other methods are sometimes evaluated.
4. It becomes more representative of universe, as the sample size increases.
5. This method is least costly and easily assessable of accuracy.

Disadvantages of simple random sampling

1. Complete and up-to-date catalogued universe is necessary.
2. To establish reliability requires a large sample size.
3. The study of sample item has larger cost and greater time, when the geographical dispersion is wider.
4. May cause wrong results due to Unskilled and Untrained Investigator.
5. A great deal of time must be spent in listing and numbering the members of the population, in case the population size is large.

6.3.1.2 Systematic Sampling

Systematic sampling is another method of nonprobability sampling plan.

Systematic sampling involves the selection of every k^{th} element from a sampling frame, where k represents the skip interval and is calculated using the following formula:

$$\text{Skip interval}(k) = \frac{\text{Population size}}{\text{Sample size}}$$

Often used as a substitute to simple random sampling, it involves the selection of units from a list using a skip interval (k) so that every k^{th} element on the list, following which a random start between 1 and k is included in the sample.

For example, if $k = 6$, and the random start were 2, then the sample would consist of 2nd, 8th, 14th, 20th,..., elements of the sampling frame. If the skip interval is not a whole number then it is rounded off to the nearest whole number. In systematic sampling, the sample units are selected from the population at an equal intervals in terms of time, space, or order. The selection of a sample using systematic sampling method is very simple. From a population of "N" units, a sample of "n" units may be selected by the following steps:

1. Arrange all the units in the population in an order by giving serial numbers from 1 to N.
2. Determine the sampling interval by dividing the population by the sample size. That is, $K = \frac{N}{n}$.
3. From the first sampling interval (1 to K), select the first sample unit at random.
4. At an equal regular intervals, select the subsequent sample units.

For example, if we want to have a sample of ten units from a population of 100 units. Steps are as follows:

1. First, arrange the population units in some serial order by giving numbers from 1 to 100.
2. The sample interval size is $K = \frac{N}{n} = \frac{100}{10} = 10$. Select the first sample unit at random from the first ten units, i.e., from 1 to 10.
3. Suppose the first sample unit selected is 5, then the subsequent sample units are 15, 25, 35,..., 95. Thus, in the systematic sampling, the first sample unit is selected at random and this sample unit in turn determines the subsequent sample units that are to be selected.

Advantages of Systematic Sampling

1. Since the time taken and work involved is less than in simple random sampling, it is more expeditious to collect a sample systematically. For example, it is frequently used in exit polls and store consumers.
2. Even when no formal list of the population units is available, this method can be used.

Disadvantages of Systematic Sampling

1. If there is periodicity in the occurrence of elements of a population, then using systematic sampling, the selection of sample could give a highly unrepresentative sample.
2. Every unit of the population does not have an equal chance of being selected and the selection of units for the sample depends on the initial unit selection. Regardless of how we select the first unit of sample, subsequent units are automatically determined lacking complete randomness.

6.3.1.3 Stratified Random Sampling

In stratified sampling, the entire population is divided into various mutually exclusive and collectively exhaustive strata or groups. Mutually exclusive means that, if an element of a group belongs to one strata, then it doesn't belong to any other strata. By collectively exhaustive, we mean that all the elements of various strata put together completely cover all the elements of the population.

The groups or strata are created on the basis of a variable or criteria known to be correlated with the variable under study. The possible criteria for stratification of a population could be Income of the individuals, Age, Sex, Purchasing frequency, Household size, Retail store size, Region of the country, etc.

Also, on the basis of more than one variable, the stratification is possible. This increases the number of stratum; the cost of stratification may come as a constraint in increasing the number of stratum or group.

A variable that is considered to be good in stratification of one population may not be so in the case of other. However, one aspect that must be kept in mind is that stratification should be done in such a way as to minimize the variability among sampling unit within strata, i.e., more homogeneous and maximize the variability among strata, i.e., more heterogeneous.

The stratified sampling method is used when the population is heterogeneous rather than homogeneous. The process of grouping members of the population into relatively homogeneous subgroups before sampling is called as stratification. It should be ensured that each element in the population is assigned a particular stratum only. The strata should also be collectively exhaustive to ensure that no population element is excluded.

Separate simple random sample of various sizes is selected from each stratum, once the population has been divided into various strata. Then random sampling is applied within each stratum independently. This often improves the representativeness of the sample by reducing sampling error.

A heterogeneous population is composed of unlike elements such as male or female, rural or urban, literate or illiterate, high-income or low-income groups, etc. In such cases, the use of simple random sampling may not always provide a representative sample of the population.

There are two approaches to decide the sample size from each stratum, namely, proportional stratified sample and disproportional stratified sample. With either approach, the stratified sampling guarantees that every unit in the population has a chance of being selected. These two approaches of selecting samples are discussed below.

6.3.1.3.1 Proportional Stratified Sample

The number of members selected from each stratum is proportional to its share of the total population in proportionate stratified sampling.

We say the sample is a proportional stratified sample if the number of sampling units drawn from each stratum is in proportion to the corresponding stratum population size.

For example, suppose we want to draw a stratified random sample from a heterogeneous population on some characteristics consisting of rural, urban, and male, female, and other respondents.

So, we have to create six homogeneous subgroups called stratums as follows (Table 6.3):

To ensure each stratum in the sample will represent the corresponding stratum in the population, we must ensure each stratum in the sample is represented in the same proportion to the stratums as they are in the population.

Let us assume that we know or can estimate the population distribution as follows: 65% male, 30% female, 5% other, 20% urban, and 80% rural. Now we can determine the approximate proportions of our six stratums in the population as shown below (Table 6.4).

Thus, a representative sample would be composed of 13% urban-males, 6% urban-females, 1% urban-other, 52% rural-males, 24% rural females, and 4% rural other. Each percentage should be multiplied by the total sample size needed to arrive at the actual sample size required from each stratum. Suppose we require 100 samples then the required sample in each stratum is as follows (Table 6.5).

6.3.1.3.2 Disproportional Stratified Sample

The number of members selected from each stratum is not proportional to its share of total population, in case of disproportionate stratified sampling. The choice of proportionate or disproportionate sampling method among strata depends on whether the variances in each group or stratum are equal or not. One should go for proportionate stratified sampling if the variance of each stratum is almost equal. A large sample should be taken from the stratum with large variance if variances are not equal.

In a disproportional stratified sample, sample size for each stratum is not allocated on a proportional basis with the population size, but by analytical considerations of the researcher such as stratum variance, stratum population, time, financial constraints, etc.

For example, disproportional sampling should be used if the researcher is interested in finding differences among different stratums.

TABLE 6.3

Homogeneous Subgroups or Stratum of Heterogeneous Population

Urban			**Rural**		
Male	Female	Other	Male	Female	Other

TABLE 6.4

The Approximate Proportions of Six Stratums in the Population

Urban			**Rural**		
Male	**Female**	**Other**	**Male**	**Female**	**Other**
$0.20 \times 0.65 = 0.13$	$0.20 \times 0.30 = 0.06$	$0.20 \times 0.05 = 0.01$	$0.80 \times 0.65 = 0.52$	$0.80 \times 0.30 = 0.24$	$0.80 \times 0.05 = 0.04$

TABLE 6.5

Required 100 Sample in Each Stratum

Population Distribution	Sample in each Stratum
Urban-Male	$0.13 \times 100 = 13$
Urban-Female	$0.06 \times 100 = 6$
Urban-Other	$0.01 \times 100 = 1$
Rural-Male	$0.52 \times 100 = 52$
Rural-Female	$0.24 \times 100 = 24$
Rural-Other	$0.04 \times 100 = 4$
Total	100

Consider the income distribution of households. There is a small percentage of households within the high-income brackets and a large percentage of households within the low-income brackets. The income among high-income group households has higher variance than the variance among the low-income group households. A disproportional sample is taken to avoid under-representation of high-income groups in the sample. This indicates that as the variability within the stratum increases sample size must increase to provide accurate estimates and vice versa. Suppose in our example of rural, urban, and male, female, and other stratum populations, the stratum estimated variances $\left(S^2\right)$ are as follows (Table 6.6).

The above figures are, normally, estimated on the basis of previous knowledge of a researcher. Then the allocation of sample size of 100 for each strata using disproportional stratified sampling method will be as summarized in Table 6.7.

TABLE 6.6

Stratum Estimated Variances $\left(S^2\right)$

Population Distribution	Stratum Estimated Variances $\left(S^2\right)$
Urban-Male	2.0
Urban-Female	4.5
Urban-Other	0.2
Rural-Male	1.5
Rural-Female	0.75
Rural-Other	0.1

TABLE 6.7

The Allocation of Sample Size of 100 for Each Strata Using Disproportional Stratified Sampling Method

Stratum	Stratum Population Proportion (P_i)	Stratum Variance $\left(\sigma_i^2\right)$	Stratum Standard Deviation (σ_i)	$(P_i)\times(\sigma_i)$	Sample Size $\left(\frac{P_i \times \sigma_i \times 100}{\sum P_i\ \sigma_i}\right)$
Urban-Male	0.13	2.0	1.41	0.1833	15.69
Urban-Female	0.06	4.5	2.12	0.1272	10.88
Urban-other	0.01	0.2	0.44	0.0044	0.37
Rural-Male	0.52	1.5	1.22	0.6344	54.31
Rural-Female	0.24	0.75	0.86	0.2064	17.46
Rural-other	0.04	0.1	0.31	0.0124	1.06
Total				1.1681	$99.77 \cong 100$

Advantages of Stratified Random Sampling

1. Stratified sampling is more representative as the samples are drawn from each of the stratum of the population, and thus more accurately reflect the characteristics of the population from which they are chosen.
2. Formation of stratum and random selection of item from each stratum make it hard to exclude in strata of the universe and increase the sample's representation to the population or universe.
3. It is more precise and avoids bias to a great extent.
4. In this method, sample size is less, due to which it saves a lot of time, money, and other resources for data collection.

Disadvantages of Stratified Random Sampling

1. To determine the homogeneous groups that lie within it, stratified sampling requires a detailed knowledge of the distribution of attributes or characteristics of interest in the population. If we cannot accurately identify the homogeneous groups, then it is better to use simple random sample since improper stratification can lead to serious errors.
2. As the stratified lists may not be readily available, preparing a stratified list is a difficult task.
3. Improper stratification may cause wrong results.
4. Greater geographical concentration may result in heavy cost and more time.
5. For stratification trained investigators are required.

6.3.1.4 Cluster Sampling

Cluster sampling involves grouping the elements in a population into various clusters and then for further study selecting a few clusters randomly. In other words, in cluster sampling, we divide the population into groups having heterogeneous characteristics called clusters and then select a sample of clusters using simple random sampling.

The criteria for dividing the population into mutually exclusive and collectively exhaustive clusters is that the elements in the clusters should be as heterogeneous as possible and elements between clusters should be as homogeneous as possible.

The researcher should ensure that clusters are homogeneous based on some characteristic of the units, in nature and the elements within each cluster are as heterogeneous as possible, i.e., each cluster should be similar to the population. We assume that each of the clusters is representative of the population as a whole. Once the clusters are formed, the researcher can form either one-stage, two-stage, or multistage cluster sampling.

In a one-stage cluster sampling, all the elements from each of the selected clusters are studied.

In two-stage cluster sampling, the researcher uses random sampling to select a few elements from each of the selected cluster.

For example, if we are interested in finding the attitudes of consumers residing in Mumbai toward a newly launched product of a company, the whole city of Mumbai can be divided into ten blocks.

We assume that each of these blocks will represent the attitudes of consumers of Mumbai as a whole, then we can use cluster sampling by treating each block as a cluster.

We can then select a sample of two or three clusters and obtain the information from consumers covering all of them.

The basic principles to the cluster sampling are as follows:

1. The differences or variability within a cluster should be as large as possible. As far as possible the variability within each cluster should be the same as that of the population.
2. The variability between clusters should be as small as possible. Once the clusters are selected, all the units in the selected clusters are covered for obtaining data.

Advantages of Cluster Sampling

1. Since traveling costs are smaller, cluster sampling provides significant gains in data collection costs.
2. Since the researcher need not cover all the clusters and only a sample of clusters are covered, it becomes a more practical method that facilitates fieldwork.
3 This method of collecting data is cheaper since collection of data from nearby units is easier, faster, and more convenient than collecting data over units scattered over a region. For example, it would not only be cheaper but also convenient to collect data on all households in a sample of few villages or clusters than to survey a sample of the same number of households selected randomly from a list of all households.
4. For conducting research studies, cluster sampling is suitable as it covers large geographic areas with respondents scattered all over. This sampling is widely used for geographical studies of many issues.

Disadvantages of Cluster Sampling

1. The cluster sampling method is less precise than sampling of units from the whole population since the latter is expected to provide a better cross section of the population than the former due to the usual tendency of units in a cluster to be homogeneous.
2. With the decrease in cluster size or increase in number of clusters, the sampling efficiency of cluster sampling is likely to decrease.

6.3.1.5 Area Sampling

Another version of cluster sampling, namely, area sampling, is used in a research study involving sampling of population, which may be grouped according to geographical areas or blocks, census tracts, communities, constituencies, etc.

The entire area is divided into various clusters. The cluster may or may not be of equal size. A sampling scheme, where sampling is done by taking into account the size of the cluster. This type of design is called Probability Proportional to Size (PPS) sampling.

6.3.1.5.1 Probability Proportional to Size

When we have to sample cluster of varying sizes, this sampling design is used.

For example, suppose a book is divided into nine volumes or blocks or clusters of varying sizes. Assume that the size of the household is 140 in one of the blocks, whereas in the other block the size of the household is 100. The block having 140 size household should have greater probability of being selected than the one with a size of 100. In fact, the probability in the first case should be 1.4 times the probability in the second case. This concept is used in this sampling design so that a greater weightage is given to the cluster with higher size.

Consider the Data Given in Table 6.8

In Table 6.8, a small area has been divided into seven blocks. It is evident from the table that the number of households in each block are varying. If a sample of size 20 is to be selected from a population of 1,000, each household should have a probability of $\frac{20}{1,000} = 0.05$ being selected in the sample.

We have identified a total of 1,000 households in seven blocks or clusters. We will assign a number from 1 to 1,000 for each household. This is shown in the last column of the Table 6.8.

Suppose we have to select three clusters randomly. We will look at the four-digit random number from the random number tables with numbers 0001 to 1,000. We select three numbers. Suppose the numbers are 190, 170, and 60. This means cluster numbering 4, 7, and 2 are selected. Now we would select a sample of size 10 from each of the clusters. The probability of selecting the required household are as follows:

TABLE 6.8

Small Area Has Been Divided into Seven Blocks (Three Selected Numbers of four-digit random number, from the random number tables with numbers 0001 to 1,000 is shown by the bold)

Block Number	Household Number	Household Cumulative Number	Associated Random Numbers
1	100	100	001–100
2	**60**	160	101–160
3	190	350	161–350
4	**170**	520	351–520
5	200	720	521–720
6	90	810	721–810
7	**190**	1000	811–1000
	1000		

Probability of household in Block A = Number of Blocks to be chosen × Block probability

× Within block household probability

$$\text{For household in block 2} = 3 \times \frac{60}{1{,}000} \times \frac{10}{60} = 0.03$$

$$\text{For household in block 4} = 3 \times \frac{170}{1{,}000} \times \frac{10}{170} = 0.03$$

$$\text{For household in block 7} = 3 \times \frac{190}{1{,}000} \times \frac{10}{190} = 0.03$$

We find that irrespective of the block size, the probability of selecting a household equals 0.03. This is the same as required by the overall sampling design. We note that in the PPS design the blocks with larger size are given more weightage selected in the sample. It can be shown that the efficiency of the estimate increases by this procedure compared to when all the blocks or clusters have equal probability of being selected.

6.3.1.6 Multistage Sampling

Multistage sampling involves selecting a sample in two or more successive stages. Here, the cluster or unit selected in the first stage can be further divided into clusters or units. A generalization of two-stage sampling is a multistage sampling.

In each stage progressively smaller (population) geographic areas will be randomly selected. A type of random sample that uses multiple stages and is often used to cover wide geographic areas in which aggregated units are randomly selected and then sample are drawn from the sampled aggregated units or cluster is called as multistage sampling.

For example, if the investigator wants to survey some aspect of second-grade, elementary school-going children. Then, the different stages of the selection of samples as follows:

a. A random sample of number of states from the country would be selected.
b. Within each selected state, a random selection of certain number of districts would be made.
c. Within district a random selection of certain number of elementary schools would be made.
d. Within each elementary school, a certain number of children would be randomly selected. Because each level is randomly sampled, the final sample becomes random.

However, selection of samples is done in different stages. This is called multistage sampling. This sampling method is more flexible than the other methods. Subdivisions at the second stage unit need to be carried out for only those units selected in the first stage.

Advantages of Multistage Sampling

1. It provides cost gains by reducing the costs on data collection.
2. It is more flexible and allows us to use different sampling procedures in different stages of sampling.
3. It is the only sampling method available in a number of practical situations, when the population is spread over a very wide geographical area.

Disadvantages of Multistage Sampling

Multistage sampling becomes less precise and efficient when the sampling units selected at different stages are not representative.

6.3.2 Nonprobability Sampling Methods or Nonrandom Sampling Methods

Nonprobability sampling or nonrandom sampling is also known as deliberate sampling and purposive sampling, and involves the selection of units based on factors other than random chance.

In this sampling method the probability of any particular unit of the population being chosen is unknown. Here the method of selection of sampling units is quite arbitrary as the researchers rely heavily on personal judgment. Usually, these sampling methods do not produce samples that are representative of the general population from which they are drawn. When the researcher attempts to generalize the results on the basis of a sample to the entire population, the greatest error occurs. Such an error is insidious because it is not at all obvious from merely looking at the data, or even from looking at the sample. To determine whether the sample is selected randomly or not is the easiest way to recognize whether a sample is representative or not. Nevertheless, there are occasions where nonrandom samples are best suited for the researcher's purpose.

The various non-random sampling methods commonly used are Haphazard or Accidental or Convenience Sampling, Quota Sampling and Purposive Sampling or Judgment Sampling.

These methods do not provide every item of population any known chance of being selected in the sample. Here no attempt is made to select a representative sample. On the convenience and/or judgment of the researcher or field interviewer, the elements of samples are selected. The selection process is subjective. It is not possible to make an estimate of sampling error, as the sample is not representative of population. Also, we cannot say whether our sample estimates are correct or not.

It may be worth mentioning that most of the research studies make use of nonprobability methods of sampling. As in these methods, the selection process of these samples is subjective, therefore one should not conclude that the results obtained from it are inferior to what one would obtain by using probability sampling methods. Also, the sample obtained through such methods need not be less representative of the population.

For example, suppose the objective of a marketing researcher is to develop an index of performance of sales force by measuring items such as sales per salesman, number of calls per day, order call ratio, number of customer complaints, and so on. Because the market researcher feels that is represents "performance," any particular item is included in the list. To achieve a representation of the population by a random selection of items representing various characteristics of performances, nonprobability method of selection is a better way.

Following are the important techniques of nonprobability sampling methods.

6.3.2.1 Haphazard, Accidental, or Convenience Sampling

In convenience sampling, the selection of units from the population-based on their easy availability and accessibility to the researcher. Convenience sampling refers to the method of obtaining a sample that is most conveniently available to the researcher. For example, to reflect public opinion on various issues of public interest, such as budget, election, price rise of petrol or diesel, fee rise of university, etc., the television channels often present on-the-street interviews with people.

It may be cautioned that the generalization of results based on convenience sampling beyond that particular sample may not be appropriate. When additional research will be subsequently conducted with a random sample, convenience samples are best used for exploratory research. Convenience sampling is also useful in testing the questionnaires designed on a pilot basis. Convenience sampling is extensively used in marketing studies.

As the name implies, under convenience sampling the samples are selected at the convenience of the researcher or investigator. Here, we have no way of determining the representativeness of the sample. This results in biased estimates. Therefore, it is not possible, in terms of both magnitude and direction to make an estimate of sampling error as the difference between sample estimate and population parameter is unknown. It is therefore suggested that convenience sampling should not be used in both descriptive and causal studies as it is not possible to make any definitive statements about the results from such a sample.

This method may be quite useful in exploratory designs as a basis for generating hypotheses. The method is also useful in testing of questionnaire, etc., at the pretest phase of the study.

6.3.2.2 Quota Sampling

In marketing research studies, the quota sampling method is commonly used. The samples are selected on the basis of some parameters such as age, sex, geographical region, education qualification, annual income, type of occupation, etc., in order to make them as representative samples. Then, meeting these population characteristics, investigators assigned fixed quotas of the sample. To ensure that various subgroups of the population are represented on pertinent sample characteristics to the extent that the investigator desires, is the purpose of quota sampling. The stratified random sampling also has this objective but should not be confused with quota sampling.

In the stratified sampling method the researcher selects a random sample from each group of the population, whereas, in quota sampling, the interviewer has a quota fixed for him or her to achieve.

Under stratified sampling, the fieldworker selects a random sample from each cell of the population, whereas under quota sampling the selection of sample is not random. It is left to the judgment of the field worker.

For example, if a city has 20 market centers, a soft drink company may decide to interview 10 consumers from each of these 20 market centers to elicit information on their products. It is entirely left to the investigator whom he or she will interview at each of the market centers and the time of interview. The interview may take place in the morning, mid-day, or evening or it may be in the winter or summer. Quota sampling has the advantage that the sample confirms the selected characteristics of the population that the researcher desires. Also, the cost and time involved in collecting the data are greatly reduced.

Advantages of Quota Sampling

1. The method has a lower cost and field workers have a free hand to select respondents for each cell to fill their quota. If the samples are selected with care then it would result in more definitive findings.
2. Quota sampling enables the researcher to introduce a few controls into his or her research plan. And because this method is more convenient and less costly than many other methods of sampling, it is popular among nonprobability methods of sampling

Disadvantages of Quota Sampling

1. In quota sampling, the respondents are selected according to the convenience of the field investigator rather than on a random basis. This kind of selection of sample may be biased.
2. It becomes difficult for the researcher to fix the quota for each subgroup, when the number of parameters on which basis the quotas are fixed are larger.
3. To obtain an accurate and up-to-date proportion of respondents assigned to each cell is difficult.
4. The total number of cells increase as the number of parameters or control characteristics associated with the objectives of the study become large. As it may not be easy to get a desired respondent, this makes difficult the task of field staff's.

5. All of the proper parameters or control characteristics related to the study in question must be incorporated while taking sample, which is very important. If any relevant parameter is omitted for one reason or the other, the results of the study could be misleading.

6.3.2.3 Purposive Sampling or Judgment Sampling

The selection of a unit from the population based on the judgment of an experienced researcher or an expert is known as judgment or purposive sampling. Here, based on the population's parameters, the sample units are selected. It is often noticed that companies frequently select certain preferred cities during test marketing of their products. This is because they consider the population of that particular city to be representative of the total population of the country.

Under this sampling procedure, a researcher deliberately or purposively draws a sample from the population that he or she believes is a representative of the population. Needless to mention, all members of the population are not given opportunity to be selected in the sample. The personal bias of the investigator has a great chance of entering the sample. If the investigator chooses a sample to give results that favor his or her view point, the entire study may be vitiated. The relevant experience and the acquaintance of the investigator with the population may help to choose a relatively representative sample from the population if personal biases are avoided. As we cannot determine how precise our sample estimates are, it is not possible to make an estimate of sampling error.

In this method of sampling the selection of sample is based on the researcher's judgment about some appropriate characteristic required of the sample units. For example, judgment sampling is also often used in forecasting election results.

Advantages of Judgment Sampling

1. Suppose, to decide about the launching of a new product in the next year, we have a panel of experts. If for some reason or the other, a member drops out from the panel, the chairman of the panel may suggest the name of another person whom he or she thinks has the same expertise and experience to be a member of the said panel. This new member was chosen deliberately—a case of judgment sampling.
2. For special situations, purposive sampling is a valuable kind of sampling. It is used in exploratory research or in field research. It uses the judgment of an expert in selecting cases or it selects cases with a specific purpose in mind. Purposes sampling is somewhat less costly, more readily accessible, more convenient, and helps in selecting only those individuals who are relevant to research design.

Limitation of Judgment Sampling

1. There is no way to ensure that the sample truly represents the population and
2. To assess the elements of population, more emphasis is placed on the ability of the researcher.

6.3.2.4 Snowball or Network or Chain Referral or Reputation Sampling

Snowball sampling is also known as Network, Chain Referral, or Reputation Sampling Method. Sampling procedure that involves the selection of additional respondents based on referrals of initial respondents are known as snowball sampling. Against low incidence or rare populations, this sampling technique is used. As the defined population from which the sample can be drawn is not available, in this case, sampling is a huge problem. Therefore, the process of sampling depends on the chain system of referrals.

Snowball sampling, which is a nonprobability sampling method, is basically sociometric. It begins by the collection of data on one or more contacts usually known to the person collecting the data. At the end of the data collection process, e.g., questionnaire, survey, or an interview, the data collector asks the respondent to provide contact information for other potential respondents. These potential respondents are contacted and provided more contacts. When there are very few methods to secure a list of the population or when the population is unknowable, snowball sampling is most useful.

Advantages of Snowball Sampling

1. Small sample sizes and low costs.
2. In studying small informal social group and its impact upon formal organizational structure, snowball sampling, which is primarily a sociometric sampling technique, has proved very important and is helpful.
3. Snowball sampling reveals communication pattern in community organization concepts such as community power, and decision-making can also be studied with the help of such sampling technique.

Disadvantages of Snowball Sampling

1. Bias is one of the disadvantages. The referral names obtained from those sampled in the initial stages may be similar to those initially sampled. Therefore, the sample may not represent a cross section of the total population. It may also happen that visitors to the site or interviewees may refuse to disclose the names of those whom they know.
2. When is large or say it exceeds 100, snowball sampling becomes cumbersome and difficult.
3. This method of sampling does not allow the researcher to use probability statistical methods. In fact, the elements included in sample are not randomly drawn. And elements are dependent on the subjective choices of the originally selected respondents. This introduces some bias in the sampling.

6.4 Steps in a Sampling Process

6.4.1 Defining the Target Population

For research, defining the population of interest is the first step in the sampling process. In general, target population is defined in terms of element, sampling unit, extent, and time frame. The definition should be in line with the objectives of the research study.

For example, if the population is defined as all women about the age of 30 years, the researcher may end up taking the opinions of a large number of women who cannot afford to buy a mixer-juicer.

6.4.2 Specifying the Sampling Frame

Once the definition of the population is clear a researcher should decide on the sampling frame. A sampling frame is the list of elements from which the sample may be drawn. Continuing with the mixer-juicer example, an ideal sampling frame would be a database that contains all the households that have a monthly income above Rs. 15,000. However, in practice, it is difficult to obtain an exhaustive sampling frame that exactly fits the requirements of a particular research. In general, researchers use easily available sampling frames such as voting list and list of ATM cards and mobile phone users.

Various private players provide databases developed along various demographic and economic variables. For sampling frames, sometimes maps and aerial pictures are also used. An ideal sampling frame is one that represents the entire population and lists the name of its elements only once, whatever may be the case.

A sampling frame error pops up when sampling frame does not accurately represent the total population or when some elements of the population are missing.

6.4.3 Specifying the Sampling Unit

A basic unit that contains a single element or a group of elements of the population to be sampled is a sampling unit.

6.4.4 Selection of the Sampling Method

The sampling design describes the procedure by which a sample is selected. The sampling method outlines the way in which the sample units are to be selected. The choice of the sampling method is influenced by the objectives of the research, availability of financial resources, time constraints, and nature of the problem to be investigated.

Types of sampling procedures are probability sampling procedures and nonprobability sampling procedures. In the probability sampling procedure, each element has a known probability of being selected in the sample. In the nonprobability sampling procedure, there is no known probability of an element of the population being selected in the sample. The selection of an element of population in the sample depends upon the judgment of the researcher or the field interviewer.

6.4.5 Determination of Sample Size

In nonprobability sampling procedures, for the determination of sample size, research has to consider the determination of the budget allocation, thumb rules and number of subgroups to be analyzed, importance of the decision, number of variables, nature of analysis, incidence rates, and completion rates.

6.4.5.1 Techniques of Determination of Sample Size

In general, sample size depends on the nature of analysis to be performed, one wishes to achieve the estimates of the desired precision, number of variables that have to be simultaneously examined, and spread of heterogeneous population.

Moreover, technical considerations suggest that the required sample size is a function of the precision of the estimates one wishes to achieve and the population variance and statistical level of confidence one wishes to use. The sample size should be larger for the requirement of the higher precision and confidence level. Typical confidence levels are 95% and 99%, while a typical precision or significance value is 1% or 5%. Depending on the plan of the study, there are several formulas he or she can use to determine the sample size and interpretation of results, once the researcher determines the desired degree of precision and confidence level.

The following formula may be more useful when the researcher plans the results in a variety of ways or when he or she has difficulty in estimating the proportion or standard deviation of the attribute of interest.

$$n = \frac{NZ^2 \times 0.25}{\left[\alpha^2 \times (N-1)\right] + \left[Z^2 \times 0.25\right]}$$

where

n=Required Sample size

α=Precision level Accuracy (i.e., 0.01, 0.05, 0.10, etc.)

Z=Standardization value indicating a confidence level, Z=1.96 at 95% confidence level and Z=2.56 at 99% confidence level.

N=Known or estimated population size

For example,

$$\text{If the population size } (N) = 1{,}000,$$

$$\text{You wish a confidence level} = 95\%,$$

$$\text{Precision level} = \pm 5\%,$$

$$\text{i.e., } \alpha = 0.05 \text{ and } Z = 1.96, \text{ then the sample size } (n):$$

$$n = \frac{NZ^2 \times 0.25}{\left[\alpha^2 \times (N-1)\right] + \left[Z^2 \times 0.25\right]}$$

$$n = \frac{1,000 \times (1.96)^2 \times 0.25}{\left\{\left[(0.05)^2 \times 999\right]\left[(1.96)^2 \times 0.25\right]\right\}}$$

$$n = 277.7 = 280$$

Similarly, if we want to use sample proportion (p) to estimate the population proportion (p) and we want to be able to assert with a $(1-\alpha)\%$ confidence that the allowable error of this estimate is "*e*".Again, to achieve this, how large is a sample is the question.

We know that

$$Z = \frac{\overline{p} - p}{\sqrt{\dfrac{p(1-p)}{n}}} \tag{6.1}$$

where

$Z \rightarrow$ Standard normal variate

$n \rightarrow$ The required sample size.

The quantity $\overline{p} - p \rightarrow$ represents allowable error "*e*".

Therefore, we have

$$Z = \frac{e}{\sqrt{\dfrac{p(1-p)}{n}}}$$

$$Z = \frac{\sqrt{n}e}{\sqrt{p(1-p)}}$$

$$\therefore n = \frac{p(1-p)\, Z^2}{e^2} \tag{6.2}$$

The population p may be unknown. Therefore, in Formula 6.2, we substitute the maximum possible value of $p\,(1-p)$, which can be shown to be equal to 1/4 and occurs when $p = 1/2$.

Therefore, in Formula 6.2, we substitute $p\,(1-p) = 1/4$ and this would increase our sample size and would ensure that the allowable error lies within the prescribed limit and at given confidence level. Therefore, Formula (6.2) can be rewritten as:

$$n = \frac{1}{4}\frac{Z^2}{e^2} \tag{6.3}$$

6.4.5.2 Numerical (The Case of Means)

A large retailer wants to determine whether the income of families living within 3 miles of a proposed building site exceeds Rs. 5,500. It is given that the population standard deviation is Rs. 500. How large a sample will be required if the probability of a Type I error is to be 0.05 and the probability of a Type II error is to be 0.01 when mean income of population is Rs. 6,000?

Solution: Here, our Null (H_0) and Alternative (H_1) hypothesis to be tested in terms of population mean μ_0 and μ_1, respectively, are $H_0 : \mu_0 = 5,500, \quad H_1 : \mu_1 = 6,000$

$$\text{Given: } \alpha \rightarrow \text{Probability of Type I error} = 0.05$$

$$\beta \rightarrow \text{Probability of Type II error} = 0.01$$

For a one-tailed test the Z values for the 0.05 and 0.01 risks are found from standard normal table to be $Z_\alpha = 1.64$ and $Z_\beta = 2.33$. The population standard deviation $(\sigma) = 500$. Now, we have to calculate the sample size that will meet the α and β requirements. For that, we need to solve the following two simultaneous equations:

$$\text{Critical value} = \mu_0 + Z_\alpha \frac{\sigma}{\sqrt{n}}, \quad \text{Critical value} = \mu_1 - Z_\beta \frac{\sigma}{\sqrt{n}}$$

By equating these two equations and solving n, we get

$$n = \frac{(Z_\alpha + Z_\beta)^2 \sigma^2}{(\mu_1 - \mu_0)^2}$$

For our example, $Z_\alpha = 1.64$, $Z_\beta = 2.33, \sigma = 500, \mu_0 = 5{,}500$, and $\mu_1 = 7{,}000$. Therefore, substituting these values in the above formula we get

$$n = \frac{(Z_\alpha + Z_\beta)^2 \sigma^2}{(\mu_1 - \mu_0)^2} = \frac{(1.64 + 2.33)^2 \times (500)^2}{(7{,}000 - 6{,}600)^2} = 15.7609 \cong 16$$

6.4.5.3 The Case of Proportion

To discuss the case of estimation of sample size, let us consider the following example, it has been claimed that 30% of all shoppers can identify a highly advertised trademark. How large a sample will be required if the probability of a Type I error is to be 0.05 and the probability of a Type II error is to be 0.01 when the population proportion is 31%?

We set up the hypotheses $H_0 : P_0 = 0.30, \quad H_1 : P_1 = 0.31$

The Value $Z_\alpha = Z_{0.05} = 1.64, \quad Z_\beta = Z_{0.01} = 2.33$

To estimate the sample size, we use the following formula:

$$n = \left[\frac{Z_\alpha \sqrt{P_0 (1 - p_0)} + Z_\beta \sqrt{p_1 (1 - p_1)}}{P_1 - p_0}\right]^2$$

$$n = \left[\frac{1.64 \sqrt{0.3\,(0.7)} + 2.33\sqrt{0.31(0.69)}}{0.01}\right]^2 = \left[\frac{1.8291516706}{0.01}\right]^2$$

$$n = [182.9151670642]^2 = 33457.958342115 = 33{,}458$$

After having decided upon the size of the sample, the field workers should be given clear and accurate instruction.

6.4.6 Specifying the Sampling Plan

To collect actual data very clear and accurate instructions as to how to do this job are given to the interviewers who are going to the field. Until these instructions are prepared and handed over to the investigators, the sampling planning is not complete. In this step, outlines are the specifications and decisions with regard to the implementation of the research process.

Suppose, blocks or wards in a city are the sampling units and the households are the sampling elements. This step outlines the sampling plan in identifying houses based on specified characteristics. It includes issues such as: How is the interview going to take a systematic sample of the houses? What should the interviewer do when a house is vacant? What is the recontact procedure for respondents who were unavailable?

For the smooth functioning of the research process, all these and many other questions need to be answered. In every step of the process, these are guidelines that would help the researcher. As most of the time, the interviewers and their coworkers will be on field duty, a proper specification of the sampling plans would make their work easy. And when faced with operational problems, they would not have to revert to their seniors.

6.4.7 Selecting the Sample

This is the final step in the sampling process. Here, the actual selection of the sample elements is carried out. For the smooth implementation of the research at this stage, it is necessary that the interviewers stick to the rules outlined. To select a sample required for the survey this step involves implementing the sampling plan.

6.5 Criteria for Selecting an Appropriate Sampling Design

Common criteria for evaluating and selecting the appropriate sampling design are discussed below.

6.5.1 Degree of Accuracy

The sample is representative of the target population while drawing upon a sample. If this is not the case, sampling errors may arise that would lead to errors in the subsequent steps. However, the degree of accuracy sought by a researcher varies from one research to another.

For example, an exploratory survey may not demand a highly accurate sampling design. But the same is required for research that is more conclusive and where the researcher is willing to invest a lot of money and time. The need for accuracy also depends on the decision it is going to support. A high degree of accuracy is sought in case the stakes attached with the decision area are high.

6.5.2 Resources

Resources in the form of budget allocation and manpower also influence a researcher's choice of sampling design. For example, the researcher may choose a nonprobability sampling design that can be implemented within the time and budgetary constraints if limited resources are allocated for a research program.

6.5.3 Time

In the research process in case of a time constraint researchers are likely to opt for simple, less time-consuming sampling designs. To cluster and stratify sampling, a researcher would prefer telephone survey or convenience sampling.

6.5.4 Prior Knowledge of the Population

Prior knowledge of the population in terms of characteristics, availability, and lists is imperative for a researcher. In the case of researches where population is defined in terms of ownership, experience, or some other qualitative dimension, a population may not be accessible for sampling. The use of better sampling designs may rule out due to the lack of unavailability of data and a researcher may have to engage in telephonic survey, convenience sampling, or snowball sampling relevant data for further progress.

The geographical spread of the elements in the population also influences the selection of a sampling design, apart from the above-mentioned factors. For example, if the scope of the research covers the whole of one country say India, and the elements in the population are scattered across the country, the researcher has to opt for cluster sampling.

6.6 Essentials of a Good Sample

The sampling results must reflect the characteristics of the population. Therefore, sample from which it is drawn must represent a true picture of the population, by the sampling procedure a sample must be unbiased, sample must be taken at random so that every member of the population of data has an equal chance of selection, sample must be sufficiently large but as economical as possible, sample must be accurate and complete. It should include all the respondents, units, or items included in the sample. It should not leave any information incomplete, and considering the degree of precision required in the results of inquiry, adequate sample size must be taken.

6.7 Sampling Errors

The term error is an act, assertion, or belief that unintentionally deviates from what is correct, right, or true. Because there is the involvement of human intelligence and the use of sampling methods that may not be always accurate, there is sure to be some error in the results in research process.

Sampling error is the absolute value of the difference between an unbiased point estimate and the corresponding population parameter. It arises because the data are collected from a part, rather than the whole of the population. The sampling error can be reduced but not eliminated. And by increasing the sample size, the study findings can be assumed to be more reliable.

Survey errors are random sampling error and nonsampling error.

The quality of a research project depends on the accuracy of the data collected, data representation to the population.

We know that only a fraction of population data represents the sample data. Therefore, the problem is related to how well the sample represents the characteristic of the populations of which it is a part.

6.7.1 Random Sampling Errors

The difference between the sample results and the results of a census conducted by identical procedures is called as random sampling error or sampling error. Although a representative sample is taken, there is always a slight deviation between the true population value and the sample value. This is because the sample selected is not perfectly representative of the test population. Therefore, a small random sampling error is evident. The laws of probability are applicable to it because sampling error is the outcome of chance. The sampling error is inversely proportional to the sample size.

As the sample size increases, the sampling error decreases. Although sampling errors cannot be avoided altogether, they can be controlled through careful sample designs, large samples, and multiple contacts to assure representative.

Random sampling error represents how accurately the sample's true mean value$\overline{X}_{\text{Sample}}$, is representative of the population's true mean value $\overline{X}_{\text{Population}}$. The sampling method applied and the sample size are the principal sources of sampling errors. This is because only a part of the population is covered in the sample. Even for the same sample size, the magnitude of the sampling error varies from one sampling method to the other.

Sampling error is made while selecting a sample that is not representative of the population. It represents the difference between sample value and true value of population parameters.

A sampling error is bound to occur while selecting a sample as it is difficult, if not impossible, for a sample or a small part of the population to be exactly representative of the population. This occurs no matter how careful the researcher is randomly choosing the sample. Therefore, sampling error, is a result of chance. The sampling error usually decreases with increase in sample size and it is nonexistent in a complete enumeration survey.

6.7.2 Nonsampling Error

Nonsampling errors, also known as systematic errors, occur due to the nature of the study's design and correctness of execution. Nonsampling error includes nonobservation errors and measurement errors.

6.7.2.1 Non-observational Errors

It occur, when data cannot be collected from the sampling unit or variable. Measurement errors arise from various sources such as respondents, interviewers, supervisors, and even data-processing systems. Nonobservation error is further divided into noncoverage and nonresponse errors.

6.7.2.1.1 Non-coverage Error

In probability sampling, each element of the population has a nonzero chance of selection into the sample. Non-coverage error occurs when an element in the target population has no chance of being selected into the sample.

6.7.2.1.2 Non-response Errors

Non-response error occurs when data cannot be collected from the element actually selected into the sample. This may be due to the refusal of the element to cooperate, due to a language barrier, health limitation, and nonavailability of the element during the survey period. Selection of a faulty sampling frame may also result in a nonsampling error. Sampling frame error is said to occur when nonpotential respondents are included in the sampling frame and certain deserving respondents are rejected.

The nonsampling errors arise from faulty research design and mistakes in executing research. The sources of nonsampling errors are respondent errors and administrative errors.

6.7.2.2 Respondent Errors

The objectives of the researcher can be easily accomplished if the respondents cooperate and provide the correct information. However, in practice, this may not happen. The respondents may either refuse to provide information or even he or she may provide biased information. If the respondent fails to provide information, it is termed as nonresponse error. Although, in all types of surveys this problem is present, the problem is more acute in mailed surveys.

The researcher often seeks to re-contact with the nonrespondents if they were not available earlier, in order to minimize the nonresponse error. If the researcher finds that the nonresponse rate is more in a particular group of respondents, then for these people who are not responding to the mailed questionnaires, personal interviews may be conducted to obtain data. When the respondents may not give the correct information and try to mislead the investigator in a certain direction, response bias occurs. The respondents may consciously or unconsciously misrepresent the truth.

For example, if the investigator asks a question on the annual income or age or tax payment of the respondent he or she for obvious reasons may not give the correct information.

6.7.2.3 Administrative Errors

These errors arose due to improper administration of the research process. The types of administrative errors are Sample Selection Error, Investigator Error, Investigator Cheating, and Data Processing Error.

1. **Sample Selection Error:** To execute a sampling plan is difficult.

   For example, we may plan to use systematic sampling plan in a market research study of a newly launched product and decide to interview every forth customer coming out of a consumer store. If the day of interview happened to be a working day then we are excluding working consumers. Because of the unrepresentative sample selection, this may lead to an error.

2. **Investigator Error**: When the investigator interviews the respondent, he or she may fail to record the information correctly or may fail to cross check the information provided by the respondent. Therefore, due to the way the investigator records the information, the error may arise.
3. **Investigator Cheating:** Sometimes, even without meeting the concerned respondents, the investigator may try to fake the data. There should be some mechanism to crosscheck this type of faking by the investigator.
4. **Data-Processing Error:** Once the data are collected the next job the researcher does, for further processing and analysis is edit, code, and enter the data into a computer. By careful editing, coding, and entering the data into a computer, the errors can be minimized. One may commit tabulation errors while processing the data.

6.7.3 Error Control

The ways of reducing the errors are designing and executing a good questionnaire, selection of appropriate sampling method, adequate sample size, employing trained investigators to collect the data, and care in editing, coding, and entering the data into the computer.

Summary

Sampling is the act, process, or technique of selection a suitable sample, or a representative part of a population for the purpose of determining parameters or characteristics of the whole population. A sample is the segment of the population that is selected for investigation. A complete survey of population is called census.

There are two broad categories of sampling methods: (a) random sampling methods and (b) nonrandom sampling methods. Probability sampling involves random selection to ensure that each unit in the population has known probability of being selected into the sample. On the other hand, nonprobability sampling does not follow a random selection process and the units in the population have an unequal probability of being selected into the sample.

Probability method includes simple random sampling, systematic sampling, stratified random sampling, and cluster sampling. Simple random sampling can be beneficial when the target population size is small, homogeneous, the sampling frame is clearly defined, and not much information is available regarding the same. Systematic sampling, which just involves a skip interval for choosing sampling units, is often used as an alternative to simple random sampling. Stratified sampling involves segregating the target population into strata before sampling, to ensure that the sample is more representative of the population. Cluster sampling is similar to stratified sampling, but here the clusters are selected first and then samples are drawn from these clusters. Nonprobability sampling includes convenient, quota, judgment, and snowball sampling techniques. Convenience sampling is sampling based on convenience or easy accessibility to the sampling units. Quota sampling involves sampling based on certain predefined characteristics in a particular category. Judgment sampling is a process where the researcher chooses a sampling frame based on his or her experience. Snowball sampling deals with sampling through reference when the availability of sampling units is rare.

The determination of sample size is discussed. In general, researchers face two errors: random sampling errors and nonsampling errors. Random sampling error or sampling error is the difference between the sample results and the results of a census conducted by identical procedures. Nonsampling errors are the errors caused by factors other than sampling and include nonobservation errors and measurement errors. With an appropriate sampling plan and selection of random sampling method, the sampling error can be minimized. It may not be possible to completely eliminate the sampling and nonsampling errors.

Review Questions

1. What is sampling? Discuss its importance.
2. Point out the basic difference between stratified sampling and cluster sampling.
3. What is simple random sampling? Discuss its advantages and disadvantages.
4. What do you mean by probability sampling method? Discuss any two types of probability sampling methods.
5. Define stratified random sampling. What is stratification? Discuss its advantages and disadvantages.
6. Explain in detail the various sampling designs under nonprobability sampling method.
7. What is the difference between random sampling and nonrandom sampling?
8. List some of the situations where (a) sampling is more appropriate than census and (b) census is more appropriate than sampling.
9. What are the advantages and disadvantages of stratified random sampling?
10 What are the ways to control survey errors?
11. What are the advantages of sampling over census?
12. Discuss the method of cluster sampling. What is the difference between cluster sampling and stratified random sampling?
13. Discuss the sources of sampling and nonsampling errors.
14. What are the essentials of a good sample?
15. Define the following:
    a. Sampling unit
    b. Population
    c. Sampling frame

7

Secondary Data

7.1 Introduction

Secondary data can help the researcher in identifying the research problem, formulation of research hypotheses, and generation of new ideas that can be later authenticated by primary research. For measuring the soundness and precision of the primary data, secondary data can serve as a reference point.

Through appropriate data and their analysis the decision maker becomes equipped with proper tools of decision making.

Secondary data are those that have already been collected by some other agency and that have already been processed. Secondary data are the data that already exist, which has been collected by some other person or organization for their own use, and is generally made available to other researches free or at a concessional rate. Sources of secondary data include websites, trade associations, journals, books, etc.

Secondary data may not always answer the specific question of a researcher. Hence, within the allotted time and money, a researcher should use a judicious mix of primary and secondary data to optimize the quality of research findings. It is necessary to take help of various types of secondary data to design a proper sampling scheme, even when an organization is interested to collect primary data.

7.2 Classification of Secondary Data

Classification of secondary data is based on source, category, medium, and databases.

7.2.1 Classification by Source

7.2.1.1 Internal Sources of Secondary Data

Internal sources of secondary data are those that are available within the organization. The examples of internal sources of secondary data are departmental reports, production summaries, financial and accounting reports, and marketing and sales studies.

Sales records can be a source of valuable information with regard to territory-wise sales, sales by customer type, prices and discounts, average size of customer order, customer type, geographical area, average sales by salesperson, sales by pack, size and pack type, trends within the enterprise's existing customer groups, etc.

Financial data records have information with regard to the cost of producing, storing, transporting, and product lines.

7.2.1.2 External Sources of Secondary Data

External sources exist outside the company. External sources of secondary data occur in books and periodicals, government sources, computer-retrievable databases, trade and manufactures' associations, publications, median sources, commercial sources, syndicated services, directories, external experts, and special collections. Pertaining to research these sources can also provide valuable information.

DOI: 10.1201/9781315167138-7

7.2.1.2.1 Population Statistics

Since 1871–1872, Population statistics in India have been collected every 10 years. For estimation of consumer demand for various goods and services, it provides factual bases by furnishing data on size of population and its distribution by age, sex, occupation, and income levels.

7.2.1.2.2 Statistical Abstract of India

This publication contains the statistics of various sections of the Indian economy for the preceding 5 years.

7.2.1.2.3 Estimation of National Product

This annual publication publishes annual estimates of national income, savings and consumption, capital formation, and expenditure as well as national and public sector accounts.

7.2.1.2.4 Monthly Statistics on the Production of Selected Industries

The Central Statistical Organization (CSO) publishes monthly statistics relating to production, installed capacity, and stock positions in selected industries, to bridge the gap between the census taken and data published.

7.2.1.2.5 Basic Statistics Relating to Indian Economy

This annual publication contains basic indicators on various aspects of economy.

7.2.1.2.6 India Pocketbook of Economic Information

This annual publication includes statistics on the various aspects of the national economy. Other important publications include India, a reference manual; Agricultural situation in India; Reserve Bank of India (RBI) Bulletin, Economic Survey; Bulletin of Food Statistics; Commercial Crop Statistics; Indian labor statistics, etc.

7.2.1.2.7 Trade Statistics

It publishes data on import and export of goods in terms of their quantity and value, classified as received from or sent to centers of consignment. Also, this publication provides the Value of foreign trade, Balance of trade, Foreign trade with each country and currency area, Foreign trade in groups of commodities with each country and currency area, Foreign trade with selected countries, etc.

7.2.2 Classification by Category

Classifying secondary data by category includes Books and periodicals, Databases, Government documents, Publications, Associations, External Experts, Directories, Median sources, Commercial sources, and Special collections.

7.2.3 Books and Periodicals

Book and Periodicals are a typical source for a desk researcher procured from various sources. A researcher who located the right book pertaining to his or her research gets off to a good start.

Professional journals serve as rich sources of secondary data, which includes Journal of Business Research, Accounting Research, Marketing Research, Financial Analysts Journal, Business Week, Business World, Economies, Fortune, and Harvard Business Review. For market research information, newspapers can also be a vital source.

7.2.4 Government Publications

The publications by the Union government are the largest single source of secondary data. Marketing researches have relied on this source of data for the estimating market potential and sales forecasts, determining distribution penetration, location of intermediate and final outlets, defining sales territories, and routing schedules. The important official publications are Statistical Abstract, India-Annual, Monthly Abstract of Statistics both published by Central Statistical Organization (CSO); Indian Agricultural Statistics published annually by Ministry of Food and Agriculture; Index Number of Wholesale Prices in India published weekly by Ministry of Commerce and Industry; Reserve Bank of India Bulletin published monthly by Reserve Bank of India. The National Sample Survey (NSS) is another important source, publishes data, on an elaborate and continuing basis with regard to social economic, demographic, industrial, and agricultural statistics.

The important publications of international bodies are United Nation Organization (UNO), Food and Agriculture Organization (FAO), World Health Organization (WHO), United Nations Educational, Scientific and Cultural Organization (UNESCO), International Labor Organization (ILO), Statistical Year Book published by the Statistical Office of the United Nations, Yearbook of Labor Statistics published by ILO, Geneva, etc.

Other government agencies in the United States, which bring out useful secondary information, are the Bureau of Alcohol, Tobacco, Firearms and Explosives, Bureau of Industry and Security from the U.S. Department of Commerce, the U.S. Food and Drug Administration, etc.

7.2.5 Nongovernmental Associations

Nongovernmental associations include the marketing research agencies as well as the data services, which in addition to providing standardized data also undertake specific data collection research projects.

Other nongovernmental sources of publications are the annual statistic by the Market Research and Statistical Bureau of the Coffee Board Bangalore, an annual publication "India's Production, Exports and Internal Consumption of Coir and Coir Goods" by the Coir Board, Cochin, an annual report, the Rubber Statistics, by the Rubber Board, Kottayam, the Indian Sugar Year Book by the Indian Sugar Mills Association, Delhi, etc.

Several chambers of commerce include Federation of Indian Chambers of Commerce and Industry (FICCI), Associated Chambers of Commerce and Industry of India, etc.

In the United States, several nongovernmental associations that publish industry-related information include "The American Statistics Index" published monthly, quarterly, and annually by LexisNexis, and bring out indexes and abstracts of a wide range of statistical publications, etc.

7.2.6 Directories, Industry Experts, Special Collections

7.2.6.1 Directories

Industry-specific directories provide first-hand information about the existing players, their products, and strategies. Researchers, when they are preparing sampling frames often make use of directories. For detailed information on the corporate sector, Stock exchange directories can be a handy source.

Some Indian directories include "The City as it Happens," detailed information about major metros, people and their lifestyles, pubs, latest fashion trends, etc., "Tata Yellow Pages" guide offers classified information of products, services, and organizations in major Indian cities, etc.

Some directories in other countries include "The Thomas Register of American Manufacturers," New York Thomas Publishing Co., which has data on more than 150,000 companies, "Who Owns Whom," North American edition lists more than 6,500 parent companies and more than 100,000 domestic and foreign subsidiaries and associated companies, etc.

7.2.6.2 Industry Experts

Getting information from Industry experts is often highly useful for research, as these experts specialize in their own domain.

Some of the expertise consulting services are Price Waterhouse Coopers (PWC), which has expert financial consultants, Data monitor's Business Information Centre, which provides services to the world's largest companies in the fields of Automotive and logistics, Consumer markets, Energy and Utilities, etc.

7.2.6.3 Special Collections

It consists of Reference books, University publications consisting of master's thesis, Doctoral dissertations, and Research papers; company publications include financial reports, company policy statements speeches by eminent personalities, sales literature, etc.

7.2.7 Classification by Medium

Secondary data classified by medium include hard copy and Internet. Hard copy refers to nondatabase information. Hard copy comprises all books, magazines, journals, and special collections contained in hard-copy libraries. Browsers such as Microsoft's Internet Explorer and Netscape's Navigator make it possible to access sites and user groups of all those connected through the Net.

7.2.8 Classification by Database Content

A classification of database by content of information includes online, Internet, and offline databases. Online databases consist of a central data bank accessible by a terminal, via a telecommunications network. Internet databases are those that can be accessed on the Net and can also be downloaded if required. Offline databases are those that make the information available on diskettes, CD-ROMs, etc.

7.2.8.1 Reference Database

A bibliography of documents, abstracts, or locations of original information is provided by a reference database. They are also referred to as bibliographic databases, as they provide online indices, citations, and abstracts.

Some of the reference databases include Marketing and Advertising Reference Services (MARS), Aerospace or Defense Markets and Technology (A or DM&T), (PTS) Newsletter Database, First and Second (F&S) Index, etc.

7.2.8.2 Source Database

Numerical data, full text, or a combination of both are usually published by source databases. They include full texts of various economic and financial databases. These databases provide complete text and numerical information.

7.3 Scrutiny of Secondary Data

Secondary data are to be scrutinized before they are compiled from the source. The scrutiny should be made to assess the suitability, reliability, adequacy, and accuracy of the data to be compiled and to be used for the proposed study.

7.3.1 Suitability

The compiler should satisfy himself or herself that the data contained in the publication will be suitable for his or her study. In particular, the conformity of the definitions, units of measurement, and time frame should be checked. For example, one US gallon is different from one British gallon.

7.3.2 Reliability

The reliability of the secondary data can be ascertained from the collecting agency, mode of collection and the time period of collection. For example, secondary data are unlikely to be reliable collected by a voluntary agency with unskilled investigators.

7.3.3 Adequacy

The source of data may be suitable and reliable but for the proposed enquiry the data may not be adequate. The original data may cover a bigger or narrower geographical region or the data may not cover suitable periods. For example, per capita income of India prior to 1970s is inadequate, as it became separated into two different countries, with-considerable variation in standard of living for reference during the subsequent periods.

7.3.4 Accuracy

About the accuracy of secondary data, the user must be satisfied. The process of collecting raw data, the reproduction of processed data in the publication, and the degree of accuracy desired and achieved should also be satisfactory and acceptable to the researcher.

7.4 Advantages and Disadvantages of Secondary Data

Advantages of secondary data

1. It is cheaper, and takes less time to gather, thus saving the researchers a lot of money and time that they would have otherwise spent in gathering primary data.
2. It can help identify, clarify, and redefine the research problem.
3. It might also hold a solution to the problem.
4. It may provide alternative methods that can be used for primary research.
5. For better creativity, it generates requisite information.
6. It can be collected by the researcher from a published or compiled research at very little cost and usually very speedily.
7. It provide access to information that would not ordinarily be obtainable by an individual organization.

Disadvantages of Secondary Data

1. **Lack of Availability:** It might so happen that there are no secondary data available for special cases or that the organization holding such data is not willing to make it accessible to outsiders. For example, if a company such as Ashok Leyland would like to conduct a research for the market potential of its vehicles in particular cities in India, then, in this context, it is very unlikely that any secondary data would be available.
2. **Lack of Relevance:** Due to difference in units of measurement, use of surrogate data in the secondary sources, difference in definition of classes and time, relevance might be reduced.
3. **Inaccurate Data:** Errors that can occur in any of the steps or due to personal bias, and can make the secondary data inaccurate and therefore unusable.
4. **Data Fitness Problem:** As secondary data have been compiled for other purposes, rarely are they completely pertinent to the information needs of the problem at hand.
5. **Identify the Data Source**: When secondary sources of secondary data are used, not only are details of methodology of original source nonavailable, but also errors originating at the secondary source may affect the accuracy of the original data.
6. **Examine the purpose for which data were published:** Sources must be treated with caution, which publish to promote the interests of a particular group or for political commercial or social reasons. Similarly, data published have limited use to promote sales or to carry on a particular propaganda or to promote views of a particular interest group, unless it is carefully interpreted in light of purpose of publication.

Suspected data include published anonymously, defensive by an organization, under conditions that suggest a controversy, in a form that reveals a strained attempt at frankness, to controvert inferences from other data, etc.

Summary

The secondary data are those that have already been collected by some other agency but also can be used by the organization under consideration. Secondary data are available in various published and unpublished documents. The suitability, reliability, adequacy, and accuracy of the secondary data should, however, be ensured before they are used for research problems.

Secondary data are gathered and recorded by some other firm or individual for a purpose other than that of the current research. Researchers have a huge quantity of secondary data at their disposal and it can be classified based on source, category medium, and database type.

Secondary data can be either sourced from various departments internal to an organization or from some external source. Researchers can browse through different books, periodicals, governmental and nongovernmental publications, directories, and special collections while beginning their research. Apart from these sources, a researcher can refer to various reference and full-text databases. A reference database provides a bibliography of documents, abstracts, or locations of original information related to various topics that are of interest to the researcher. On the other hand, full-text databases provide numerical data, full text, or a combination of both.

Review Questions

1. Discuss the main sources of secondary data.
2. What is the type of data available from official publications?
3. What are the limitations associated with the use of secondary data?
4. Discuss the important sources of error in secondary data.
5. Describe the various methods of collecting secondary data and comment on their relative advantages and disadvantages.
6. Define secondary data. State their chief sources and point out the dangers involved in their use and the precautions necessary to use them. Illustrate with examples.

8

Survey Research

8.1 Introduction to Survey Research

Survey research is one of the most relevant techniques basically used for collecting data. This method involves any measurement procedures that prominently include asking questions from respondents or the subjects selected for the research study. The term "survey" can be defined as a process that may involve an investigation or an examination or assessment in the form of a short paper and pencil feedback form to an intensive one-on-one in-depth interview. The method tries to gather data about people, their thoughts and behaviors, with the help of the questionnaire or other statistical tools.

A researcher should at first explore all sources of secondary data and verify the possibility of their usage for the research at hand. But it may so happen that after a point, these secondary data, for making further marketing decisions, might prove to be inadequate or of no use to the researcher. In such cases, the researcher has to go in for primary data research employing survey research method.

A survey cannot measure the behavior of people, but can trace a certain behavior to the perceptions, feelings, attitudes, beliefs, and other personal characteristics of the respondent.

8.2 Concept and Meaning of Survey Research

The method of survey research does not involve any observation under controlled conditions, hence it is a nonexperimental. Descriptive research used for studying of large sample is one of the quantitative method. Population is referred to as the universe of a study, which can be defined as a collection of people or object, which possesses at least one common characteristic.

In a survey research, the researcher collects data with the help of standardized questionnaires or interviews, which is administered on a sample of respondents from a population.

8.3 Nature of Surveys

The method of collecting information by asking a set of preformulated questions in a predetermined sequence in a structured questionnaire to a sample of individuals drawn so as to be representative of a defined population is known as survey research.

A researcher conducting a survey has to deal with Sampling, Questionnaire design, Questionnaire administration, and Data analysis. Through interviews, these questionnaires are administered to an individual or a group of individuals.

These interviews, which have to be decided in advance can be face-to-face, over the phone, or through any other communication medium.

Typical survey objectives involve describing or learning from an ongoing activity by studying the changes in behavioral patterns of the subject of interest to the researcher. Thus, surveys tend to be descriptive in nature. Although surveys are often quantitative in nature, surveys also entail some qualitative aspects as in research concerning consumer satisfaction surveys, and new product development.

DOI: 10.1201/9781315167138-8

8.4 Classifying Survey Research Methods

Surveys can be classified on basis of the method and mode of communication, the degree of structure, the amount of disguise in a questionnaire and its structure, and the timeframe for data collection. Based on the selection of an instrument or method of data collection, the researcher can use qualitative, e.g., ask open-ended questions or quantitative, e.g., use forced choice questions.

The major types of surveys are cross-sectional surveys and longitudinal surveys.

8.4.1 Cross-Sectional Survey

In this type of survey, the total target population is divided into various segments and then using a sampling method data are collected from all these segments. Then, the collected data are analyzed to define the relationship among the various variables based on cross-tabulation. For example, a study designed to establish the relationship between ethics of parents and their views on Internet filtering is likely to bring in varied responses who are studied at the same time from different sections of society.

8.4.1.1 Advantage of Cross-Sectional Survey

1. More representative of the population, less time-consuming, and economical.
2. Can be used to study the difference in the consumption levels, trends in income, job changes, and buying behavior of individuals from various groups and subgroups of the population.
3. When he or she wants to collect data from varied or different types of groups, which may be in terms of age, sex, group, nation, tribes, and so on, at a single time.
4. Can be a study on the effect of socialization of children of different age groups of a particular country.

8.4.1.2 Disadvantage of Cross-Sectional Survey

Cross-sectional studies cannot be used when it comes to defining the same research objectives over a period of time. Here, longitudinal studies are required.

8.4.2 Longitudinal Survey

Longitudinal studies use multiple surveys to gather data over a period of time. It is used only when the subject wants to study the same sample for a longer period of time. It may be used to study the behavioral changes, attitude changes, and religious effects or any event or practice that may have a long-time effect on the selected sample or population. They help in monitoring the behavioral changes, which is of interest to the researcher taking place in the population. This type of survey is flexible and can over a period of time interview different respondents provided the new subjects are also from the same group or subgroup originally interviewed. Hence, longitudinal surveys are essential not only to learn about current social situations but also to measure their variation over a time period. A number of different designs are available for the construction of longitudinal surveys. They are Trend Studies, Panel Surveys, and Cohort Panels.

8.4.2.1 Trend Studies

Longitudinal surveys consisting of a series of cross-sectional surveys conducted at two or more points in time, with a new sample on each occasion, are known as trend studies. When the researcher needs to analyze a trend of a phenomenon in a population, he or she conducts trend studies. But it should be ensured that, as trend studies focus on the changing patterns of a particular population, the new sample is from the same category or segment of population originally surveyed.

Since, at a particular point of time, each survey brings out the existing trend, data from several cross-sectional studies of the same population can be integrated and a time trend analysis can be established into the longitudinal survey. This can be done in each of the cross-sectional studies, by using consistent questions.

The sample of the selected population might not be the same, as over a period of time, for various reasons, they might have shifted or not available but they belong to the same population. This selected population is sampled and examined regularly. Since it is a type of longitudinal research, it may not be started as well as ended by just one researcher or research project. For example, a yearly survey of number of graduates, postgraduates, and PhD scholars actively using books and journals from the library of a university.

8.4.2.2 Cohort Studies

Those people within a geographically or otherwise delineated population, who experienced the same significant life event within a given period of time, is defined as a Cohort. Cohort panels can be considered as a specific form of panel study that takes the process of generation replacement explicitly into account. Thus, over their life course one or more generations are followed.

The focus of this type of longitudinal study is also on a particular population, which is sampled and studied more than once within a time gap.

The study usually probes into the long-term changes and the individual development processes. If the same sample people are investigated, in each particular generation, then a cohort study consists of a series of panel studies. If in each generation, in each period of observation, a new sample of respondents is drawn, then a cohort study consists of a series of trend studies.

For example, this study can be an investigation of the number of graduates, postgraduates, and PhD scholars of the last year, who have been actively using the library, and 4 years later, the researcher may examine the same issue on another sample of the graduates, postgraduates, and PhD scholars and investigate whether after the time gap there has been any difference in the attitudes toward the importance of the library within the members of the same class. Wherein, in the trend study, the research scholar would study such an attitude within the graduates of different batches of the same university.

8.4.2.3 Panel Studies

Panel studies is a longitudinal survey, which involves collecting data from the same sample of individuals or households across time. The researcher in a panel study uses the same sample of people every time and that sample is called as a "panel." This means that the selected sample is called a panel. Such a study is used in order to investigate the changes in attitudes, behavior, or practices of the same panel within a period of time. They are more specific and focused as the researcher studies a particular change in the attitude, behavior, belief, or practice of the same group.

Panel surveys enable the researcher to detect and establish, over a period of time, the nature of changers occurring in the population. As the surveys are conducted on the same panel over a period of time, these changes can be traced to the level of the individual. A particular sample of interviewees might respond or react to an impulse in a certain way, which might differentiate them from others over a period of time. The very basis of longitudinal surveys lies in detecting these changes.

For example, a researcher may study the library usage trends amongst the graduates, postgraduates, and research students and ask them questions related to their frequency of library usage habits. Thereafter, the researcher may ask the same group or panel similar questions and also the reasons behind the changes in their habits, if it has occurred. The study is difficult enough as it faces a greater trend of attrition rates, difficulty in availability of the same people.

8.4.2.3.1 Advantages of Panel Studies

They provide highly specific information.

8.4.2.3.2 Disadvantages of Panel Studies

They are time consuming, expensive, and are known to have high attrition rates as people often drop out of the study.

8.5 Survey Methods

Surveys conducted through interviews are generally classified based on the method of communication used in the interview. Surveys should be designed in such a way that helps in making accurate decisions. Predominantly there are three major ways that can be used as an instrument in collecting data with the help of survey research.

8.5.1 Sampling

A representation of the population or universe selected for the study is known as a sample. In survey research, the technique of sampling in collecting data can itself act as an instrument.

For example, if the objective of researcher is to study the level of job satisfaction amongst the employees of an organization. Then, the researcher can select at least five to ten persons of each department of the organization and study their attitude. The sampling can be done with the help of randomization, a method of sampling that provides an equal chance for each subject to be involved in the study, in order to avoid any bias. It can be done with the help of the Lottery Method, Fish bowl technique and stratification, or a method of sampling that categorizes the population into various categories and subcategories and then conducting the research.

8.5.2 Questionnaire Design

Questionnaire design is a vital issue in interviewing. From the respondent a properly designed questionnaire can tap the necessary information. Therefore, researchers always design a tactful set of questions to give useful answers to probe and prompt the interviewee.

The categories of questionnaires are Structured, Unstructured, Disguised, and Undisguised.

Questionnaires, in which the individual needs to select the most suitable alternative, are basically a kind of paper pencil and multiple choice test. With the help of a questionnaire, the researcher at a single time may collect data from a large number of samples.

Questionnaires can be administered to the sample by Mail survey, Group-administered questionnaire, and Household drop-off survey.

8.5.2.1 Mail Survey

Through mail the researcher may forward a soft copy of the questionnaire to a large number of respondents and can get the data collected from them at a single time. Mail survey is one of the methods of obtaining responses that are relatively less time consuming, convenient, and inexpensive. Yet, the questions that require on-the-spot response or detailed answer is difficult to be achieved through mail survey.

8.5.2.2 Group-Administered Questionnaire

This questionnaire is one of the traditional methods of administering questionnaire. The researcher calls for a large number of respondents to be present as a group at a stipulated time period. Under such group settings, the respondents are asked to respond to a structured sequence of questions written in paper or questionnaire. The greatest advantage of this method is that with regard to any question, the respondents can clarify their doubt that has been asked by the researcher instantly.

8.5.2.3 Household Drop-off Survey

In this method, the researcher goes door to door to the respondents and personally hands over as well as collects the questionnaire from them. It is a kind of pick-and-drop facility that is provided by the researcher so that the researcher can answer the questions according to their convenience.

8.5.3 Personal Interviews

Based on the respondents to be interviewed and the means to contact them, the different methods of personal interview are door-to-door interviewing, executive interviewing, and mall intercept surveys. Personal interviews are characterized by the researcher, interviewer, interviewee, and interview environment. The researcher, interviewer, and interviewee have specific to each of them some inherent and acquired characteristics. As such, in some way or the other, they are able to influence the interviewing process. Based on the type of data to be collected the choice of the interview environment is chosen by the researcher.

Advantages of Personal Interviews

1. **Feedback Opportunities:** Personal interview provides the opportunity to clarify the doubts of the interviewee. A respondent hesitant to provide sensitive information can be assured of the confidentiality of the information provided.
2. **Probing:** In a personal interview, the interviewer has the advantage of probing the respondent for complex answers. For example, a respondent might reveal his or her likes or dislikes for a certain ice-cream flavor, which is of no use to the researcher. But with the interviewer present, the actual reason can be traced back to any of the product attributes. By asking further questions, the interviewer can probe respondents' specific product attribute that they like or dislike. This kind of information is more useful to the researcher.
3. **Length of Interview:** The length of an interview is appreciably better compared to other survey in personal interviews. This is because for a reluctant respondent it would be easy to hand up the phone, not respond to an e-mail, avoid someone in a face-to-face interview, etc.

Hence, as compared to other nonpersonal survey methods, the chance of the respondent answering all the questions of the researcher is greater. Some respondents, though reluctant to participate in a nonpersonal survey method, feel comfortable with an interviewer present right in front, about sharing information. In the case of personal interviews, this leads to an increase in the length of the interview and an improvement in the quality of response.

Disadvantages of Personal Interviews

1. **Cost:** Personal interviews are generally expensive as compared to e-mail, internet, and telephonic surveys. The costs are directly related to the number or quantity of interviewee, quality of the workforce employed, reach of respondents, length and complexity of questionnaires, nonavailability and ignorance, and extent of nonresponse.
2. **Lack of Anonymity of Respondents:** As respondent's identity is known to the interviewer, in a personal interview respondents may hesitate to provide the right information. For example, questions such as smoking habits during driving and extramarital affairs are sure to fetch falsified answers. Thus, the interview suffers from social desirability bias. To overcome this issue, interviewers spend a lot of time in framing questions in the best possible way, so as to be able to prompt the true responses from the interviewees even for sensitive issues.
3. **Necessity for Call-Backs:** The characteristics of those who remain at home such as housewife, nonworking women, and retired people are different from those who go to work. Hence, it becomes necessary to recontact people who were unavailable at the first call. This requires a systematic procedure and often turns out more costly than interviewing the individual in the first call itself.

8.5.3.1 Structured Interview

Structured interviews are those interviews in which the questions that are to be asked from the respondents are prepared and preplanned in advance by the researcher. The researcher serially imposes those prepared questions on the respondents and note down the answers given by them.

Structured interviews are from the most part orally administered questionnaires. Such questions restrict the interviewee from giving his or her own answers and require him or her to choose from among the alternatives given. This saves a considerable amount of time as the respondent is quick to choose from among the options given to him or her. Thus, rather than going off the track, the interviewer takes the interview in the required direction. The structured questionnaire makes the interview somewhat "funnel' shaped. Without any influencing factors, the interviewer consciously guides the interviewee through a sequential, preformulated set of tactful questions to extract the "factual" responses. This leads to accomplish the goal of the interview. The common features of structured interviews include a common vocabulary for all interviewees, question formats have the same meaning for all, in exactly the same way all respondents are interviewed, in advance with their order the questions are set and for all respondents the range of possible responses are the same.

Advantages of Structured Questionnaire

1. They are easy and the interviewee can answer them quickly.
2. Similar questions and a uniform format make the answers easy to decode and analyze.
3. The factual information has a high degree of reliability and reduces the possibility of any interviewer bias.

Although structured questionnaires help the researcher in eliciting programmed responses, they fail to probe into the actual motives of the respondent. By including some unstructured questions in the questionnaire, this drawback can be overcome.

It might happen that sometimes a questionnaire has a set of personal and sensitive questions to which the respondent might give incorrect answers. These are a set of questions to which the interviewee might take offence or questions that might threaten his or her ego or prestige. In such situations, the interviewee may knowingly give the wrong information. To nullify such instances of deliberate falsification, the interviewer frames the questions in a disguised manner. To elicit the right information from the respondent in an indirect manner, these disguised questions help in framing in a tactful manner thus leading to the accomplishment of the research objective.

Depending on the degree of structure and disguise involved, questionnaires can be further categorized as Structured-undisguised, Unstructured-undisguised, and Structured-disguised.

Disadvantages of Structured Questionnaire

1. Due to the variance in the degree of structure and disguise in the questions makes them less straightforward and liable to misunderstanding by the respondent.
2. As the number of responses is limited, interviewees feel forced to choose one even if it does not divulge their true feelings.
3. They tend to adopt a hybrid style of questionnaire format including structured, unstructured, disguised, and undisguised questions because surveys have a mix of personal and general questions. This leaves no alternative other than response bias to creep into the research data.

8.5.3.2 Unstructured Interview

Unstructured questionnaires are usually open-ended. Unstructured questionnaires try to probe into the mind of respondent, allowing the interviewee to express his or her own thoughts rather than restricting him or her to the available response options.

When the researcher conducts an interaction with the respondent in an informal atmosphere, interviews are said to be unstructured. Nothing is preplanned in advance. The response of the sample gives a clue to the researcher to ask the next question.

8.5.3.3 Telephonic Interview

Telephonic interviews, once thought of as "quick and dirty," providing less reliable or valid data, have finally come of age, and are currently judged as one of the best cost-effective alternatives to face-to-face interviews and mail surveys.

The reasons for shift of focus to telephonic are in certain urban areas, plunging response rates in face-to-face interviews, lower cost of telephonic interviews as interviewer travel time and mileage are eliminated. Also, the introduction of random digit dialing and adoption of new technology in telephonic interviewing in the form of Computer-Assisted Telephone Interviewing (CATI) and Computer Voice Activated Telephonic Interviewing (CVATI).

The use of random digit dialing eradicated many problems associated with telephonic interviews, as a sampling procedure. Instead of sampling from existing telephone directories, it used sampling through a random number procedure. This ensured that even those individuals in the sample who had shifted or changed their telephone numbers could be included. But the sampling frame for telephonic interviews is not restricted only to directories.

Researchers are also known to make use of the sampling frames for telephonic interviews such as Student Registers, Hospital and Clinic Records, Census Tract Information, and Employee Lists of Corporations.

The researcher may call the subjects or sample through telephone and ask them questions to collect data, in order to save time and money. This method helps in saving time and energy. But, the sample gets limited to only that part of the population who have at their residences or offices the facility of telephones.

Advantages of Telephonic Interview

1. Speed in data collection, making call-backs is easier.
2. Through improved techniques, potential to produce a high-quality sample.
3. As individuals reluctant to respond to face-to-face interviews feel more comfortable with telephonic interviews, thereby increasing cooperation and quality of data.
4. Ability to interview respondents in high-crime areas, which is a limitation for face-to-face interviews.
5. Facilitation of collection of socially undesirable responses, which is a drawback in face-to-face interviews.

Disadvantages of Telephonic Interview

1. The inability of the interviewer to display products, concepts, and advertisements, or to judge the respondent on demographic characteristics due to the absence of face-to-face contacts.
2. Time length of interviews is less and it is easy for a reluctant respondent to hang up the phone rather than avoid someone in a face-to-face interview.
3. Interviews on sensitive topics, although they may exceed the expected length of time give rise to doubts regarding the quality of data.
4. Uses of screening devices such as Caller ID and answering machines have increased the nonresponse rates for telephonic interviews. Respondents are more willing to participate in a legitimate survey rather than entertaining callers who wish to sell products.
5. When the interest group consists of the general population, and directories are used as sampling frames, samples are usually not representative.

In telephonic interviews, adoption of advanced techniques has helped interviewers to overcome many of the problems associated with this method. These advanced techniques are discussed below:

8.5.3.3.1 Central Location Telephone Interviews (CLTI)

For central location telephone interviews, the interviewers make calls from a centrally located marketing research facility to reach the interview respondents. For making the calls Wide-Area

Telecommunication Service (WATS) lines are used. These lines, facilitated at fixed rates, unlimited long distance calls throughout the country or geographical area. The superiority of CLTI can be attributed to one factor, i.e., control. Using special monitoring equipment, the whole interviewing process can be monitored by supervisors. This means that interviewers can be either be corrected or removed, who do not conduct the interview properly. For on-the-spot quality, this also facilitates editing and checking interviews. Interviewers can be appraised of any deficiencies in their work. Finally, since interviewers report in and out of the workplace, it helps to scrutinize and control their work hours.

8.5.3.3.2 Computer-Assisted Telephone Interviewing (CATI)

Computer-Assisted Telephone Interviewing is the process in which the telephonic interview responses can be directly entered into the computer. Here, while interviewing qualified respondents, the telephonic interviewer is seated at a computer terminal. Usually close-ended questions are used. The questions appear on the computer screen one at a time, in front of the interviewer, along with their possible response options. The interviewer reads out the questions and enters the corresponding answers of the interviewee into the computer. The computer automatically skips to the next question, as soon as the answer to the question is entered by interviewee. The questionnaire needs to be highly structured because the interview consists of close-ended questions with possible options for each.

With the use of the latest technology, the processing of the CATI has become much easier. This technology includes Telephone Management Systems, which take care of everything, starting from selecting telephone numbers at random to dialing them.

Another call management feature is automatic callback scheduling where the computer is programmed to make the necessary recalls as per the desired timings. Thus, timings can be set to recall "Busy Numbers" after 10 minutes and "no-contacts" after 2 hours.

The computer can also be programmed to fill a certain quota and to deliver daily status reports according to the quota.

Advantages of CATI

1. As data can be edited with their subsequent data entry, a separate step of editing is not required.
2. In the traditional way, tabulations that would require a week or more to compile, using CATI, it can be done at the click of a button.
3. In indicating clearly whether certain questions need to be deleted or added to the existing questionnaire to make it more specific, speed in tabulations also proves to be advantageous.

8.5.3.3.3 Completely Automated Telephone Surveys (CATS)

This process, which combines computerized telephone dialing and voice-activated computer messages, makes use of Interactive Voice Response (IVR) technology to record the responses of the interviewees. Since CATS involves a voice-synthesized module controlled by a microprocessor, the need for an interviewer is eliminated. The questions are highly structured, and close-ended with response options.

Functioning of CATS

1. To ask the questions, the computer uses the recorded voice of a professional interviewer.
2. By choosing from the options available, the interviewees are required to answer.
3. Then, pressing a number button on their telephone sets to mark their choice of options.
4. The options selected are thus recorded by the computer.
5. The system is so designed that, if a respondent does not answer the first couple of questions, the computer moves on to dial the next respondent.

For short, simple questionnaires, the use of CATS is handy. CATS technology is known to produce quality data at good speed and is also considered to be much economical compared to other telephonic methods. CATS shares the same advantages as CATI, because the computer handles the entire interview.

8.5.3.4 Door-to-Door Interviewing

This survey method involves consumers being interviewed in their homes.

Advantages of Door-to-Door Interviewing

1. Door-to-door interview involves a direct, face-to-face contact with the interviewee. Therefore, it has the inherent advantages of instant feedback and explanation of complex and difficult tasks.
2. To improve data quality, special questionnaire technique requiring visual contact, can be used in this method.
3. Door-to-door interviewing is an obvious choice, where complex product concepts are to be explained to the customer.
4. Since, as the customer being at home is at ease and is likely to reveal factual information, it is also helpful to the interviewer.
5. It provides a sample that is more representative of the population as compared to mail questionnaires. Even people who do not have a telephone or whose numbers are not listed in the telephone directory can be reached by door-to-door interviewing.
6. It is the best possible way for in-home product tests, which require either establishing facts about the product or explaining complex product features to the customer.
7. It reduces the chances of nonresponse error, as it involves direct, face-to-face interaction.

Disadvantages of Door-to-Door Interviewing

1. The number of potential respondents is low in a population where both adults work outside the home.
2. Unsafe areas, distance, and lack of accessibility pose a hindrance in reaching the desired sample.
3. Dearth of qualified interviewers.
4. Factors that might pose a hindrance to reaching the target samples, such as fluctuations in weather conditions, vehicle breakdown, sickness, etc.
5. Individuals who reside in high-rise apartments or are too busy to entertain personal interviews, it might not be possible to interview. Hence, these individuals have to be excluded from the list.

8.5.3.5 Executive Interviewing

Executive interviewing is similar to door-to-door interviewing with the only difference that it is specific to workplace respondents. It is concerned with the finding out information related to some industrial product of service, required the interviewing of people who use these products in their offices, etc.

The process is expensive but it is worth it. This is because the users more often than not make time for the interview as they too are interested in expressing their opinions and learning more about the products and services they use at work.

The interviewer should ensure that he or she reaches the venue on time. Often, the interviewees are busy at work and the interviewer might be required to wait for the meeting, at other times, the appointment might be postponed due to time constraints.

8.5.3.6 Self-Administered Interviews

In a self-administered interview, the questionnaire is filled out by the respondent without the intervention of an interviewer. Such interviews are not assisted by interviewer or the computer. These self-administered interviews are mostly conducted in the shopping malls, supermarkets or Big Bazaar or D-Mart, hotels, theatres, and airlines as these locations provide captive audiences, etc.

Passengers and regular customers are given brief questionnaires to enquire about their views of the quality of service offered in an airline or hotel. However, the absence of the interviewer results in a limitation, namely, that clarifications on responses to open-ended questions cannot be obtained. A customer might just indicate his or her liking as a reason for buying a particular product or brand, which is of no utility from a managerial perspective. The absence of the interviewer thus makes it difficult to race the buying decision of the customer to any of the product or brand attributes. Even the limited quantity of information is generated. However, the absence of the interviewer proves to be a boon in disguise as it eliminates the possibility of interviewer bias.

The use of Kiosks is another recent improvement in self-administered interviews. Kiosks are multimedia, touch-screen computers contained in freestanding cabinets.

The capacity of these preprogrammed computers to administer complex surveys is enhanced by their ability to display full-color scanned images and play stereo sound clips on any show videos.

Due to their numerous applications, these kiosks having been successfully tested at trade shows are now being tried in retail stores. Kiosk interviewing is less expensive and is known to derive more honest results than methods that involve an interviewer.

8.5.4 Mall-Intercept Surveys

The concept of mall interviewing, a predominant type of personal interview, in the United States today has become a popular way to collect survey data. The technique gained popularity in the early 1960s when big, enclosed shopping centers attracted a large number of people from various sections of society. Something of an ideal sample for researchers. Mall-intercept interviews are often viewed as an inexpensive substitute for door-to-door interviews.

Shopping mall-intercept interviewing involves the mall-intercept, as the name implies, stopping or intercepting shoppers in a mall at random, qualifying them if necessary, asking whether they would be willing to participate in a research study, conducting the interview right on the spot, taking them to the research agency's interviewing facilities located in the mall.

Prior to the mall intercept, surveys were conducted in other places having a high concentration of people, such as supermarkets, discount stores, theatres, and railway stations.

Since its interception, mall intercept surveys have come a long way. The present period is witnessing huge developments and advancements in mall intercept surveys, with enterprising researchers opening permanent offices and test centers in malls.

Today, some mall research facilities are equipped with complete food preparation, storage facilities for conducting taste tests, focus group facilities, video tape equipment, etc.

Since each mail has its own customer characteristics, the chances of deriving biased information is more as compared to door-to-door sampling. When the chances of demographic influences are negligible or the target group is a special population, mall-intercept interviews are useful. It comes in handy for surveys that require coordination and timing such as cooking and tasting food products and for products that need to be demonstrated. A special case of mall-intercept interviewing is Purchase Intercept Technique. This technique involves an in-store observation and in-store interviewing, where consumers are intercepted and interviewed while buying a specific product. The interviewer then probes into the reasons for selecting the particular product.

Mall-intercept surveys screen over other modes of survey interviews in respect of cost of research and degree of control, time taken for execution, and the quality of information collected.

Advantages of Mall-Intercept Surveys

1. As compared to any other face-to-face interview, depth of response is greater for mall-intercept interviews.
2. The interviewing environment is controlled by the researcher.
3. To the nonverbal indications of the interviewee, interviewer can notice and react.
4. To analyze the responses, various types of equipment are available.
5. The memories about the shopping experience are fresh, and hence the situation is conductive for studying purchase behavior.

Disadvantages of Mall-Intercept Surveys

1. From respondents getting personal information is not easy and involves many problems.
2. Social desirability effect and Interviewer bias.
3. Shoppers and respondents, who are in a hurry, might respond carelessly leading to wrong information. The respondents might be in a hurry to leave the mall.
4. Samples drawn may not be representative of the population and inapplicability of probability-based sampling techniques lower completion rates of questionnaires.

8.5.5 Mail Surveys

In a mail survey, questionnaires are sent to qualified respondents by mail or e-mail. In research, mail surveys used are ad-hoc mail surveys and mail panels. The only difference between the two is that in the case of ad-hoc mail surveys there is no prior contact. A questionnaire is just sent to a sample selected from an appropriate source and responses are awaited. The selected sample is used only for a single project.

Functioning of Mail Panel Surveys

1. The process starts with, from various sources obtaining mailing lists, after ensuring that they have the current, complete address of potential participants. It should be ensured that the list of participants is closely related to the group under study.
2. The next step involves contracting the sample participants through mail, postcards, letters, or telephone. The purpose of their participating is explained to the participants, in the panel survey. If the participants contacted agree to take part, they are required to fill in an initial questionnaire pertaining to their background and demographic details, which may be used to determine whether the participant qualifies for inclusion in the survey.
3. The panel participants are sent questionnaires from time to time, on successful selection.
4. Thereafter, participants are contacted by various means, to remind them, to mail back, the completed questionnaire.

An essential feature in mail panel surveys is that it is a type of longitudinal study where the same respondents are surveyed at different point of time to note specific changes pertaining to the topic of research.

The advantages are similar to those of self-administered interviews. The method is cost-effective as the need to recruit, train, monitor, and pay the interviewers is eliminated. The questionnaire can be administered from a single location for better control. It is even possible to contact respondents who are hard to reach. Respondents can spend as much time as they like answering the questionnaires and can complete them at their convenience. Thus, the respondents tend to give more detailed responses.

However, the absence of a qualified interviewer gives rise to the same limitations in mail surveys as for self-administered interviews. Mail surveys are however characterized by a high rate of nonresponse.

Typical ways to cope with nonresponse in mail surveys are monetary incentives, stamped, self-addressed return envelope with a persuasive covering letter, premiums such as pen, pencil, and other small gifts, promise of contributions toward charity, entry into drawings for prizes, emotional appeals, reminder that the respondent participated in previous surveys, etc.

8.6 Steps in Conduction of Survey

Step 1: Determination of the Aims and Objectives of Study: At the outset, the researcher must analyze and assess the relevant areas or issues that need to be studied. The basic aims and objectives have to be clearly specified by the researcher, once the research area is selected. These have to be focused and analyzed so as to make the purpose of research relevant and understandable. The researchers have to come up with the basic aims and objectives that would be focused and analyzed in their overall research.

Step 2: Define the Population to be Studied: The researcher also needs to define the target population, which would be studied by him or her after selecting the theme of the research. The population or universe would be a collection of people or object who would possess at least one common characteristic, which is going to be helpful and which would also provide direction in the process of conducting the research.

Step 3 Design and Construct a Survey: Once the target population is defined by the researcher, he or she needs to design a survey research. On the basis of the framed design, the research decides to conduct a survey, selects instrument for survey, for example, telephonic interview, with the help of which data will be collected. After the selection of the instrument, the researcher conducts a pilot study, i.e., a small survey taken in advance of a major investigation or research. The pilot study helps the researcher to analyze for the present research the significance and relevance of the instruments selected by the researcher.

Step 3: Select a Representative Sample: The process of construction of the survey instruments gives a way to the selection of the sample from the target population. The researcher selects a sample, which represents nearly maximum characteristics of the whole universe or population. The results or the findings of the survey conducted on the sample can be easily generalized on the population as a whole, if the selected sample is a good representation of the population.

Step 4: Administer the Survey: The researcher conducts the survey, after the selection of the sample, by administering the survey instrument or tool on the selected sample. This step helps in the collection of the required data or information from the sample.

Step 5: Analyze and Interpret the Findings of the Survey: The researcher analyzes the data; once the data have been collected, with the help of required statistical tools and then on the basis of the information revealed, interprets the findings. This step involves several processes such as coding the data and then processing it.

Step 6: Prepare the Report of the Survey: The researcher prepares a report of the overall research conducted, on the basis of the analysis and interpretation of the results.

The report contains the details of aims, objectives, data analysis, interpretation, and discussion of the results. In this step, the researcher tries to evaluate how the findings meet the proposed aims and objectives of the research.

Step 7: Communicate the Findings of the Survey: The most important step of conducting the survey research is to disseminate the survey findings. The researcher needs to communicate the findings to the target population and for the future research to be done on a similar field, it is an equally important record. The impacts of the survey results are also assessed on them, on the basis of which the researcher may also recommend certain policies on decision making.

8.7 Constructing a Survey Research

In the process of conducting a survey research, the researcher needs to design a framework of the instruments and processes of data collection, on the basis of which the overall research would be done.

The researcher needs to decide the Survey Research content, format, and wordings that would be included in the survey instruments.

No doubt the researcher selects any one of the kinds of instruments that is a questionnaire or an interview, he or she needs to frame questions. The questions should be so worded that there is clarity in what is being asked and should have the capability of eliciting response. The survey instruments are the backbone of research and that is why the statements or the questions of the researcher should be short and specific as well.

The researcher constructs the survey instrument by framing questions, once the framework of the process of research is decided.

While writing the questions for the survey, the researcher needs to take care of a few basic aspects, which include:

1. Deciding the content, scope, and objectives of the question;
2. Selecting the most convenient format of response, for example, Likert-type 5-point scale, multiple choice questions, and so on;
3. Deciding on how to frame the questions that would elicit the required response;
4. Bringing out the best response and favorable conditions for the survey formatting the series of questions; and
5. While preparing the questions and taking full care of the moral values and ethics of the respondents to get the best results, the researcher needs to be very sensitive.

8.8 Advantages and Disadvantages of Survey Research

Advantages of Survey Research

1. For the researcher, it is convenient, less time taking, and economical.
2. The survey can be conducted for a longer period of time, which gives a chance of knowing about the latest changes or advancements that might have taken place in the agenda under study.
3. The researcher gets a full chance to well organize and present the reasons of the study to get full and honest answers from the respondents.

Disadvantages of Survey Research

1. Under a group interview, maintaining the privacy of responses of each respondent is questionable and that may also restrict full and honest answers from them.
2. High attrition rate of the respondents might hinder the longitudinal based studies.

8.9 Difficulties and Issues of Survey Research

There are certain issues that the researcher might have to understand and take full care, if he or she plans to go for a survey research. They are as follows:

1. **Issues on Selecting the Type of Survey:** Selecting the kind of survey that might be most appropriate or suitable for his or her study is one of the most critical decisions for a researcher. The researcher should be aware of the kind of population that would be suitable for the study. Again, they should also be comfortable with the language of the selected population. The researcher should also analyze the geographic restrictions and try to find out for a dispersed population which method can be most feasible.
2. **Issues on Survey Instruments:** The researcher should have complete knowledge of the suitability of the questions that would be asked to the respondents, while constructing the survey. Within a survey research some of the controversial issues are the type of questions, clarity, specificity of the questions, the length of the questions, etc.
3. **Bias Issues**: On the findings of the survey research, the researcher's biases and prejudices might have a significant influence, so they should be fully aware of the repercussions of their biases. Their behavior should be socially desired ones, so he or she should not lose track and also avoid false reports. In such cases, issues of bias are really difficult but an essential agenda in a survey research.
4. **Administrative Issues:** Important aspects that need to be preplanned even before the advancement of the research are cost, mode of survey, feasibility of the area selected, required time period, etc.

Summary

In this chapter, a brief classification of the surveys based on the amount of structure and disguise in a questionnaire, and the time and mode of communication, were explained. Structured questions are usually close-ended, whereas unstructured ones are usually open-ended. Cross-sectional studies are single surveys, which involve data collection from different segments of the population, and analysis of them to gather information on a population at a single point of time.

Longitudinal studies, on the other hand, are multiple surveys that tend to reveal the differences in the samples responses over a period of time.

Classification of survey methods based on the mode of communication can be broadly grouped under personal interviews, telephonic interviews, self-administered surveys, and mail surveys.

Telephonic interviews that involve contacting the respondents over the telephone can be similarly subgrouped into central location telephone interviews (CLTI), computer-assisted telephone interviewing (CATI), and completely automated telephone surveys (CATS).

Self-administered interviews are those in which the respondent due to his or her involvement and interest in the research objective answers a preformulated set of questionnaires designed by the researcher.

Mail surveys require the mailing across of the questionnaire to qualified respondents who return it to the researcher after completion.

We also discussed the steps involved in conducting survey research. We discussed how to construct a survey research and what are the ways in which questions should be asked in a survey and how to avoid biases. We dealt with the many precautions while designing instruments and discussed the advantages and disadvantages of the survey research. The methods and kinds of survey research that are most common and are frequently used by a number of research scholars were also mentioned.

Review Questions

1. Describe the different methods or techniques of the survey research method.
2. Explain the different types of conducting survey research.
3. Explain the different issues of survey research method.
4. What are the different types of questions that can be designed for a survey instrument?
5. Explain the different types of interviews that can be used for conducting a survey research.

9

Questionnaire

9.1 Introduction

A questionnaire may include a series of questions to be answered by an individual or group, with the aim of obtaining relevant data on the topic of research. This could be self-administered or could be administered by an interviewer. Broadly survey questions can be classified as structured and unstructured. In research, both structured and unstructured questionnaire are in use.

9.2 Definition of Questionnaire Method

Questionnaire is a data collection instrument. The researchers for collecting data most commonly use this method. The researcher lists the questions to which he or she requires answers, in order to gather data on a particular research topic. The list of questions grouped in some order is either given personally, or mailed to the target population.

9.3 Construction of Questionnaire

Caution must be taken in the selection of questions and variables, so that the researcher can receive accurate answers that he or she wants to explore. The purpose of this type of data-gathering technique is to obtain valid and reliable information so that smooth investigation can be conducted and hypothesis can be tested. A clear understanding of the problem under study is essential for the researcher. Hence, before finalizing the contents of the questionnaire, he or she needs to review the related literature.

The covering letter explains the identity of the researcher, objectives of the research, need for questionnaire, tell the respondents what use will be made of the results and precisely what will happen to their answers, request the respondent for cooperation, explain the purpose of the questionnaire, and ensure the confidentiality of respondent's answers. This assurance will motivate the respondents to express their views freely.

9.3.1 Steps in Questionnaire Construction

1. Determining the scope of the questionnaire
2. Deciding the type of questions to be asked, i.e., close ended or open ended
3. Preparing the draft questionnaire
4. Pretesting the questionnaire with a sample population
5. Revising the questionnaire, according to the suggestions received
6. Distributing the questionnaire
7. Sending reminders to the population under study
8. Receiving the responses
9. Analyzing and interpreting the data received and
10. Writing the research report.

DOI: 10.1201/9781315167138-9

9.3.2 Length of the Questionnaire

For a standard questionnaire, there is no prescribed length. Due to its cost effectiveness, researchers want to ask a maximum number of questions in one questionnaire. Length of the questionnaire depends on the topic of the research problem and size of the target population.

9.3.3 Guiding Principles to be Followed for Questionnaire Construction

1. It should be self-explanatory.
2. As open-ended questions receive vague and incomplete responses, which are difficult to interpret, questions should be restricted to close ended.
3. Less number of questions in the questionnaire helps in receiving high response rate.
4. Attractive layout of the questionnaire helps in obtaining completed questionnaires.
5. In filling up the questionnaire, proper instructions should be provided to the respondents.

9.4 Structured Questions

Those questions that pose definite and concrete questions, the format of which is preplanned and predefined in advance, are called as structured questionnaires. When there is a need to clarify vague or inadequate replies by respondents or when further details are needed, only then additional questions are asked. Some of the types of structured questions are Dichotomous questions, Level of measurement-based questions, and Filter or Contingency questions.

9.4.1 Dichotomous Questions

Dichotomous question has only two possible responses. For example, Yes or No, True or False, On or Off, Right or Wrong, and so on. The layout of these questions appears in the questionnaire:

Q: Does the library of your College/School have an electronic database system?
Yes No

Q: Please mention your gender:
Male Female Other

9.4.2 Level of Measurement-Based Questions

The basic levels of measurement are Nominal, i.e., based on names, classification of persons, objects, and groups; Ordinal, i.e., based on ranks and preferences and Interval measurements, i.e. based on ratings, for example, a nominal question may have numbers before each response, which may only represent the serial order, as follows:

Please state the category to which you belong:
General/Open OBC SC/ST NT/VJ

The question based on interval scale may be based on rating the choices, out of which the most commonly used scale is Likert response scale, which has a rating of 1–5, or 1–7, or 1–9. For example,

The university has a well-equipped and fully furnished computer-laboratory:

1	2	3	4	5
Strongly agree	Agree	Cannot say	Disagree	Strongly disagree

9.4.3 Filter or Contingency Questions

When a question framed is design in such a way that it is followed by succeeding questions, which are subparts of the main question, such types of question frame design is known as filter or contingency questions. For example, if a researcher wants to ask whether the respondent has ever attended the library of the college and if the researcher also wants to know how many times the respondent has attended the library, then the format of the question will be as follows:

Q: Have you ever regularly attended the library of your college?
Yes No

Q: If yes, then how many times?
Once in a month Every week of the month Every day in a month

To get the subsequent answers, the researcher should use multiple filter question responses. But he or she should take full care that in order to maintain the interest of the respondent, they should not exceed more than two to three levels for any question.

9.5 Unstructured Questions

The chief advantage of the unstructured questionnaire is flexibility. Unstructured questions are usually used in interviews, where either the researcher does not prepare a list of questions and the series of questions might depend upon the response of the subjects or they ask questions in an informal atmosphere. The researcher should take full care and should give a silent probe, verbally encourage, ask for clarification, and have full empathy with the respondent, in order to get adequate and required information.

9.6 Designing a Questionnaire

Step 1: Preliminary decision: What information will be sought after a thorough scanning of secondary sources of data will be assessed. Determine the target respondent.

Step 2: Decide on the type of questionnaire and the method of administration.

Step 3: Evaluation of question content: Before including a question in the schedule, examine the question is really essential, respondent can understand the question, i.e., for the target respondent, it may not be too technical, ambiguous, or advanced, and the respondent can answer the question. Say, the respondents possess sufficient knowledge. As such, it is better not to ask too much of factual data or about history specially, if it invades into one's privacy or usually they refuse to cooperate, when it requires too much effort to answer.

Step 4: Check question phrasing: Like words have ambiguity in meaning, any implied alternatives in the question, any assumptions to be made to answer the question, etc.

Step 5: The type of response format will depend on the objective of the research, nature of data to be collected, and analysis to be performed.

Step 6: Determine sequence of questions: Use simple, easy, and interesting opening questions; design branching questions with utmost care; place the questions in a proper sequence and logical manner, etc.

Step 7: The physical layout of the questionnaire must be assessed. The questionnaire must be printed properly and design must be attractive; presented in an elegant form; and facilitate handling.

Step 8: Pretest or Pilot test of the questionnaire is essential. On a small number of target respondents, it is a good practice to pretest a questionnaire. To assess both individual questions and their sequence of response pattern, the pretest is done. Accordingly, a researcher must revise the questions that cause problems.

While developing a questionnaire, a researcher must do's and don't the following:

1. Use simple and easy words,
2. Ambiguous questions should be avoided,
3. Implicit alternatives should be avoided,

4. Questions that require too much memory recall and calculation should be avoided,
5. Double-barreled questions should be avoided,
6. A questionnaire should first secure some basic information to get the respondent's cooperation, interest, and gradually try to collect more information about the phenomenon of interest,
7. It is easier to administer multiple choice response categories that require one simple tick.

9.7 Questionnaire Format

Questionnaire format is normally used when the data are collected from a large population about the Awareness, Attitudes, Opinions, Past and Present behavior.

Situation and experience of the researcher greatly influence the process, as there is no standard procedure to construct a questionnaire.

A questionnaire format depends upon the amount of structure and disguise required during data collection.

9.7.1 Structure

A highly structured questionnaire is one in which the question to be asked and the responses permitted are explicitly prespecified. In a nonstructured questionnaire, the questions to be asked are kept flexible in their own words. Also, the respondents are allowed to answer the questions in a manner they like. The response pattern may vary from open-ended to close-ended.

In an open-ended question, the respondent is free to choose the possible response. In a close-ended question, the researcher prespecifies certain options and from the given options, the respondent is allowed to choose the alternatives. For example,

Open-ended: What brand of hair oil do you use?

Close-ended: Mention the brand of hair oil you use from the list given below:

( )Parachute ( )Indulekha ( )Dabur ( )Kesh King ( )Hair & Care

9.7.2 Disguise

In disguised questions, the purpose is not made obvious to the respondents and is asked in an indirect manner. Questions that are direct and the purpose of asking them is known clearly to the respondents are termed as nondisguised questions. When the issues concerned are such that respondents may not give correct answer to direct questions, in such situations disguised questions are used.

Questionnaires could be classified into categories such as structured–nondisguised, structured—disguised, nonstructured–nondisguised, and nonstructured–disguised questionnaire.

9.7.3 Structured–Nondisguised Questionnaire

In marketing research studies, structured–nondisguised questionnaires are very popular. These are more applicable when sample size is large.

9.7.4 Structured–Disguised Questionnaire

When responses are required toward certain sensitive issues such as attitude toward aids patients, abortion, etc., Structured–Disguised Questionnaires are more appropriate.

9.7.5 Nonstructured–Nondisguised Questionnaire

When a freehand is to be provided to the respondents, so that an indepth information on the subject could be solicited, then in such situation Nonstructured–Nondisguised questionnaires are used. For example, in industrial marketing research wherein number of respondents would be low.

9.7.6 Nonstructured–Disguised Questionnaire

These are mainly used in Motivation research, Word association test, Sentence completion test, Thematic appreciation test, Cartoon test, etc.

9.8 Questionnaire Administration

Depending "on the way it is administered," the questionnaire method may vary. These could be broadly classified into following categories:

1. Personal interview, wherein there is a face-to-face interaction between interviewers and respondents.
2. Telephonic survey, in which survey is conducted over telephone, i.e., unlike personal interview there is only a voice contact.
3. Mail survey, which is conducted through mail and as such there are no interviewers.

9.9 Preliminary Decisions

Before framing the actual questionnaire, a researcher has to take many decisions. These decisions are related to the information required, target respondents, and choice of interviewing techniques.

9.9.1 Required Information

The researcher is expected to know and understand the survey's objectives before he or she can take further steps. The researcher must ensure that, in framing a questionnaire, the questions are designed to draw information that will fulfil research objectives. A researcher should go through the secondary data and research studies that are similar to the current research. This helps in planning current research related to the topic under study, based on existing research findings. To understand the nature of the problem and the information that would help managers in solving a problem, the researcher, with the prospective target audience, can conduct informal interviews.

9.9.2 Target Respondents

The researcher must make sure of the target population for the survey before conducting the actual survey. For example, a researcher has to decide whether to include both users and nonusers of a product or service, in case of market research. As the sampling frame would be drawn after the target respondents are defined, this is a crucial step. As the task of developing a questionnaire that will be suitable to all cross-sectional groups of a diversified population, defining the target respondents becomes vital.

9.9.3 Interview Technique

A lot depends on the choice of interviewing technique, in developing a questionnaire. The format and type of questions will be different for Personal interviews, Focus groups, Telephonic interviews, and Mailed questionnaires. A questionnaire designed for direct interviewing cannot be used for a survey through mail. In personal interviews, the respondent should be clearly explained about the details and the form of answers the questions require. In telephonic interviews, it is prudent for questionnaires to be brief and to the point. Mail survey questionnaires should give clear instructions about the type of details that are desired, as an interviewer does not mediate these interviews.

9.10 Question Content

The question content decides the general nature of the questions, the information they are supposed to elicit, etc. In this process, things become easier because there are some set standards that can be followed. Irrespective of the type of research, while deciding the question content, a researcher has to find answers to the following major questions:

1. What is the scope and utility of the data collected?
2. How effective is a questionnaire in producing the required data fulfilling the objectives of the researcher?
3. Is the respondent willing to answer the question accurately and error free?
4. Is the respondent cooperating to answer the question and feels interesting?
5. What is the chance of the responses being influenced by external events and disturbances in environment?

9.10.1 Utility of Data

Each question in the questionnaire contributes to the survey must be ensured by the researcher. For this, before adding the question to the questionnaire each question needs to be screened.

9.10.2 Data-Producing Effectiveness

Researcher should assess whether the question will be able to generate the required information. If it needs to be broken down, i.e., double-barreled questions into two specific questions, so that a researcher can elicit better and accurate answers from the respondents. Are the questions effective enough, so that the researcher can extract required information from the interviewee.

9.10.3 The Participant's Ability to Answer Accurately

It is necessary that respondents understand the question in a way that the researcher wants. This will eliminate the probability of potentially incorrect responses.

Genuine Ignorance about the Topic: It refers to respondents being unaware or uninformed about the subject of the question. As respondents will rarely admit to lack of knowledge on a topic, this can lead to respondent bias

1. **Unable to Recollect the Answer:** This happens because of recall and memory decay, the respondents forget an answer. This happens when questions overtax the respondents recall ability. For example, questions such as "What was your expenditure on Vegetable and Fruits in the last week?" To answer this question, it requires respondents to bank on their memory. While answering the above question, the respondent might not recollect the purchases made in the last week and hence might fail to give the actual and correct data.
2. **Telescoping:** When an interviewee thinks that an event that occurred sometime in past occurred more recently. In other words, the respondent may report purchases made a fortnight ago as done in the last week.
3. **Creation:** When an interviewee feels that the incident or event did not occur at all. In other words, total forgetfulness.

The above three aspects of forgetting increase with the length of the recall period. This means that an interviewee should be asked questions that need only recalling of incidents and events from the near past.

9.10.3.1 Unable to Verbalize the Response

This refers to the respondents' inability to verbalize factors influencing their buying motives. It is not quite possible to answer the questions such as "Why did you buy that Bike?" or "What made you buy that brand of Mobile?" This is because many times people buy things for reasons other than what they admit to themselves. Behind the purchase there might be the reasons such as Habit, Vanity, Taste, etc., but people are generally unable to articulate reasons when asked "Why?" as they are not conscious of what is in their subconscious mind. Through effective projective techniques, researchers can awaken the subconscious minds of the respondents.

9.10.4 The Respondent's Willingness to Answer Accurately

This refers to the researcher assessing the likelihood of the respondent answering a particular question accurately. Nonresponse results in situations such as "when a respondent is unwilling to answer a specific question, when a respondent completes the rest of the questions other than those he or she is uncomfortable with, or when a respondent refuses to complete the rest of the questionnaire in deliberate falsification."

Responses are virtually sure to attract stereotype responses or refusals from participants to questions such as "Were you involved in any extra-marital relationship in the 5 years of your marriage?" or "Would you resort to stealing things in a Big Bazaar or D-Mart, if you knew there were no hidden cameras?"

This refusal can be because of asking the types of questions too personal and embarrassing, offending, and reflecting on prestige when the respondents decide that the topic is irrelevant to their interests, etc.

9.10.5 Effect of External Events

Sometimes, because of the interference of external events, the respondent's answer to a particular question is exaggerated, for example, external events such as weather or time.

9.11 Response Format

The format usually deals with issues relating to the degree of freedom that should be given to respondents while answering a question. Following are the two popular response formats.

9.11.1 Open-Ended Questions

Open-ended question is a type of question that requires participants to respond in his or her own words without being restricted to predefined response choices. They are also called infinite response or unsaturated type questions. Open-ended questions are structured in themselves.

Although they probe for unstructured responses, there is a definite structure in the arrangement of questions in the questionnaire. They help establish rapport, gather information, and increase understanding. Open-ended questions at times require the respondent to recollect past experiences, hence these questions act as memory prompts. Therefore, the interviewer should refrain from making suggestions. He or she should rather invite the participant to use his or her own choice of words to answer. The interviewer should make the respondent to talk as much as possible and record answers in the same words used by the respondent.

Open-ended questions are useful when the respondent is able to provide a narrative answer, when the researcher is uncertain what answers are needed or wants to conduct an exploratory research. Such questions can be subdivided into Free Response, Probing, and Projective.

Free response questions, which the interviewer gives to the interviewee, typically fluctuate in the degree of freedom. Consider the following question:

Q. What do you think of the performance of the Indian cricket team in the recent matches?

9.11.1.1 Probing

Probing open-ended questions are those where the actual open-ended questions are reached a little later in the process. Consider the following example:

Q. Which brand of soft drink do you like? Limca or Pepsi?
A: Limca

Q. Why do you prefer Limca to Pepsi?
A: I like the taste.

Q. What aspect of its taste do you like? (Probe)

This is where the interviewer starts probing to get to the specific product attributes linked to the interviewee's liking of Limca and the role that the subconscious mind of the interviewee plays in influencing the buying decisions.

9.11.1.2 Projective

A vague question or stimulus used by the researcher to project a person's attitudes from the responses is known as a projective open-ended question. Such questions are primarily used in projective techniques.

9.11.1.3 Advantages of Open-Ended Questions

1. It can discover uncommon but intelligent opinions of which the surveyor would otherwise have remained unaware.
2. Respondents have freedom to qualify their answers.
3. The respondent has greater freedom of expressions.
4. Researcher can have real views of the respondents.
5. Respondents can give their views in their own language reflecting creativity, self-expression, and richness of detail. Such answers reveal the logic of the respondents.

9.11.1.4 Disadvantages of Open-Ended Questions

1. Coding open-ended questions is difficult and time-consuming.
2. As the questions require more thought and time on the part of the interviewee, it reduces the number of questions that can be asked within a specified time span.
3. There are chances that a researcher or interviewer might misinterpret a response as it becomes difficult in pooling an opinion across the sample.
4. Different answers may be received from the respondents on the same question.
5. The responses to open-ended questions are difficult to analyze but not impossible. It is possible that some of the answers may be put forcibly into one of the categories. The researcher has to carefully study the responses and categorize them into different categories. However, it is very time consuming.

9.11.2 Close-Ended Questions

Questions, which restrict the interviewee's answers to predefined response options, are called close-ended questions. Close-ended questions give respondents a finite set of specified responses to choose from. These questions are common in survey researches. Such questions are deemed appropriate when the respondent has a specific answer to provide. For example, gender, when the researcher has a predefined set of answers in mind, when detailed narrative information is not needed, when there is a finite number of ways to answer a question, etc.

The major structures exist for close-ended questions are Binary, Ranking Questions, Multiple Choice, Checklist, etc.

9.11.2.1 Binary Questions

As binary questions permit only two possible answers, these are also known as dichotomous questions. The respondent has to choose one of the two permissible answers. Binary questions are helpful in collecting simple, factual data to record classification data about the interviewee, i.e., demographic data, etc. These questions may have the response options such as "Yes" or "No," "True" or "False," "Agree" or "Disagree," "Right" or "Wrong," etc.

The respondent might be compelled to give an answer whether or not they represent their true feelings. This tends to affect the survey's accuracy.

9.11.2.2 Ranking Questions

These questions require the participant to rank the response options listed on a continuum basis in order of preference. Ranking questions are used to obtain information that reveal the participants' attitudes and opinions, list several alternatives that might influence an individual's decision-making, etc. The participant assigns a rank to each option listed as per the scale mentioned. For example, the factors that influence your decision to buy from a particular supermarket are listed below. Please rank them from the most important (1) to the least important (7), listed below.

Convenient location-----, Helpful sales staff -----, Regular discounts offered-----, Instant home delivery-----, Availability of everything needed -----

Such questions make it easy to compare different alternatives at the same time.

9.11.2.3 Multiple Choice Questions

All significant degree of responses are covered by multiple choice questions. They are also known as "cafeteria" questions. The respondent has to select an option that best describes their feelings. These are mostly a variation of binary questions with more responses provided. The reasons behind the popularity of multiple choice questions are their simplicity and applicability.

9.11.2.4 Checklist Questions

These are questions where the participant has the freedom to choose one or more of the response options available. This is different from multiple choice questions in that it gives freedom to the respondents to choose one or more of the options available. For example,

Q. Which premium brand of shoes do you possess? (Tick as many of the following as apply)
A: 1. Adidas 2. Bata 3. Paragon 4. Puma 5. Other

It should be insured that options are placed in a random sequence rather than in any preferential order. An option called "others" should be provided, apart from the options selected by the researcher so that the respondent can fill it in, if he or she wants to. With all significant categories present, this method facilitates replies from the respondent and subsequent tabulations.

9.11.2.5 Advantages of Close-Ended Questions

1. Best suited for large-scale surveys and answers of the respondents can be compared.
2. They are easier, more specific, and quicker for the respondents to answer as well as for the researcher to interpret.
3. They provide a high level of control to the interviewer by obliging the interviewee to answer questions using a particular set of options.
4. The uniformity of the questions makes them easier to code, record, and analyze results quantitatively.
5. Higher response rate, less expensive, cost effective, easy to code and tabulate.

6. Performance of the respondents is more reliable, sensitive questions can be properly answered, and less number of confused answers. With limited number of answer options, the possibility of obtaining enough responses to the options or categories may be useful for analysis.

9.11.2.6 Disadvantages of Close-Ended Questions

1. While coining different choices for a particular question. If the choices are less, the desired results cannot be achieved. If the choices are too many, it may confuse the participant and appropriate answers cannot be received.
2. Close-ended questions are provided with a list of acceptable options. A respondent has to pick and choose one or more of these options. The responses chosen by respondents may be best suited to them, but they may not be the correct answers.
3. There is a possibility that the list of alternative answers may not be sufficient for the respondent, hence, limiting his or her options.
4. The options might not reveal the true feelings of the participants. Misleading conclusions can be drawn because of poor questionnaire design and limited range of options.
5. Ideas of the researchers are imposed on the respondents. Less knowledgeable person can also answer due to the availability of multiple choices. They may not give the correct answers to the question.

To decide what type of the questions will be adopted depends on the following factors:

1. For quantitative data the factual questions are required. In this case close-ended questions will suit the most.
2. For qualitative data the open-ended questions are required though they are less easy to categories and less amenable to computer techniques.
3. Close-ended questionnaires are used when categorized data are required.
   i. They include a set of questions to which a respondent can reply in a limited number of ways—"yes," "no," "no-opinion," or an answer from a short list of possible responses.
   ii. He or she is asked to put a tick $(\surd)$ mark in a space provided on the answer sheet or is requested to underline a response.
   iii. Sometimes he or she is asked to insert his or her own brief answer.

9.12 Question Wording

In effective cross-communication, designing questionnaires can be an exercise, as it tests the communication abilities of the person framing them. In gathering responses, the effective translation of the desired question content into appropriate words does the trick. Questions tend to get longer to be explicit, present alternatives, and explain meanings.

In such cases, a lack of appropriate words can result in the respondent misunderstanding the question and giving inappropriate answers or the respondent refusing to answer. A slight mistake in designing the questionnaire can be annoying and cause potential problems in data analysis, resulting in incorrect results.

While framing a questionnaire, the factors that should be considered are Shared Vocabulary, Unsupported Assumptions, Frame of Reference, Biased Wording, Adequate Alternatives, Double-barreled Questions, Generalizations, Estimates, etc.

Common Problems with Question Wording

1. Avoid objectionable and Sensitive Questions

   Objectionable: How often do you trade-in in a train without a ticket?

   Unobjectionable: How often do you forget to take a ticket while traveling by train? (disguised)

2. Avoid Biased Questions
   Biased: Do you think that playing of mobile games has a negative effect on children?
   Unbiased: What are your views about the effect of playing of mobile games on children?
3. Avoid Vague Questions
   Vague: How satisfied are you with service provided at hotel?
   Better: How would you describe the hospitality in hotel in your own words?
4. Avoid unwarranted Presumptions
   Presumptive: How satisfied are you with the Service provided in Nationalized Banks? (assume that customers are satisfied)
   Better: How satisfied or dissatisfied are you with Service provided in Nationalized Banks?
5. Avoid the use of leading questions that prompt the respondent to a particular answer
   Leading: Would you prefer a Big Bazaar nearer your home?
   Better: How often would you shop from a Big Bazaar based on its distance from your house?
6. Avoid asking negative questions
   Negative: Medical representative should not be allowed to make visits in the evening. Agree or Disagree
   Positive: Medical representative should be allowed to make visits at any time. Agree/Disagree
7. Ensure that the wording is completely unambiguous
   Ambiguous: How seldom, occasionally, and frequently do you visit Big Bazaar?
   Unambiguous: How often do you visit Big Bazaar? (a) Seldom, (b) Occasionally, (c) Frequently
8. Avoid double-barreled questions
   Double-Barreled: Do you drive or take the car every day to office? Yes or No.
   Better: How do you go to your office every day? Drive or take a car?
9. Have as narrow a reference range as possible
   Too Broad a Time Period: How many times have TV Advertisement influenced you to switch brands over the last 1 year?
   Better: How many times in the last month have TV advertisement influence you to switch brands?

Question using words that are ambiguous in context:
Sometimes the words that are fairly understandable on their own may be used in a way that renders their ambiguous. In the following question:

Q: Do you watch television news regularly?

For different people, the word `regularly may have different meanings. It does not specify whether regularly means the whole day long, seven times a week, five times a week, four times a week, three times a week, two times a week, one time a week, or certain programmes on every telecast. Therefore, in light of the information sought, the question` needs to be rephrased.

9.12.1 Shared Vocabulary

An interview of any kind is mostly an exchange of ideas between the interviewer and the interviewee. Mostly through words this exchange takes place. This makes it imperative for the interactive language to be kept simple and easily understood by both the interviewer and the interviewee. Following things are worth ensuring in this respect:

1. The involvement and usage of technical language has to be dealt with carefully. This is necessary as using highly technical language in the questions may create understanding problems for both the interviewer and the interviewee.
2. Another issue is the appropriate choice of words. To ensure that the words are simple is not enough. It also has to be seen that the words are not ambiguous or vague.

9.12.2 Unsupported Questions

For better response rates, while preparing questionnaire, the researcher must take care that in a questionnaire should avoid the use of implicit assumptions, should not contain questions framed on assumptions that are not explained in the questions, should not leave anything for the respondents and the audience to interpret, and question should be supported with valid assumptions that would make it clearer to the audience. Unsupported implied assumptions tend to procure exaggerated estimates from respondents.

For example, consider following question to a lady:

Q: How often does your man accompany you to…?

This will elicit varied responses and may even be misinterpreted. The question assumes that every lady has a spouse or a boyfriend, which is obviously not the case.

9.12.3 Frame of Reference

Under different situations, a single work can have several connotations. Words such as "often" and "regularly" can mean different time frames for different individuals. The word "capacity" can mean very different things to an industrialist, a businessman, and an educator. But the framework of social desirability makes the interviewer extend a common frame of reference to the participants. The interviewer assumes that in its denotative terms, the interviewee has understood the question and qualifies the answer as valid. This is a mistake, as the respondent might have answered the question using an individual frame of reference rather than from the interviewer's point of view.

9.12.4 Biased Wording

Biased and loaded words tend to be too emotionally colored, eliciting automatic feelings of approval or disapproval. They make participants aware of the desired response, thereby taking the focus away from the actual response. For example, consider a question to a bank employee.

Q: Would you favor the replacement of manual staff of bank by computers?

It is sure to receive a negative response. A way of asking the question to read the subconscious mind of the employee would be, "How do you think the introduction of computers would affect manual staff in a bank?"

9.12.5 Adequate Alternatives

An ample number of alternative answers to each question. Alternatives should be explicit rather than implicit. This gives respondents the freedom to choose among alternatives rather than delve into their own mind to recollect responses. To gather responses, it is a faster way. For example, consider a question:

Q: How often do you purchase an Ice Cream?
(a) Seldom (b) Occasionally (c) Frequently (d) Rarely

9.12.6 Double-Barreled Questions

Questionnaires should avoid asking double-barreled questions. For example, consider a question: Q: "Do you like fuel-efficient cars with comfortable seats?" Actually, this is a combination of two questions. It does not distinguish between people, either who prefer cars due to their fuel efficiency and people who prefer a car for its comfortable seats or other competing reasons. Such questions can be easily divided into two different questions. Because two or more ideas are included, answers to double-barreled questions will be ambiguous.

9.12.7 Generalizations and Estimates

To avoid generalizations and estimates, questionnaires should be structured. When respondents are asked for the frequency of a particular activity over a longer period, it is seen that they tend to provide generalizations and estimates rather than the actual figures. By changing the time reference point to a more specific base, this trend can be reduced. Answers that require calculations by the respondent should also be avoided. Minimal necessary information can be gathered and then the calculations should be done by the interviewer.

9.12.8 Length of the Question

Long questions become incomprehensible to respondents when it is comprising of complex and compound sentence structures. The longer the question, the greater will be possibility of its being misunderstood, as all words are potential sources of ambiguity. For example, consider a question that was actually used in a study on distribution network for LPG Gas Agency:

Q: Under the new system, do you think LPG Gas Agency dealers would be independent like business appliance dealers and furniture merchants who own their outlets, or they would be employees of the companies?

This sort of question would pose problems of comprehension among most respondents. It could easily be rephrased as:

Q: Under the new system, do you think LPG Gas Agency dealers would be owners of their business or employees of the companies?

9.12.9 Unfamiliar Vocabulary

As far as possible, the questions should consist of words that are a part of the normal vocabulary of the respondent. For example, consider a question: Do you think the pasteurization process interferes with the lactogenic balance of milk? This question if put to doctors, chemists, biochemists, pharmacists, or medical representatives would not match the vocabulary of this researcher to most of the members of the general public.

9.12.10 Combined Questions

Sometimes, poor question construction results in two questions being asked as one.

For example, consider a question put to housewives: Q: What do you think is a healthier, nutritious, and economic medium for your cooking: Soya bean oil or other oil? It is clear that the housewife who thought that one medium was healthier, nutritious and the other more economical would not be able to respond logically to this question: The simpler and more effective way to get this information would be to break this question into two, one dealing with healthier, nutritional value and the other with economy.

9.13 Questionnaire Sequence

The sequencing tends to drive the interview through a "funnel-shaped" process, i.e., starting with general questions and progressing to more specific ones. Before moving to sequential steps, the interviewer gives a brief introduction about the basic purpose of survey and about client confidentiality.

Steps for questionnaire sequencing are Lead-in Questions, Qualifying Questions, Warm-up Questions, Specifics Questions, and Demographic Questions.

Questions in the questionnaire should be arranged or grouped in logical sequences. Questions should be arranged in such a way that they do not scare the respondent but make him or her comfortable in answering or responding. Questionnaire should start with general but relevant questions and then move to the specific ones. It helps to set the logical flow in the questionnaire. The same type of questions can be put together in a group or section.

9.13.1 Lead-in Questions

This is the introductory phase of the interview and consists of tactfully designed ice-breakers. In gaining the participant's confidence and cooperation these can prove crucial. At this stage, the questions should be simple, nonthreatening, and not too personal. By asking a "ringer or throw away" question or a dichotomous question with two responses is the best way to start the session. The respondent's interest and willingness to respond are measured through these questions. The questions can be about the main headlines of the day, where responses are of little importance to the survey. For example, consider a typical lead-in question Q: It is often said that there are some things money can't buy. Do you agree with this? Ans: Yes/No.

9.13.2 Qualifying Questions

These are questions that slowly lead to the survey's objective. This stage is characterized by questions that evaluate the respondent and qualify him or her for further questioning. The interviewer directs the interview toward a relevant set of questions, depending on the responses. Prior to this, the interviewees related to the survey in some meaningful terms should be ensured. For example, a survey for estimating market potential for a newly launched brand of Bath Soap should ask the following qualifying question.

Q: Which type of Bath Soap do you like?
(a) Bubbly (b) Cleansing (c) Hardness (d) Conditioning

Depending on the interviewee's response, the interviewer can further give directions to the next questions.

9.13.3 Warm-up Questions

This stage, related to the survey questions plays on the respondent's mind by making him or her think of certain facts. For example, consider a question: Q: "When was the last time you bought tooth-brush?"; "Was it comfortable in shape, size, handle, and bristles you need to choose?"; "Looking back, can you recollect how many times you might have used tooth-brush over the six months?" tend to make the respondent think and recollect past experiences.

A person who is straightway asked such questions may not be interested in answering or providing details, but the resistance slowly decreases and gives way to cooperation, after a series of lead-in and qualifying questions.

9.13.4 Specific Questions

This stage consists of questions that are specific to the research objectives. As such, they are asked of participants who show a favorable response or are end users of the product. For example, consider a question:

Q: Which type of Bath Soap do you like?
(a) Bubbly (b) Cleansing (c) Hardness (d) Conditioning Respondent: Bubbly

In above case, Bubbly Bath Soap. These questions tend to estimate the usage pattern and influential factors in using Bubbly Bath Soap. In data collection and analysis, these specific questions play a major role. This section can probe to gain insight into sensitive issues, after ensuring that enough rapport has been established.

9.13.5 Demographics Questions

Demographic questions usually consist of a set of questions related to age, sex, educational qualification, location, occupation, etc. To avoid interviewee resistance and to prevent the interviewee's attention from being diverted, these questions are kept at the end.

9.14 Questionnaire Pretest, Revision, and Final Draft

For gathering data, Survey research questionnaire acts as an instrument. It should be pretested before putting it to actual use. Pretesting refers to testing the questionnaire on a small sample of respondents selected on a convenient basis that is not too divergent from the actual respondents. The aim is to identify and eliminate flaws and problems. Testing every aspect of the questionnaire starting from the question-content to question-sequence is included in pretesting.

Once the final questionnaire is printed, there is no room for corrections and improvement. If the researcher tries to make corrections it will be expensive as well as difficult. To pretest the questionnaire, it has to be circulated to the sample population to receive useful comments and the researcher can revise accordingly. For example, questions on frequency of use, visit the library, visit college canteen, etc., should not use the terms such as "Frequently," "Often," and "Occasionally" as the users may interpret these terms differently. Pretest also includes verbal communication with sample population about Confusing, Difficult, Over lapping categories questions, etc.

Pretesting helps the researcher in recording, simplifying, and transforming some of the questions. The process generally involves drafting the questionnaire, discussing it with colleagues, and circulating it among the small sample of the population for whom the questionnaire is designed. Pretest also indicates the time required to fill the questionnaire. It is a practice not to include the pretest sample to the actual population.

Pretesting helps to reveal incomprehensible meanings, wrong order of questions, leading questions, awkward responses, etc. No matter what the final mode of administration is, pretesting should be done by personal interviews. This will facilitate the interviewers in the following:

1. To observe respondent's reactions and attitudes,
2. Giving respondent's a firsthand experience of the potential problems and the data that can be expected from a questionnaire,
3. To facilitate analysis the responses gathered from pretesting are coded;
4. By identifying flaws and eliminating any ambiguous questions pretesting enables the researcher to revise the questionnaire;
5. Pretesting helps in further improving the questionnaire and works such as a measuring yardstick seeking perfection, etc.

9.14.1 Final Draft

After the revision, the research instrument is ready for its final draft, which is to be used for the actual survey. While drafting close-ended questions the researcher has to decide the inclusion of negative answers, like not sure, not at all, do not know, etc. There is a possibility that the target population finds it easy to just strike these choices without trying out other options. If such a choice were not given the respondent would definitely choose some other option. In obtaining relevant responses, terminology and jargon play a very important role. The researcher who is immersed in the subject so much sometimes forgets the target population's limited knowledge to understand the terms.

9.15 Advantages and Disadvantages of Questionnaire Method

9.15.1 Advantages of Questionnaire Method

1. Easier and quicker to collect data. Through this method, in a short span of time, large amount of data can be generated.
2. Direct responses can be obtained, empirical data can be collected, cost-effective.
3. From educated population, high rate of response can be generated.
4. Before finally answering the questionnaire, respondents get a chance to prepare and revise their answers.
5. Through pretest, researchers also get a chance to revise the final questionnaire.
6. Responses are easy to analyze and tabulate. Questionnaire method provides the respondents to express their views on any given topic freely.
7. Questionnaire has a fixed format with a given number of questions. This helps in eliminating variation in questioning process.

9.15.2 Disadvantages of Questionnaire Method

1. To receive adequate response rate is difficult. People are in the habit of not filing and returning the questionnaire in time.
2. Reliability of data can be questioned. Truthful answers cannot be ascertained.
3. It is time-consuming activity, which includes preparation, pretest, revision, distribution, sending reminders, etc., adding to time and cost of data collection.
4. During analysis stage, incomplete answers also cause problem. At the time of filling up of the questionnaire, a researcher cannot observe the respondent's reaction.
5. The questions can be wrongly interpreted, which can affect the analysis. For certain things, assumptions of the researcher may prove opposite to the respondent's perceptions.
6. Technical jargon or professional terminology may play havoc for the respondent. It may be possible that the questionnaire is returned with a substantial number of unanswered questions.
7. Complex worded questions also fetch poor results. There is a possibility that the questionnaire may be biased. It may not have included certain important questions though are very useful.
8. Verification of the accuracy of the responses received from questionnaires might be difficult. A questionnaire cannot be used with children and illiterate people.

Summary

Questionnaire is an effective tool to gather both quantitative and quantitative data in survey research quickly and is considered the most popular research method.

Steps involved in constructing questionnaires have been explained. A good questionnaire is imperative for good survey results.

Questionnaires consist of a series of questions dealing with psychological, social, educational, and other related issues. Questionnaires are either structured or unstructured.

The first step in questionnaire designing is arriving at preliminary decisions with regard to the issues of required information, the target respondents, and the interview techniques to be adopted. The next step is to determine the questionnaire content, so that it deals with identifying the need for data, the question's ability to yield data, the participant's ability to answer without generalizations and estimates, and the willingness to answer sensitive questions. Knowing how each question should be phrased requires familiarity with the different types of questions. This leads to the next step of the questionnaire designing, i.e., questionnaire response format. This deals with issues of using open-ended or close-ended questions.

Open-ended questions require the respondent to do most of the talking while close-ended questions restrict the respondent's responses to the variable options.

The questions should be free of implicit assumptions and biased and loaded words; the merits and demerits of asking open and close-ended questions have also been discussed.

Questionnaire sequencing is very important to elicit required information from the participant. The questions are sequenced in the following manner: lead-in, qualifying, warm-up, specific, and demographic.

A good questionnaire is specific in purpose: simple in language, logical in arrangement, and moderate in length. Proper attention has to be paid to carry out a questionnaire survey. If not done properly, it may lead to misleading results.

Review Questions

1. What is meant by a questionnaire?
2. What are the merits and demerits of a structured questionnaire?

3. In what sort of marketing studies will you use an unstructured questionnaire?
4. Explain the meaning of open-ended and close-ended questions.
5. Describe briefly the characteristics of a good questionnaire.
6. Differentiate between an open-ended and close-ended questionnaire.
7. List the advantages and disadvantages of questionnaires.
8. Why is covering letter necessary in a mailed questionnaire?
9. Why pretesting is required before finalization of questionnaire?
10. Explain why the length of a questionnaire determines its response rate.

10

Qualitative Research

10.1 Introduction

Qualitative research can be defined as a type of scientific research that tries to bridge the gap of incomplete information, systematically collects evidence, produces findings, and thereby seeks answer to a problem or question. It is widely used in collecting and understanding specific information about the behaviors, opinions, values, and other social aspects of a particular community, culture, or population. Qualitative research helps in providing an in-depth knowledge with regard to human behavior and tries to find out reasons behind decision-making tendencies of humans.

10.2 Rationale for Using Qualitative Methods

A major drawback in using quantitative methods is the problem of deliberate falsification, where the respondents knowingly fabricate the answers to private and sensitive questions. Therefore, it becomes difficult to probe into the subconscious mind of the respondent to collect factual data. This is where the need for qualitative data arises.

The exquisiteness of qualitative research lies in its flexibility to adapt to different situations. Not only it can help in probing the subconscious mind of the respondent, but it also finds extensive use in brainstorming sessions that often pave the way for embarking on product development or solving marketing problems.

A quantitative research generally has a predesigned set of responses. A respondent has to choose from the limited answers irrespective of whether or not they represent his or her true feelings.

Qualitative researches involve group dynamics where participants can interact with one another. These interactions have the inherent tendency to draw out responses that may not have been obtained in a one-to-one confrontation with the interviewer.

10.3 Types of Qualitative Research

There are certain approaches of qualitative research attempting to understand human nature, market research purposes, current trends, changing tastes, and preferences of people.

10.3.1 Case Study

With the help of case study method, a case of an individual, group, event, institution, or society is studied. It helps in providing an in-depth knowledge of the nature, process, or phenomena of a specific case under study. Multiple methods of data collection are often used in case study research, for example, Interviews, Observation, Documents, and Questionnaires.

10.3.2 Ethnography

This approach mainly focuses on a particular community. It is more of a kind of close field observation and basically tries to study a socio-cultural phenomenon. For example, judging others based on the

DOI: 10.1201/9781315167138-10

researchers' cultural standards. Ethnography can be used for comparative analysis of cultural groups. Eating habits of North Indians and South Indians, also known as "Ethnology." Further, it can also be used to analyze the cultural past of a group of people, for example, Harrapan civilization, also known as "Ethno history."

10.3.3 Historical Method

Historical method helps in understanding and analyzing the causal relationships. With the help of this technique, the data related to the occurrence of an event are collected and evaluated, in order to understand the reasons behind the occurrence of such events. It helps in testing hypothesis concerning cause, effects, and trends of events that may help to explain present events and anticipate future events as well.

10.3.4 Grounded Theory

This approach involves an active participation of the researcher in the activities of the group, the culture, or the community under study. With the help of observation, the data with regard to the required information are collected. It is generally used in generating or developing theories. This means that the ground theorists cannot only work upon generation of new theories, but also can test or elaborate previously grounded theories. The clear and understandable theory is generated by grounding. For further analysis or generating more theories, the theory provides much information and scope. The theory generated is valid as it has been analyzed under controlled conditions.

Grounded theory helps in identifying anchors or codes that allow the key points of the data to be gathered. It helps in making implicit belief systems, explicit with the help of researchers' questions and analysis. It consists of a set of steps whose careful execution is thought to "guarantee" a good theory as the outcome. Data collection and analysis continue throughout the study.

10.4 Comparison between Qualitative and Quantitative Research

The basic conceptual difference between both of the research techniques is as follows (Table 10.1):

10.5 Qualitative Research Methods

Qualitative research methods can be subdivided into individual Depth or Intensive Interviews, Focus Groups, Discussion and Projective Techniques.

10.5.1 Individual 'Depth' or 'Intensive' Interviews

Depth interview is a qualitative approach, in which a trained moderator conducts interviews with individuals, rather than with groups, to obtain information about a product or brand. These interviews are primarily conducted on a one-to-one basis. Depth interview is simply the routing of an ordinary conversation that permits both the researcher and the interviewer to interact and explore an issue.

Depending upon the amount of guidance extended by the interviewer, an individual in-depth interviews can be subdivided into Nondirective or Unstructured Interviews, Semi-Structured or Focused Interviews and Standardized Open-ended Interviews.

10.5.2 Nondirective or Unstructured Interviews

Here, the respondent is given maximum freedom to respond in a manner that he or she wishes to, within a reasonable limit of relevancy to the topic under discussion.

Unstructured interviews, during the course of the conversation, take the form of a natural conversation and the interviewer brings up various topics that are of interest to him or her. While expressing his or

TABLE 10.1

Comparison between Qualitative and Quantitative Research

Sl. No.	Points of Distinction	Qualitative Research	Quantitative Research
1	**Dependent**	Qualitative research is based on observation and experiences	Quantitative research is based on numerical or graphical representation of data
2	**General framework**	Seeks to explore phenomena using some structured methods such as in depth interviews, experiences, and participant observation	Seeks to confirm hypothesis related to phenomena using highly structured methods such as questionnaires, surveys, and structured observation
3	**Objectives**	It aims to describe variation, explain relationships, and describe behavior, experiences and norms of individuals and groups	It aims to quantify variation, predict causal relationships
4	**Questions**	The questions used for data collection are open-ended ones	The questions used for data collection are close-ended ones
5	**Representation of data**	Data are represented in the form of notes, recordings, and video tapes	Data are represented in the form of numbers and graphs
6	**Research design**	The research design allows some flexibility in certain situational aspects. The questions used for the data Collection differ individually and depend on the response of the participants	The research design is predetermined and stable from the beginning. The questions used for data collection are structured and same for all the participants
7	**Use**	Exploratory or diagnostic in nature—used to understand behavior and generate hypothesis	Once hypothesis has been generated, used to test out hypothesis
8	**Sample size calculation**	No calculation of sample size possible—size of sample depends on time available to conduct research, cost, and variation in the population	Generally a probabilistic approach is used to calculate sample size, using the sample size formula
9	**Represents**	Sample selected is such that it represents different sections of the population	Random selection of respondents to be part of research work—may or may not represent different population segments, depending upon the sampling method utilized
10	**Conclusions**	It is dangerous to generalize conclusions for the entire population	Conclusions are generalized to the universe, of which the sample is to be representative

her opinions or narrating his or her experiences relating to a topic, the respondent is given the freedom to decide the direction of the conversation.

This unstructured characteristic of the interview enables the interviewer to develop a rapport with the interviewee and understand him or her better. As there are no preformulated set of questionnaires and no predetermined paths to route the interviewee responses, unstructured interviews are also known as nondirective interviews.

10.5.3 Semi-Structured or Focused Interviews

In a semi-structured or focused interview, the initiative is retained by the interviewer, and the interview has to cover a specific list of points, which has been decided in advance. To maximize data collection, there is also a tighter control over the interview, and also in collecting data relevant to the topic under consideration. The interviewer should be determined as to which questions are to be asked. For example, consider the chat shows that take place on television. Even though the participants are given maximum freedom with respect to his or her answers, the initiative is retained by the interviewer. And he or she has decided in advance the questions that would be asked in the course of the interview.

These types of interviews are more structured than the nondirective interviews. While allowing some amount of flexibility in the interview, the interviewer ensures that he or she keeps the interview limited to the topics that are essential to the research.

At this stage, to provide details for relevant responses, probing techniques can be used to encourage the respondents. This technique is primarily used to interact with busy executives, technical experts, and thought leaders.

As the technique calls for interactions with experts, the interviewer must possess knowledge of the latest trends in technology, market demand, legislation, competitive activity, etc. This will enable him or her to apply the probing techniques better. Therefore, the inherent skills of the interviewer are crucial.

Disadvantages of Semi-Structured or Focused Interviews

1. The technique is known to have no provision to permit the interviewer to probe into unanticipated issues cropping up during the interaction, which were not a part of the basic checklist.
2. Even the flexibility with regard to the choice of words of the interviewer may lend bias, leading to different responses from different individuals.

10.5.4 Standardized Open-Ended Interviews

Here, the questionnaire contains a set of sequential ordered, carefully worded, open-ended questions. When two or more interviewers conduct the interviews, this technique is appropriate. It minimizes the variation in the questions posed by them to different interviewees. These types of interviews enable the evaluator to collect data systematically, thus facilitating comparison of responses collected from different respondents. This method, however, limits the use of substitute questioning to probe into individual differences.

Advantages of In-depth Interviews

In-depth interviews are appropriate in the following situations:

1. When detailed probing of an individual's behavior, attitude and needs is required.
2. When the subject matter is highly confidential in nature, e.g., how do you plan your investments required for annual tax planning.
3. Attitudes and emotions of the test persons can be explored in detail and are close to reality as there is no social pressure to conform to group responses, as is the case of focus groups.
4. When interviews are conducted with highly qualified professionals, e.g., statistician, on the usage of statistical techniques for analysis or analysis and interpretation of problems with various statistical techniques, a normal questionnaire would not suffice for obtaining information, and a detailed probing is required, which would come out only through an in-depth interview.

Disadvantages of In-Depth Interviews

1. The skill of the interviewer is very critical in drawing out the respondent's true feelings.
2. In an interview, there is not only verbal communication but also nonverbal communication, and the interviewer should also keep track of the respondent's voice tone, facial expression, movement of hands, i.e., gestures.
3. The sample size cannot be large, as conducting in-depth interviews takes a longer time.
4. Generalizations to the entire population cannot be done because analysis and interpretation of data are highly subjective processes.
5. To find and employ skilled interviewers is difficult and expensive.
6. Lack of structure in questionnaires in nondirective and semi-structured interviews introduces interviewer bias.
7. The cost and the length of the interviews combined do not permit more number of interviews to be conducted.

10.6 Focus Group Discussion

Group discussion can be conducted by the following methods.

10.6.1 Brain Storming

In brain storming, there is no moderator for the group and the group freely expresses its ideas on the given topic. The ideas could be absolutely abstract, but then this would help in generating new product ideas and also better ways of conducting a particular business, etc.

In this method, videotaping of the proceedings is also done in order to record the group discussion, to record the facial expressions of the participants and also the intensity of their feelings.

10.6.2 Focused Group Discussion

A focus group is defined as group of individuals selected and assembled by researchers, to discuss and comment on, from personal experience, the topic that is the subject of research. Here, the group is given a topic and asked to discuss the topic. A moderator would also be involved in order to ensure that the group 'discussion remains relevant and does not go off the track'.

The moderator could stop the discussion between time intervals to find out what conclusions are being drawn by the group after each time interval. A focus group consists of a group of anywhere between 6 and 12 members. This size of the group encourages the participants to express their views on a specific issue. The very essence of the focus group as a technique lies in tapping the unexpected findings that result from an interactive session between the members of the group. These members take part in the discussion for about 2 hours, which is the normal time for a focus group interview. These members are selected from a planned sample. It should be ensured that the participants have ample knowledge and experience of the issue or topic to be discussed.

Prior to conducting the focus group discussion, the participants are updated over the phone regarding the purpose of the focus group and the confidentiality of the members and their information. At the onset of the discussion, the moderator reiterates the same things in addition to introducing any co-moderators and explaining how and why these group members were invited to participate and stating the purpose of note taking and recording.

It should be ensured that the focus group is homogeneous with the participants having common interests, experiences, or demographic characteristics. This would facilitate proper blending among the members resulting in a productive discussion.

Due to lack of representativeness, it is not possible to compare the results from different groups in a strict quantitative sense. However, in doing so, it should be ensured that people who know each other or are in some sort of command chain are not recruited into the same sessions.

For this purpose, the researchers segregate the participants into different groups based on the difference in views for or against or some other parameters.

The effectiveness of a focus group depends on the person who moderates the discussion. A moderator should balance a directive role with that of a moderator, which calls for the moderator to be skilled in establishing and upholding group dynamics and being able to provoke intense discussion on relevant issues.

This is important, as the quality of data collected is directly proportional to the effectiveness. With which, the moderator monitors the discussion by asking questions and keeps the discussion targeted on the research objectives.

A moderator external to the research organization but with sufficient expertise can also be invited to preside over and facilitate the discussions. If there are different groups then the moderator is expected to be flexible enough to customize his or her style to each of them.

The questions should be open-ended, clearly formatted, neutral, and sequential. Close-ended questions and leading questions, i.e., questions that favor a particular response from the participants, should be avoided.

The responses of members of the group are also affected by the immediate physical surroundings. Therefore, the provision of a relaxed, informal atmosphere should be ensured.

Advantages of Focus Groups Discussion

1. In a group discussion setting, the interaction among members acts as a stimulant to generate new ideas, which may never be possible in an in-depth individual interview situation.
2. The group discussion setting leads to spontaneity in conversation, which cannot come in individual interview situation.
3. If the group setting is emotionally provocative, the conversation may set off a thinking process, and one might recall older memories and conversations that is relevant to the discussion on hand.

Disadvantages of Focus Groups Discussion

1. To conduct group discussion, qualified and trained personnel are required.
2. Analysis and interpretation is highly subjective.
3. A few members in a group could dominate the entire discussion.

10.7 Projective Techniques

When a researcher is conducting an in-depth interview or conducting a survey through the questionnaire method, he or she might face some problem with the respondent in the form of language barriers, illiterate respondent especially in social research and rural research, social barriers, i.e., respondent is embarrassed to talk about a topic, tries to avoid certain questions, cannot answer, etc.

To overcome such barriers faced during an interview process, the researcher may replace the questionnaire with projective techniques. The basic underlying concept behind projective techniques is that in certain situations, it is impossible to obtain correct information about what a person thinks or feels by asking him or her to describe his or her feelings. But this information can be obtained by making the respondent to project his or her feelings on to some other person or object.

Every individual has a subconscious mind that even the individual may not be aware of, which holds a lot of attitudes and motivations. Use of direct questions to unravel these attitudes and motivations are least effective. Therefore, researchers use special techniques to venture into the private worlds of a subject to uncover their inner motives. These special techniques are known as projective techniques.

The projective technique is an unstructured, indirect form of questioning that encourages respondents to project their underlying motivations, beliefs, attitudes, or feelings with regard to the issue of concern.

The respondents are exposed to various scenarios and asked to interpret them. A close observation of the way the respondents describe a situation or a scenario reveals their own motives, attitudes, values, and motivation.

Projective techniques find applications in various fields and are not limited to the exclusive study of consumer motivation.

To tap the feelings into the subconscious minds of the subjects, some of the projective techniques used by researchers are Association Techniques, Completion Techniques, Construction Techniques, Expressive Techniques, and Sociometry.

10.7.1 Association Techniques

10.7.1.1 Word Association Test

In this method, the respondent is presented with a list of stimulus words. And for each word, the respondent is asked to respond with what he or she thinks about the word. The respondent is not given time to think of the responses. The ideas is that the "first thought" responses are likely to reveal the true feelings of the respondent about the stimulus. This is known as free word association where the subject has to share his or her first word of thought. Successive word association is a slight variation in which the subject shares a series of words or thoughts that strike his or her mind in response to the stimulus.

The researcher records the responses and the time taken to respond to each word. This helps in analyzing the frequency in which the subject gives a particular word or thought in response to the stimulus. These responses are analyzed by calculating the frequency of the words given in response, time elapsed before the response, and number of nonrespondents.

Uses of Association Test

1. Association tests are especially useful in consumer research used for discovering brand image or product attributes. Brand personification is one such area.
2. Association tests can also be extended to measure attitudes about specific brands, their attributes, packaging, and even advertisements.

10.7.1.2 Sentence Completion Test

This is an extension of the word association test. In this method, the respondent is asked to finish an incomplete sentence with the first thought that comes to his mind. The idea is that the respondent projects his or her own feelings into the sentence. For example, I like to drive car because…; Today I am very happy because…

As with the word association, the frequency of responses are taken into consideration while analyzing the results of the sentence association test.

10.7.1.3 Fantasy Situation

Here, the respondents are asked to imagine that they are converted into a product itself, e.g., car, box of chocolate, train, toy, ball. This leads the respondent to imagine himself or herself as the product itself and give human characteristics to the product. This method is used for developing brand perception, brand personality, etc.

10.7.1.4 Cartoon Completion

In this method, the respondent is shown a cartoon that is similar to a comic strip, with "balloons" indicating speech. Usually, two people are shown talking to each other about a particular product or service or situation, but only one balloon contains the speech. The situation that is shown in the cartoon is obviously of special interest to the researcher and is part of the research project under study. The respondent has to fill the other "balloon" with his or her answer to what the other person is saying. With this method, one tries to measure the attitude toward a product or service. Analysis and interpretation of these results are highly subjective.

10.7.1.5 Picture Interpretation (Thematic Apperception Test)

Thematic Apperception Test (TAT), along with the Rorschach Inkblot test, is probably the most widely known and used projective technique in Clinical Psychology. Here, the respondent is shown a picture or either a line drawing, illustration, or photograph, which is rather ambiguous, and is asked to describe what is going on or is asked to tell a story about what is illustrated.

For example, a retail outlet, such as a saree shop, is interested in knowing what a shopper thinks while buying a saree, how does he or she look at the various display sarees, what he or she expects during the shopping experience, etc.

For such a research project, a researcher might show the photograph of a person entering the saree shop to the respondent, and ask him or her to develop a story on what he or she thinks could be happening inside the shop. Normally, a respondent would project his or her feelings while developing the story from the photograph. This gives an idea to the researcher as to what the respondent expects on entering the saree shop.

The major drawback with this method is that there is a great deal of subjectivity in interpretation of the responses to these projective tests. These projective techniques are used in exploratory research, whose output acts as input for developing a hypothesis and full-fledged quantitative research. In this technique, subjects are presented with a stimulus, and are asked to reveal the first word, image, or thought elicited by the stimulus.

10.7.2 Completion Techniques

Completion techniques are of two types, that is, sentence completion and story completion.

10.7.2.1 Sentence Completion

In sentence completion, a subject is asked to fill up the blank in a sentence. The subject uses his or her intuitive ideas to do the job and in the process leaves clues that are traceable to his or her underlying attitudes, thought process, and feelings.

This technique scores over word association in that the subjects can be given more directive stimulus. These statements are usually in the third person and are somewhat ambiguous. Interpretation is usually qualitative, rather than quantitative. Sentence completion is useful when time is limited, but depth of feeling still has to be trapped. A slight modification in this technique is the paragraph completion where the subjects are required to fill in an incomplete paragraph.

10.7.2.2 Story Completion

In story completion, the respondent is required to fill in the conclusion of the story. The story contains enough clues to direct the responses of the subject, but gives no hints at the ending. The choice of words and the way the respondent concludes the story helps the researcher to form an idea about the feelings and personality of the respondent.

10.7.3 Construction Techniques

In construction techniques, the subject is asked to construct his or her responses in the form of a story, description, dialogue, or picture. In construction techniques, the respondent is supplied with less initial structure, which in turn requires more complex and controlled intellectual activity on his or her part. Construction techniques are of two types, that is, picture response technique and cartoon technique.

10.7.3.1 Picture Response Techniques

Picture response techniques originated from thematic apperception test (TAT), which is based on Henry Murray's Personality Theory. The participants in this technique are given one or more pictures and asked to interpret the background situation, discussion, or direction of the story from the moment the picture is captured. For this reason, it is also known as the picture interpretation technique. It is used to identify the unspoken thoughts of the characters.

10.7.3.2 Cartoon Technique

Cartoon techniques also serve a similar purpose. Here, the statements or a thought of one of the cartoon characters is given in the box above his or her head. This is supposed to evoke certain responses in the mind of the other cartoon. This is exactly what the subjects have to identify.

This evokes varied responses from different subjects, which in turn help the researchers to understand their unique mindset and personalities. Other variations in this technique are the third-person techniques, fantasy scenarios, and personification.

10.7.4 Expressive Techniques

This is a technique that involves role playing, where the respondent is given a verbal or visual situation and is asked to play the role of a specific character such as a sales executive, a manager, a teacher, a political leader, a businessman, or an income tax commissioner.

For example, the respondent is asked to assume that he or she is a professor of a college, who has been invited to an interactive session with the Students of Final Year MSc Statistics or PhD Scholars. The students tend to pour out their difficulties and doubts about the subject. The manner in which the respondent copes with the situation, tackles the grievances, and makes statements reveals a lot about the personality of the subject.

10.7.5 Sociometry

Sociometry is a method that was devised by Jacob L. Moreno for assessing group structure. Moreno defined sociometry as the mathematical study of psychological properties of populations, and by application of quantitative methods, the results are obtained.

Sociometry is based on the fact that people make choices in interpersonal relationships. Whenever people are in a group, they choose where to sit or whom they speak to, etc. It studies patterns of affection and loyalty that bind some group members more closely than others and can be applied to situations involving the study of group behavior in research.

Advantages of Projective Techniques

1. As the respondent is aloof of the purpose of the study, he or she tends to give responses that would otherwise have not been possible.
2. Respondents do not perceive right or wrong answers to the exercise and are encouraged to respond with a wide range of ideas. This results in an increased amount of rich and accurate data collection.
3. Projective techniques with regard to why consumers behave as they do help in generating hypotheses.
4. In focus group discussions, they are useful in "breaking the ice."

Disadvantages of Projective Techniques

1. Requires trained interviewers and skilled analysts who are difficult to find, to analyze the responses, due to the complexity of the techniques.
2. It is difficult to administer the techniques because the employment of highly skilled staff is expensive.
3. As with the idea, all subjects may not feel comfortable, getting subjects for role-playing is difficult.
4. It is difficult to establish the measure of reliability.

10.8 Observation Methods

Another very powerful tool for getting information about the consumer is observation method. This method is used for recording behavior of people, objects, and events. Informal observations are extensively used for observing customer buying patterns, impact of competitive advertisement on buying, product availability, etc.

Observation technique is always used in conjunction with other research techniques. As it is highly subjective, and a lot depends upon the observer's perception of a situation, the inherent danger in this method is that one could draw wrong conclusions.

Unlike the methods discussed earlier, observation methods do not involve any verbal communication with the respondents. Observation methods involve recording the behavioral patterns of respondents without communicating with them.

Classification of Observation Techniques

1. **Disguised versus Undisguised Observation**

   Disguised Observation: Disguised observation means consumer is not aware of being observed. For example, two-way mirrors, hidden cameras in shops, observers dressed as sales clerks, observers dressed as waiter, etc.

   Undisguised Observation: Undisguised observation means the consumer is aware of being observed.

2. **Structured versus Unstructured Observation**

   Structured Observation: In a structured observation, the decision problem has been clearly defined. Information needs are clearly defined, and so this reduces observer bias and the reliability of observed data increases.

   Unstructured Observation: In an unstructured observation, the problem is yet to be defined and formulated, and is ideal for exploratory research.

3. **Human versus Mechanical Observation**

   Human Observation: By replacing humans with mechanical devices, the accuracy of observation increases, observer bias is reduced, and lower costs are incurred.

   Mechanical Observation: Special mechanical devices used for observation could be video cameras for recording shopping behavior, audiometer or people meter for recording TRP ratings, eye-cameras, etc.

Some of the most popular observation methods used by researchers are discussed below.

10.8.1 Direct Observation

Direct observation is a method where the observer tries to gain an insight into the behavior of a shopper in a tactful manner so as not to be noticed. This has applicability in studying merchandising effects in a Big Bazaar, D-Mart,Super Bazaar, compliance to traffic rules by motorists, etc.

In tracking the behavior of a shopper in a Big Bazaar, etc. the observer can either remain in a passive state as a silent observer or structured or disguise himself or herself as another shopper and engage in a shopping spree in close association with the subject or unstructured. In both cases, the observer note down certain specific behaviors related to the subject.

This makes it possible for the observer to find the appealing factors in the buying behavior and service problems faced by the subject.

This is a highly subjective task and requires the observer to record certain noticeable behavioral features useful for the study. It can often be a rapid and economical way of obtaining basic socioeconomic information on households or communities.

Be it structured or unstructured, it is imperative for the observer to ensure that he is not identified; else it would lead to an alteration in the behavior of the subject and introduce subject bias.

Various ways that facilitate in direct observation are one-way mirrors and disguised and hidden cameras. However, it should be ensured that there is no invasion into the privacy of the subject while using one-way mirrors or hidden cameras. It is possible to identify the exact timing and length of continuation of an activity by direct observations. There is instantaneous recording of the observations, which eliminates the necessity of having to recall later. This method is, however, prone to observer bias where the observed may wrongly assign specific demographic characteristics to the subject.

10.8.2 Natural and Contrived Observations

10.8.2.1 Natural Observations

An observation in which the subject under study is unaware of being scrutinized for specific behavior is known as a natural observation. The subjects under study have little knowledge that they are being observed for specific behavioral aspects and demographic characteristics. This method uses more of a disguised observer, who inconspicuously records the specific behavior he or she has to scrutinize. This method of natural observation has little relevance for researchers who desire to analyze special behavior, which may be rare among individuals operating in natural circumstances.

10.8.2.2 Contrived Observations

Here comes in the concept of contrived observation. The subjects in this case have some advanced knowledge of being participants in the observation study. Although the subjects are aware of their involvement in the study, they still have no idea as to which aspects of theirs are being scrutinized and observed. However advantageous it may be, the artificial setting and the awareness of the subject that he or she is being observed can bring in respondent bias.

A corollary concept to contrived observation is mystery shopping. Here, the main motive of the observer is to analyze the behavioral aspects of participants, primarily in the service sectors. The following are the situations where this concept is used.

Pizza Hut claims to deliver orders for home delivery within 30 minutes. The company may authorize any person to pose as a customer and place an order to observe the timeliness in the delivery process. Similar procedures can be applied to analyze the quality of service offered in hotels and banks.

Advantages of Observation Technique

1. This method does not rely on the willingness of the respondent to cooperate and provide information.
2. Behavior patterns of which a respondent is not aware of can be recorded by observation only, e.g., facial expression of a consumer while examining a new product display in a store.

Disadvantages of Observation Technique

1. By observation method, one cannot observe a consumer's beliefs, feelings, awareness, etc. However, other research techniques such as focus groups, in-depth interviews are required.
2. The observed behavior pattern must be of short duration, should occur frequently, in order to qualify for observation, and act as input to other research techniques.

10.8.3 Content Analysis

Another form of discourse analysis is content analysis. It is one of the methods that is used in summarizing any form of content only after having a deep study of the actual content. This enables the researcher to more objectively evaluate and understand the situations. For example, an impressionistic summary of a movie cannot help in analyzing the overall aspects of the content of the movie. Content analysis tries to analyze written words. The results of content analysis are numbers and percentages. It starts with the process of selecting content for analysis, then preparing the content for coding. After the content is coded, it is counted and weighed. Later, conclusions are drawn on the basis of the weighing. After doing a content analysis, the researcher can make a statement such as "47% of programs on FM Radio in present year mentioned at least one aspect of anti-dowry, compared with only 7% of the programs in previous year."

The content analysis helps in removing much of the subjectivity from summaries, in detection of trends in an easier and simpler manner, etc. Content analysis enables the researcher to make links between causes, e.g., program content and effect, e.g., audience size. The content analysis is used to evaluate and improve the programming of the media world. It also helps in increasing awareness and summarizing the various notes or documentaries that focus on a specific issue.

Written materials such as advertising copies and news articles, and Television and Radio programs have many implicit and explicit meanings. Therefore, their content has to be thoroughly analyzed for any mismatch or misrepresentation in communications. This is where the technique of content analysis comes into play. These written materials need to be analyzed based on words used, themes, characters, and space to enable the smooth flow of the intended communicational aspects. This helps the management to introduce the required changes in the communication process, as may be deemed necessary to generate a better response rate.

10.8.4 Physical Trace Measures

Physical trace measures refer to exposure to advertisements, computer cookie records, records of credit card usage, dirt on the floor to determine store traffic patterns, etc.

In other words, it is the process of looking systematically in the immediate surroundings for any evidence of human interaction with one another or the environment. This method usually helps in unraveling the space usage patterns of people. Following are the types of traces observed and measured. They are as follows:

1. Erosion traces are shown by deterioration or wear and tear that provides a look at the usage pattern. This refers to the traces of selective wear and tear of certain parts or things in a space that shows evidence of being used.
2. Accretion traces are a build-up of a residue or an interaction. Traces of lumps of dirt in proximity reveal the piling up of shoes. Similarly, a number of glasses together reveal their use of drinking purposes.

10.8.5 Participant Observation

A process in which a researcher establishes a many-sides and long-term relationship with individuals and groups in their natural setting, for the purposes of developing a scientific understanding of those individuals and groups is known as Participant Observation.

At the first look, it may seem as a process concerned with looking, listening, experiencing, and recording the same.

However, in reality, it is more demanding the analytically difficult. This method of observation requires the researcher to be involved in the day-to-day activities of the subjects or the social settings that are under investigation. Depending on the degree of involvement of the researcher, this involvement can be categorized into three types. These are as follows:

1. **Complete Participant:** The researcher immerses himself or herself fully in the activities of the group or organization under investigation. It supposedly produces accurate information. As the intentions of the researcher are not disclosed to the subjects or social settings under investigation and are least likely to guide researchers to enforce their own reality on the social world they seek to understand.
2. **Participant as Observer:** The research in this case keeps the group informed about his or her intensions, but does not actively involve himself or herself in the social settings.
3. **Complete Observer:** The researcher is uninvolved and detached, and merely passively records behavior from a distance. The presence of the researcher can cause some initial sparks of discomfort. Language and Cultural dissimilarities can pose barriers in this method.

The compatibility of observation and interviewing in this method makes it highly flexible. Apprehensions about observations pave the road to questions that are later clarified during interviews to understand the significance of the observations. The interview in this case is highly unstructured.

10.8.6 Behavior Recording Devices

Human observation is prone to deficiencies or errors. To overcome such errors, machine observers in the form of behavior-recording devices are used. This sort of mechanical observation include

1. Onsite cameras in stores and at home for eye-tracking analysis while subjects are shopping or watching advertisements using oculometers to identify what the subject is looking at and pupilometers to measure how interested the viewer is.
2. Electronic checkout scanners that record the universal product codes on the products as those used by A.C. Nielsen and INTAGE. These are used to record purchase behavior of the subjects under investigation or in general.

3. Nielsen People Meter for tracking television station watching.
4. Voice pitch meters that serve to measure emotional reactions.
5. Psych galvanometer that measures galvanic skin response.

It may be easier for these machines to record the behavior of the subjects, but measuring the precise level of arousal and reaction through them is questionable. Therefore, calibration and sensitivity is a limitation with the mechanical devices.

10.9 Importance of Qualitative Research

1. Exploratory or diagnostic in nature.
2. It involves a small number of people who are not sampled on any probabilistic basis.
3. Used to generate hypothesis for further research.
4. Used to get better insights into consumer behavior, and to understand the underlying behavior of the consumer in the buying process.
5. Through qualitative research, one can get subtle clues about products or brands or services that very few quantitative studies can replicate.
6. No attempt is made to draw hard and fast conclusions about facts that emerge.

10.10 Uses of Qualitative Research

1. It is used to define the problem areas more fully—in marketing research one normally starts with qualitative research, which is validated further by quantitative research.
2. It is used to formulate hypothesis for further investigation or quantification.
3. It is used to obtain a large amount of data about beliefs, attitudes, etc., as data input for developing questionnaires, attitude scales, which would be used as input for multivariate analysis studies.
4. It is also used to conduct post-research study, i.e., to amplify or explain same points that emerge from a major study, without having to repeat on a large scale.
5. In studies of distribution channels, sales, pricing strategies quantitative approach is most suitable, whereas in concept development, product development, i.e., needs of consumer, advertising research, i.e., qualitative approach, is more suitable.

10.11 Ethical Guidelines in Qualitative Research

The respondents and their responses should be respected by the researcher. The researcher must show respect and belongingness to the community he or she is studying. The respondents must be made aware of what is being analyzed by the researcher. The researcher must ensure and maintain the confidentiality of the researcher. The researcher should be aware of the expected risks and benefits including the psychological and social aspects while performing the research.

Summary

Qualitative research may be used to generate hypothesis for further research, is exploratory in nature, and involves a small number of people who are not sampled on probabilistic basis. There are various methods of conducting qualitative research. These are depth interviews, focus groups, observation, and projective techniques such as word association test, sentence completion test, cartoons, TAT.

A research can be better analyzed with qualitative information that unravels the underlying motives that act as driving factors for the subjects rather than only quantitative measurements. This underlying limitation of quantitative means is the very reason for the development of qualitative techniques for business researches. Designed tactfully, qualitative techniques help in overcoming another limitation of quantitative techniques, that is, respondent bias. As the qualitative techniques are designed in a disguised manner, they help to elicit factual responses from the subjects. This chapter is dedicated toward providing a thorough knowledge of the characteristics and functioning of different qualitative techniques.

Depth interviews involve interviewing on a one-to-one basis with a single moderator. These techniques help to determine an individual's underlying perception opinions, facts, and his or her reactions to initial findings and potential solutions.

Focus groups, on the other hand, consist of a group of individuals selected and assembled by the researcher to discuss and comment on from personal experience the topic that is the subject of the research.

Projective techniques make use of techniques that are disguised and help in eliciting the feelings, beliefs, attitudes, and motivation, which many consumers find difficult to articulate with direct questioning methods. Projective techniques are of different types: association, completion, construction, and expressive. Each has unique features and tactics to understand the subconscious feelings of the respondents.

Review Questions

1. Define qualitative research.
2. Differentiate between qualitative and quantitative research. Do you think that qualitative research is advantageous over quantitative research? Give reasons.
3. Elaborate the different types of qualitative research.
4. Explain the concept and significance of content analysis.
5. With the help of examples, discuss the areas where qualitative research can be used in marketing.
6. What are the various methods of conducting qualitative research? Discuss the advantages and limitations of each of these methods.

11

Experimentation

11.1 Introduction

An experiment refers to the process of manipulating one or more variables and measuring their effect on other variables, while controlling external variables. The variable, which is manipulated, is called the independent variable, and the variable whose behavior is to be measured after experimentation is called the dependent variable. For example, if a company wants to test the impact of advertising frequency on product sales, in a particular region. Researcher conducts an experiment by manipulating the advertising frequency to study its impact on product sales. Here, the variable, which is being manipulated, is advertising frequency and therefore it is the independent variable. The impact of change in advertising frequency on product sales is measured and analyzed. Thus, product sales is the dependent variable.

To establish and measure the causal relationship between the variables studied is the aim of experimentation. By controlling extraneous variables, a well-executed experiment can depict the casual relationship between variables.

11.2 Experimentation Issues

A researcher has to take decisions with regard to various aspects, to make an experiment successful. While conducting an experiment, a research has to consider the key issues such as Treatment of Independent Variable, Experimental and Control Groups, Selection and Measurement of the Dependent Variable, and Control of Extraneous Variables.

11.2.1 Treatment of Independent Variable

A variable over which the researcher is able to exert some control for studying its effect on a dependent variable is called as an independent variable. Experimental treatment refers to the manipulation of the independent variable. For example, consider a company planning to test changes in package design, in terms of its impact on product sales. To test the relationship between package design and sales, it has decided to expose customers to packs of three different designs, A, B, and C. These pack are placed on the shelves of select outlets. The consumer's response is measured. Here, package design is the independent variable, which is manipulated, and there are three treatment levels A, B, and C of the variable.

11.2.2 Experimental and Control Groups

In a simple experiment, a researcher uses experiment and control groups.

A group of test units that are not exposed to the change in the independent variable is called as the control group. On the other hand, the experimental group is exposed to a change in the independent variable.

In the package design example discussed above, a group of supermarkets, i.e., experimental group, are selected and each package design is displayed for a month. Another group of supermarkets, i.e., control group, continue to carry the regular package design for that particular period. Then, in each case, the

DOI: 10.1201/9781315167138-11

sales of the product are measured and the difference between the measurement of sales, in the experimental group and the control group is analyzed to determine whether the design change has affected sales or not.

11.2.3 Selection and Measurement of the Dependent Variable

The dependent or Response Variable is the variable whose behavior is the result of an experiment. Dependent variable is the variable that measured may changes due to the manipulation of independent variable. Using the same example of package designs, the Sales Volume of the product is considered as the dependent variable. Selecting a dependent variable in all cases may not be easy. For example, if the objective of a company is to do research, to evaluate the effectiveness of various advertising programs, the dependent variables can be the brand image, brand awareness, and product sales. Depending on the purpose for which the experiment is being conducted, the researcher has to select the dependent variable. Proper problem definition will help the researcher select the appropriate dependent variables.

11.2.4 Control of Extraneous Variables

Other extraneous variables, which influence the dependent variable, have to be controlled, to determine the real effect of manipulation in the independent variable on the dependent variable. The presence of these variables in the experiment will put the researcher in dilemma. As to whether the change in dependent variable is due to the change in the independent variable or due to extraneous variables. This is why extraneous variables are also called Confounding Variables. Researchers use various methods to control extraneous variables. They are Randomization, Physical Control, Matching, Design Control, and Statistical Control.

11.2.4.1 Randomization

It is the most popular method to control extraneous variables. Randomization refers to the process of assigning test units randomly to experimental treatments and assigning experimental treatments randomly to test units. This process helps researchers to spread the effects of extraneous variable equally over the test units.

11.2.4.2 Physical Control

Another approach to control the extraneous variables is Physical Control. This is achieved by keeping the level of extraneous variables constant throughout the experiment.

11.2.4.3 Matching

Matching is another variant of the physical control approach. Here, the researcher adopts judgmental sampling to assign test units to both the experimental group and the control group. This ensures that both experimental and control groups are matched in terms of characteristics of test units.

11.2.4.4 Design Control

In this method, selecting appropriate experimental designs to conduct the experiment helps researchers control particular extraneous variables that affect the dependent variable.

11.2.4.5 Statistical Control

Here, extraneous variables that are affecting the dependent variable are identified and measured using appropriate statistical tools such as analysis of variance. Then, the effects of extraneous variables on the dependent variable are adjusted statistically, canceling out the effects of extraneous variables.

11.3 Experimental Validity

Validity is the extent to which a research process is accurate and reflects actual market conditions, i.e., it is free from systematic error. Types of validity that are considered in an experimentation are Internal Validity and External Validity.

Internal validity measures to what extent the change in a dependent variable can be explained by the independent variable. External validity measures to what extent the inferences derived from experiments can be generalized to the real environment.

11.4 Internal Validity

Internal validity indicates to what extent the change in dependent variable in an experiment is caused by the manipulation of independent variable or due to extraneous variables.

If extraneous variables have an influence on the dependent variable, then establishing the causal relationship between the dependent and independent variable becomes difficult.

Any findings or conclusions drawn from experimentation in the absence of internal validity will be superficial and deceptive. Hence, while developing experimental research designs, researchers should take adequate care to include the influence of extraneous variables to improve the experiment's internal validity.

Various types of extraneous variables that are sources of threat to internal validity are History, Maturation, Testing, Instrumentation, Selection Bias, Statistical Regression, and Mortality.

11.4.1 History

It refers to a specific event in the external environment that is historic or rare occurrence in nature and occurs at the same time an experiment is being conducted. Such events may impact the dependent variables. For example, an experiment aimed at assessing the impact of a new promotional campaign for a Bike. It may be influenced by the steep spurt in petrol prices due to a historic event such as the Gulf Countries War. An occurrence of an event, which is beyond the control of the researchers, will have an impact on the dependent variable, i.e., in this case sales.

11.4.2 Maturation

The change in the test units, not due to influence of independent variable but due to passage of time, is referred to as the maturation effect. During the course of the experiment, people may become older, hungrier, or tired. For example, if a pharmaceutical company is conducted during trails on a sample of patients over a longer period of time, there may be some difference in the effect of the drug on patients due to physiological changes in them. This impacts the experiment's internal validity.

11.4.3 Testing

Another extraneous variable that affects experimental results is the testing effect. This refers to the subjects becoming alert when they are exposed to experimentation. For example, when employees are made to answer a questionnaire that tests their knowledge and skills, before attending a training program they are alerted that they are being studied. This prompts them to pay more attention to the training modules. Thus, they obtain better scores in the test conducted after the training program. Thus, there will be change in experimental results between the first test and the second test.

11.4.4 Instrumentation

To minimize the test effect, a researcher can vary the measuring instrument used for pretesting and posttesting. However, this may lead to the introduction of a new effect called the instrumentation effect.

This refers to the effect on experimental results due to change in the measurement instrument, measurement values, or the researcher's process of recording measurements, during the course of the experiment.

For example, from morning till evening, a researcher has to record observations. During the morning hours, the researcher will record observations enthusiastically and the measurements will be accurate. In the evening, due to fatigue, he or she may not show the same interest in recording the observations. Such an instrumentation effect will impact experimental results.

11.4.5 Selection Bias

The improper assignment of respondents to treatment conditions is referred to as Selection Bias. Selection bias occurs due to the wrong selection of test units in an experimental group. As a result, it does not represent the population from which the test units are drawn. Also, when the test units assigned to experimental groups differ from test units assigned for the control group, the difference will result in selection bias. For example, a company may have included more heavy users of the product in the experimental group and moderate and light users in the control group. As a result, the outcome of the experiment may be favorable to the company.

11.4.6 Statistical Regression

Statistical regression is the phenomenon where extreme values of the sample tend to converge near the mean value of the sample during the course of the experiment. This can be either positive extreme values or negatively extreme values.

For example, consider an experiment aimed at ascertaining consumer perception on the customer service levels of a financial institution. In a pretest measurement, some consumers may rate the customer service as highly exceptional, and some may rate it as very poor. However, in a post-treatment, i.e., a pilot launch of a new customer service imitative, measurements, these extreme scores tend to get closer to the mean of the sample. This is known as the statistical regression effect. This can be attributed to a continuous change in consumer attitudes. Thus, subjects who display extreme attitudes may change their perception during the course of the experiment. This will affect experimental results as the change in scores is due to statistical regression and not due to the treatment, i.e., a pilot launch of new customer service imitative.

11.4.7 Mortality

Mortality refers to the loss of subjects or test units in experiments, thus affecting experimental results. For example, suppose an educational researcher is conducting an experiment on the impact of Mobile game viewing on IQ scores on 50 students. In the course of the experiment, five students have dropped out from the experiment. Such a reduction in subjects or test units may impact experimental results.

11.5 External Validity

External Validity refers to the approximate validity, with which, we can infer that, the presumed causal relationship can be generalized to and across alternate measures of the cause and effect and across different types of persons, settings, and times. It examines to what extent the experimental findings can be generalized to the population, from which test units are drawn. We can infer from the definition that experiments that are conducted in natural settings offer a greater external validity compared to those conducted in controlled environment. So, a field experiment provides greater external validity compared to a laboratory experiment.

11.6 Experimental Environment

Experiments are conducted either in a laboratory environment or a field environment. In a laboratory environment, the experiment is conducted under artificial conditions. Field environment refers to conducting an experiment in real conditions. The researcher has to analyze, which environment will suit his or her requirements.

11.6.1 Laboratory Environment

It refers to experiments conducted in controlled conditions. For example, showing advertisements or products to select consumers in controlled conditions, and blind taste tests.

Advantages of conducting experiments in Laboratory Environment over Field Experiments

1. The conditions can be controlled. Thereby, the effect of extraneous variables on dependent variables can be minimized.
2. A controlled environment is also effective in eliminating the history effect.
3. The isolation achieved in laboratory settings will also help researchers achieve similar results.
4. If the experiments are repeated number of times with the same test units in similar conditions. Laboratory experiments, therefore, provide more internal validity.
5. As the test units and resources required for laboratory experiments are less, it also helps researchers conduct the experiment in shorter time and cost effectively. This is why companies conduct laboratory experiments during the initial stages of product development, as costs and risks associated with experiments can be minimized.
6. It also help a company to lessen the risk of information about products or ideas being passed on to competitors. The risk is more in field experiments.

Disadvantages of conducting experiments in Laboratory Environment over Field Experiments

1. Laboratory experiments are conducted in artificial conditions, and the results may not hold up well in actual conditions, i.e., in the market. So, these experiments provide less external validity.
2. The results of laboratory experiments are influenced by the testing effect, where the test units are aware that they are being tested, and so may not respond naturally.

11.6.2 Field Environment

It refers to experiments conducted in natural settings. For example, launching products in selected regions, observing consumer behavior with regard to a Point of Purchase (POP) displayed in supermarkets, and analyzing customer response to trial offers.

As field experiments are conducted in natural settings, they have a high degree of external validity. The disadvantages in field experiments are that as the researcher has no control over external variables, these experiments will have a low degree of internal validity. Field experiments also require greater time and effort and are expensive.

11.7 Types of Experimental Designs

Pure experimental research is not always possible in behavioral and social sciences as controlling all the variables is difficult. Only in a laboratory situation influences from outside of and inside the individuals is possible. In experimental situations, the control groups may not be accurate and the control of the extraneous variable is also not possible. In experimental situations, experimenter can manipulate the independent variables and has liberty to assign subjects randomly to the treatment groups. Hence, a researcher has to take such designs in which to the extent possible randomization and control of variances are possible. With complete control over all variables and all subjects, experimental design is generally conducted in the laboratory. In this research design, a researcher can assign subjects randomly to the treatment groups, manipulate the independent variable, study the pure effects of the manipulation on the dependent variable, complete control over the scheduling of independent variables, use high-level advanced statistical methods for the analysis and interpretation of the data, etc. Examples of statistical methods are *Z*-test, *t*-test, *F*-test, chi-square, correlation and regression analysis, multiple regression analysis, partial correlation, factor analysis, discriminant analysis, and analysis of variance (ANOVA).

Experimental designs are classified into pre-experimental designs, true experimental designs, quasi-experimental designs, and statistical experimental designs.

There is no proper control mechanism to deal with threats to internal and external validity in pre-experimental designs.

The true experimental designs allow researchers to randomly select test units for experimental groups and also assign treatments randomly to the experimental groups.

Quasi-experimental designs do not allow researchers to fully manipulate the independent variable, but provide a limited flexibility in assigning the treatments randomly to experimental groups.

Statistical experimental designs have statistical control mechanisms to control extraneous variables.

The following are the notations and common symbols used in explaining experimental designs:

$X \rightarrow$ Exposure of a group to an experimental treatment or independent variable.

$O \rightarrow$ Observation or measurement of the dependent variable on the test units.

$O_1, O_2, \ldots \rightarrow$ Various observations or measurements of the dependent variable taken during the course of the experiment.

$R \rightarrow$ Random assignment of test units to experimental groups.

$EG \rightarrow$ Experimental group, which is exposed to the experimental treatment.

$CG \rightarrow$ Control group of test units involved in the experiment. However, this group is not exposed to experimental treatment.

11.7.1 Pre-Experimental Designs

Pre-experimental designs lack proper control mechanisms to deal with the influence of extraneous variables on an experimental results. The prominent pre-experimental designs used by researchers are One-Shot Design or After Only Design, One-Group Pretest–Posttest Design, and Static Group Design.

11.7.1.1 One-Short Design or After Only Design

It involves exposing the experimental group to treatment X after which the measurement (O_1) of the dependent variable is taken. This can be shown symbolically $EG: XO_1$. For example, a company may launch a sales promotion initiative in selected supermarkets in a city for a month, to ascertain the impact of sales promotion on sales. Then it might measure the sales registered in that particular month. The higher sales may prompt the company to extend the sale promotion offers to other cities, where it has a presence.

Disadvantages of One-Short Design

1. The test units are not selected randomly. Instead, their selection is based on the researcher's judgment.
2. The results might not reflect the experimental treatment's impact completely, as various extraneous variables influence the dependent variable including history, maturation, and mortality.
3. As this study lacks proper control mechanisms, to deal with extraneous variables, the internal validity of the experiment is affected.
4. Moreover, we cannot infer results based on the measurement (O_1), as there is no other measurement against which (O_1) can be compared with.

Due to these limitations, one-shot design is not used for conclusive research. It is used more for exploratory research.

11.7.1.2 One-Group Pretest–Posttest Design

It involves exposing an experimental group of test units to experimental treatment (X). Measurements are taken before and after experimental treatment. This can be symbolically expressed as $EG : O_1XO_2$.

$O_1 \rightarrow$ Represents the measurement of the dependent variable before the experimental group is exposed to the treatment.

$O_1 \rightarrow$ Represents the measurement of the dependent variable after the experimental group is exposed to the treatment. So, the difference between O_1 and O_2 will be the impact of treatment on the dependent variable.

For example, an HR manager may plan a training program for employees and measure the productivity change. First, he or she may measure the productivity of employees. Then the training program will be conducted. After the training, employee productivity is again measured. However, just like the one-shot design, this experimental design too lacks proper control mechanisms to limit the influence of extraneous variables. These include history, maturation, testing effect, statistical regression effect, selection bias, and mortality effect.

11.7.1.3 Static Group Design

Here, two groups of test units, the experimental group and the control group, are involved in the experiment. The experimental group is exposed to the experimental treatment. The control group is not exposed to the experimental treatment.

The measurements are taken for both groups after the experiment. This can be symbolically expressed as:

$$EG : XO_1$$

$$CG : O_2$$

$O_1 \rightarrow$ The measurement of the dependent variable of the experimental group after exposing it to the treatment and

$O_2 \rightarrow$ The measurement of the dependent variable of the control group, which is not exposed to the treatment.

The difference between these two measurements. i.e., $O_1 - O_2 \rightarrow$ will be the effect of treatment. Various extraneous variables do influence experimental results being primary selection bias. The non-random selection of test units may result in differences between the units assigned to the experimental group and the control group. Another extraneous variable that will influence the results is the mortality effect. Some test units may drop out from the experiment. So, this is more for the experimental group, if the treatment is strenuous.

11.7.2 True Experimental Designs

To control the influence of extraneous variables, true experimental designs use randomization. Randomization refers to the assignment of test units to either experimental groups or control groups at random. Such selection of test units will reduce, the differences between the groups, on whom, the experiment is being conducted. True experimental designs also use one or more than one control groups to reduce the effect of extraneous variables.

The prominent true experimental designs widely used in research are pretest–posttest control group design, posttest-only control group design, and Solomon four-group design.

11.7.2.1 Pretest–Posttest Control Group Design

Here, two groups of test units, i.e., experimental group and control group, are considered for the experiment. The test units are assigned to these two groups randomly. Pretest measurements of dependent variable are taken for the two groups. Then, the experimental group is exposed to the treatment. The posttest measurements of the dependent variable are taken for the two groups. It can be shown symbolically as follows:

$$EG : RO_1XO_2$$

$$CG : RO_3O_4$$

O_1 and O_2 → Are the pretest and posttest measurements of dependent variable of the experimental group.

R → Represents that the assignment of testing units to each group is done on a random basis.

O_3 and O_4 → The pretest and posttest measurements of dependent variable of the control group.

We know that the control group is not exposed to experimental treatment. The treatment effect (TE) can be calculated as follows:

$$TE = (O_2 - O_1) - (O_4 - O_3)$$

For example, a fertilizer company is launching a new pesticide. To test its efficacy, the company has decided to conduct an experiment. For this, it has divided pesticide into a few parts. These parts are randomly assigned to the experimental group and the control group.

Then the pretest measurements, i.e., productivity of the fields, are taken. The parts in the experimental group are treated with pesticide and the parts in the control group are not exposed to the pesticide treatment. The posttest measurements are taken. The differences between the pretest and posttest measurements are analyzed.

This design addresses most of the extraneous variables. Hence, it provides accurate results. However, this design may not control the testing effect. This is because pretest measurements are taken, and such measurements will sensitize test units. This may have an impact on posttest measurements.

11.7.2.2 Posttest-Only Control Group Design

Here, both the experimental and control groups participate in the experiment. The first is exposed to the experimental treatment, and the second is kept unexposed. The posttest measurement of the dependent variables is taken for both groups. This can be shown symbolically:

$$EG : RXO_1$$

$$CG : RO_2$$

The *TE* can be obtained as $TE = O_2 - O_1$.

To illustrate a personal product company has claimed that the use of its new hair-oil formulation will reduce hair-fall by 70% compared to other hair-oils. To support this claim, the company has conducted an experiment by randomly assigning consumers, who use a competing Coconut Oil brand to both the experimental group and the control group.

The experimental group consumers are provided with the company's hair-oil formulation for 7 months, while the control group continues to use the competing hair-oil brand. Measurements are taken after 7 months. This type of design will address most of the extraneous variables.

11.7.2.3 Solomon Four-Group Design

Here, the sample is divided randomly into following four groups. They are experimental samples, Experience no experimental manipulation of variables, receive a pretest and a posttest and receive only a posttest.

Solomon four-group design controls for the effect of the pretest, hence it is improvement over the classical design. This type of design involves conducting an experiment with four groups, two experimental groups and two control groups. Six measurements are taken, two pretest and four posttest. This study is also known as the four-group, six-study design. The design can be symbolically represented as follows:

$$EG : RO_1XO_2$$

$$CG : RO_3O_4$$

$$EG : RXO_5$$

$$CG : RXO_6$$

The design addresses all extraneous variables; it is expensive and consumes more time and effort. The design provides various measures, which can be analyzed. They are

$$O_2 - O_1$$

$$O_4 - O_3$$

$$O_2 - O_4$$

$$O_6 - O_5$$

11.7.3 Quasi-Experimental Design

Here, all experimental situations in which the researcher or experimenter does not have full control over the assignment of experimental units randomly to the treatment conditions or the treatment cannot be manipulated. The plan of experimental situations, in which the experimenter does not have full control over the situations constitutes the quasi-experimental design.

Using this design, researcher can conduct the experiment in natural and real life setting. And hence, it has certain amount of realism and the information so gathered, can also be to quite an extent generalized. This design also provides answers to several kinds of problems, about past situations and also to those situations, which cannot be handled by employing Pure Experimental Research Design.

When it is not possible to assign test units randomly to experimental treatments or assign experimental treatment randomly to test units, quasi-experimental designs are used.

In such cases, quasi-experimental designs help control extraneous variables, though not as effectively as true experimental designs. It is better than pre-experimental designs. Prominent quasi-experimental design used by researchers is Time-Series Design.

11.7.3.1 Time-Series Designs

A series of measurements are taken before and after the test unit is exposed to the experimental treatment, in time-series designs. This can be symbolically represented as follows:

$$EG : O_1O_2O_3XO_4\ O_5O_6$$

Time-series designs are used for experiments performed over a longer period; for example, if a company wants to determine the impact of price changes on the sales of a product. The company takes a series of observations, before the price is changed and trends are identified.

Another series of observations are taken after changing the price. The trends after the treatment, i.e., post-price change are compared with trends before the treatment, i.e., pre-price change, to determine whether they are similar or not.

If there is an increase in sales levels after the price change, the researcher can conclude that the treatment had a positive effect on the dependent variable.

However, because of threats to internal validity, these experiments may not give absolutely accurate results. Key threats to internal validity are the history and instrumentation effects.

The simultaneous occurrence of events such as boom or bust in global economy or any calamity might affect experimental results. Another threat is the instrumentation effect, where there can be change in measurement units or the process followed by researcher to make measurements.

Advantages of Time-Series Design

1. Aids in identifying permanent trends and temporary trends.
2. This helps to design long-term and short-term business strategies.

11.7.4 Statistical Designs

It aids in measuring the effect of more than one independent variable. It helps researchers to conduct a single experiment to analyze the effect of more than one independent variable, instead of conducting a series of experiments for each independent variables. Also, it is helpful in isolating the effects of most extraneous variables, thereby providing better experimental results.

The prominent experimental designs in this category are Completely Randomized Design (CRD), Randomized Block Design (RBD), Latin Square Design (LSD), Factorial Design (FD), etc.

11.7.4.1 Completely Randomized Design

When the researcher has to evaluate the effect of a single variable CRD is used. The effects of extraneous variables are controlled using the randomization technique. The key difference between CRD and other statistical experimental designs is that the other statistical experimental designs use the blocking principle, which CRD does not. This design involves randomly assigning test units to treatments. For example, there are "n" test units and "k" experimental treatments. Then the "n" test units are assigned to "k" treatments randomly. Later, the posttest measurements are evaluated. This can be shown symbolically,

$$EG_1 : RX_1O_1$$

$$EG_2 : RX_2O_2$$

$$EG_3 : RX_3O_3$$

EG_1, EG_2, and, EG_3 → Experimental groups, which are exposed to various experimental treatments.

X_1, X_2, and X_3 → Experimental treatments assigned to experimental groups.

For example, a researcher at a pharmaceutical company plans to evaluate the efficacy of a Blood Pressure Control Drug made by a particular company. For this, the researcher has selected a sample of 30 consumers. These consumers are assigned randomly to two treatment levels:

20 consumers to treatment 1, and 10 to treatment 2.

Consumers under treatment 1 are asked to take the drug for 2 months. Consumers with treatment 2 are not given any drug. After the experiment, measurements are taken for both groups. Differences, if any, are analyzed to see whether the drug is effective in Blood Pressure Control.

Disadvantages of CRD

1. The design is applicable only when test units are homogeneous.
2. It can be used only in situations when a single variable is being evaluated.
3. This design can be used only when extraneous variables can be controlled.

11.7.4.2 Randomized Block Design or Matched Group Design

Here, all subjects are first tested on a common task and then they are formed into groups. The groups thus formed are known as equivalent groups. Subsequently, the different values of the independent variable are introduced to each group, and the mean scores of the dependent variable are taken for both the groups. The matching variable is usually different from the variable under study but is, in general, related to it. The two groups are not necessarily of the same size, although there should not be large differences in the number of subjects of two groups. When we use the randomized group design, the most important factor is the identification of the variables on which matching has to be done.

The matching variable should have the high correlation with dependent variable. Sometimes the dependent variable itself is used as matching variable. Sometimes an independent measure may be used as matching variable. But the variable selected should be somewhat related to the dependent variable.

For example, in a study the researcher wants to see, the effect of praise on subject's performance on Quantitative Aptitude Test. We have two groups: one group is praised for its performance on the test and urged to try to better its scores, and the second group does not receive any comment.

For assigning the subjects into two groups, the researcher may find the scores on form A of the Quantitative Aptitude Test and obtained the set of scores. On the basis of the obtained scores on form A, subjects can be paired off. Those subjects who scored 100 were selected for the study. They were divided into two groups randomly and form B of the same test was administered to see the effect of incentive on subject's score. Suitable statistical test is used to find out the significant difference in the mean scores of two groups. In RBD, we may use two methods of matching.

11.7.4.2.1 Matching by Pairs

In this type of research, matching is done initially by pairs so that each person in the first group has a match in the second group.

For example, a researcher wants to study the effect of two teaching methods on statistical achievement of the postgraduate students. Subject's Knowledge and Academic Performance were taken as matching variable.

All subjects were administered statistical academic performance test and scores were obtained. If for instance two subjects scored 80, then one subject is placed in one group and another is placed in another group. In this way, two groups are formed.

One group is taught by one method, and the other group is taught by another method and academic performance scores of both the groups are compared.

11.7.4.2.2 Matching in Terms of Mean and SD

When it is difficult to set up groups, in which subjects have been matched individual to individual, researchers often resort to matching of groups in terms of Mean and Standard Deviation. The matching variable is somewhat related to the dependent variable. For example, quantitative aptitude is a matching variable and the researcher obtained the mean and SD of academic performance scores of two groups. In the RBD, the subject may be matched on Gender, Educational Qualification, Age, and so on. However, one should be very careful in choosing the matching variables.

11.7.4.2.2.1 Within Subject Design Here, the same individual is treated differently at different times. And after they have been subjected to different treatment conditions, we compare their scores. Hence, it is also known as Repeated Measure Design.

For example, let us say a researcher wants to study the effect of colors on reaction time. The investigator selects three colors say Red, Green, and Yellow. And, let us say that 30 subjects are selected for the experiment from the population of interest. After exposing them to different colors, their reaction time is noted and compared. Within Subject Design is further divided into two categories: conditions within subject design, and multiple condition within subject design.

11.7.4.2.2.2 Two Conditions within Subject Design It is the simplest design. The two conditions are labeled as "Condition 1" and "Condition 2." All subjects experience both the conditions. Let us say that the researcher wants to compare the reaction time of Red and Green colors. We select 30 subjects from the population of interest, and the reaction time of all the subjects is noted down for the two colors. Since many experimenters involved more than two conditions and there is possibility of carryover effect from one condition to the other, despite its simplicity, this design is not used as often as one might expect.

11.7.4.2.2.3 Multiple Conditions within Subject Design The reasons for conducting multiple conditions experiment is that the researcher wants to compare several variables and ascertain their effectiveness, and to determine the shape of the function that relates the Independent and Dependent variables.

For example, the objective of researcher is to know how the sensation of brightness increases with the intensity of a light. For this, the researcher may present each of several intensities of the light to a group of subjects. From the responses to the various intensities, the researcher can plot the relation between intensity and brightness.

11.7.4.2.2.4 Controlling for Order and Sequence Effects in Within Subjects Design Within Subject Experiment, the possibility exists that order effect and sequence effect may influence the result of the repeated testing because a subject experiences more than one experimental conditions. Order effects are those that result from the ordinal position in which the condition appears in an experiment, regardless of the specific condition that is experienced. On the other hand, according to them, the sequence effect "depends on an interaction between the specific conditions of the experiment."

For example, in an experiment when judging the heaviness of lifted water cane of 10 L weight, there is possibility that a light weight of an water cane of 1 L will feel even lighter if it is followed by a heavy one, and vice versa.

By randomization, controlling for order and sequence effect is possible, which can be used when each condition is given several times to each subject.

11.7.4.2.2.5 Comparison of between Group Design and within Subject Design RBD is used when the researcher feels that there is one major extraneous variable that will influence experimental results. In this design, the test units are blocked or grouped based on the extraneous variable, which is also called the Blocking Variable (Table 11.1).

For example, drug stores may register higher sales compared to supermarkets and hypermarkets. Therefore, we can apply RBD in such situations. The retail outlets are segregated according to store type.

TABLE 11.1

Comparison of Between Group Design and Within Subject Design

Points	Within Subject Design	Between Group Design
Definition	In within subject design different subjects are treated differently at different times and we compare their scores after subjecting them to different treatment conditions. In within subject design, control of order and sequence effect is achieved through randomization or counterbalancing	In between subject design we have two groups, one is experimental group and other is control group. The allocation of subject in experimental and control groups is made randomly
Number of treatment received	Each subject in the experiment receives a number of treatments or conditions	A subject receives only one treatment
Repetition of measurement	The experimenter repeats the measures on the same group of subject and this increases the precision of the experiment by eliminating intersubject differences as a source of error	The experimenter does not repeat the measures on the same group
Preference of design	When we have small number of subjects, and they are available for extended period of experimentation and number of treatment is small then we should prefer within subject design	When there are chances of practice or carry over effect of one treatment to the subsequent task, then between group designs should be preferred

TABLE 11.2

Examples for Randomized Block Design

	Store Type		
Price Change	**Drug Stores**	**Supermarkets**	**Hypermarkets**
Rs. 8	Store 1	Store 6	Store 7
Rs. 12	Store 3	Store 2	Store 8
Rs. 16	Store 4	Store 5	Store 9

Store 1, Store 3, and Store 4 are drug stores. Store 6, Store 2, and Store 5 are supermarkets, and Store 7, Store 8, and Store 9 are hypermarkets. Then treatment levels, i.e., price are assigned randomly to each test unit, i.e., retail outlets. This is summarized in Table 11.2:

Using the design, two effects can be determined: the main effect and the interaction. The average effect of a particular treatment on the dependent variable, regardless of extraneous variables is referred to as the main effect. The influence of the extraneous variable on the effect of treatment is refers to as the interaction effect. In this example, the main effect is the direct effect of price change on the product sales. This can be achieved by determining the average impact of each treatment on each block. The interaction effect is the influence of the store type on the effect of price change. This can be obtained by determining the customer response to each price change for each store type.

11.7.4.3 Latin Square Design

Where the researcher has to control the effect of two noninteractive external variables; other than the independent variable, in situations LSD is used. It is done through the blocking technique as used in random block design.

For example, a researcher wants to examine the impact of three different ads on sales. However, the researcher feels that pricing and income levels of consumers will also impact sales. So, a researcher wants to isolate the correct of the two extraneous variables-pricing and consumer income levels. In this design, the blocking or extraneous variables, i.e., price and income levels are divided into an equal number of levels and so is the independent variable, i.e., advertising program.

Table 11.3 is then developed with levels of one extraneous variable representing the rows and levels of the other variable representing the columns. The levels of the independent variable or treatments are exposed to each cell on a random basis, so that there should be only one treatment in each row and column. Then the TE is determined. Based on the results, it can be analyzed which treatment level influences the dependent variable more.

In the advertising program example, we have created a 3×3 table, where each extraneous variable has three blocks and so does the independent variable. The advertisements programs that are to be tested are Ad-A, Ad-B, and Ad-C. The pricing levels are Rs. 12,000, Rs. 14,000, and Rs. 16,000. The income levels are low-income, middle-income, and high-income groups. In the table, income levels are represented in columns; the pricing levels are represented in rows. The advertising programs Ad-A, Ad-B, and Ad-C are assigned to each cell. Table 11.3 depicts the example for LSD.

TABLE 11.3

Examples for Latin Square Design

	Income Levels		
Pricing Levels	**Low Income**	**Middle Income**	**High Income**
Rs. 12,000	Ad-B	Ad-A	Ad-C
Rs. 14,000	Ad-C	Ad-B	Ad-A
Rs. 16,000	Ad-A	Ad-C	Ad-B

However, there are some assumptions in this design. They are:

1. There is negligible or no interaction effect between the two extraneous variables. As a result, we cannot examine the interrelationships between pricing, income levels and the advertising programs.
2. The number of levels of all three variables are equal.

Disadvantages of LSD

1. The assumption that all variables should have the same number of levels, i.e., the two extraneous variables and the independent variables, is not possible in all cases. So, in situations where any of the variables does not have the same number of levels as that of the other two variables, this design is not valid.
2. This design also assumes that there is no interaction effect between the extraneous variables. Interaction effect refers to measurement of the amount of influence the level of one variable has on another variable. The interaction effect exists between two variables when the simultaneous effect of two variables is different from the sum of the individual effects of both the variables. In situations where there are interrelationships between the variables, this design cannot be applied.

11.7.4.4 Factorial Design

In FD, an additional samples are used, as compare to classical design. Each group is exposed to a different experimental manipulation. All the above designs of research can be used in experimental research for analyzing the data. These designs are not suitable for conducting field experiments though one could use them with certain modifications.

11.7.4.4.1 Single Factor Design

When we have only one independent variable, Single Factor Design is used.

Single factor design can be classified into categories between group design and within subject design.

11.7.4.4.1.1 Between Group Design Here, subjects are assigned at random to different treatment conditions and the effect of different conditions on the subjects are computed. Here, we have two randomized group design and multigroup design.

11.7.4.4.2 Two Randomized Group Design

We assign the subjects randomly into two groups. Here, the researcher first defines the independent variable, the dependent variable, and the research population.

For example, a research investigator wants to observe the effect of knowledge of result on the rate of learning of School of Mathematics & Statistics Students of MIT WPU, Pune, Maharashtra, India. Then, the researcher randomly selects a sample of 200 students. Then the researcher will divide these 200 students randomly into two groups with 100 students in experimental group and 100 students in control group.

The assignment of the subjects into two groups can be done by various methods randomly. The most common method of assigning the subjects randomly into two groups is to use the random number table. The researcher may write down the name of all the students in alphabetical order on a paper for dividing the subjects into experimental and control groups.

After that, the researcher assigns the first student in experimental group, the second in control group, and the third in experimental group, and so on.

The researcher also may write on separate slips the name of the subjects and place them in a box; after folding them and pick the slip one by one. The experimenter may place first slip in one group and second in the next group. It is expected that at the start of an experiment these two groups will not differ significantly.

Now, the knowledge of result of their performance will be received by the students of experimental group and the feedback of their performance will not be received by the students of control group.

Then, the scores of all subjects of experimental and control groups will be recorded and subjected to statistical analysis.

If the statistical test reveals that two groups differ significantly on dependent variable, then it can be concluded that the difference in rate of learning is due to the manipulation of independent variable.

If the rate of learning of experimental group is more than that of the control group, then we may conclude that knowledge of result facilitated the learning.

In two randomized group design, "t" test or Mann–Whitney U-test is most commonly applied statistical techniques.

11.7.4.4.2.1 Multigroup Design Here, there are more than two or three experimental groups and one control group. In more than two randomized group designs, sometimes we have three or four experimental groups only.

For example, an experimenter wants to study the effect of Four Teaching Methodologies on learning of a particular method of statistical analysis. Suppose for this, the researcher randomly selects 100 students and assign 25 subject randomly in each group. These groups are supposed to be equivalent groups after random assignment. In the experiment, one group will be taught by method A, the second by method B, the third by method C, and the fourth by method D.

All subjects were administered a particular task and the scores are obtained on dependent variable. Through appropriate statistical technique, we can find out which method of statistical analysis is most effective.

In multigroup design, the most commonly applied statistic used are one-way ANOVA and Z-test.

Advantages of FD

1. It overcomes the drawbacks of the LSD regarding the interaction effect.
2. It can be used in cases where there is interrelationship between the variables.
3. They are used to examine the effect of two or more independent variables at various levels.

FD can be depicted in tabular from. In a two-factor design, the level of one variable can be represented by rows and the level of another variable by columns. Each test unit is assigned to a particular cell. The cell is exposed to a particular treatment combination randomly. This design enables a researcher to determine the main effect of each independent variable as well as the interaction effect between them.

For example, a market researcher plans to study the effect of in-store promotions on the sales of a product and the impact of price change too. The researcher has decided to use two types of in-store promotions, POP-display and trial packs, and three price levels: Rs. 60, Rs. 70, and Rs. 80.

Six stores, namely, A, B, C, D, E, and F, have been selected for the experiment. By using FD, we develop Table 11.4, with row containing in-store promotions variable and columns containing pricing variable. We assign test units, i.e., supermarkets to each cell randomly. The test unit, i.e., supermarket in each cell is then exposed to POP-display and Rs. 70 price level and supermarket B is exposed to trial packs and Rs. 80 price level, posttest measurements are taken.

TABLE 11.4

Effect of In-Store Promotions on the Sales of a Product and the Impact of Price Change

	Pricing Variable		
In-Store Promotion Variable	**Rs. 60**	**Rs. 70**	**Rs. 80**
POP-display	C	A	F
Trial packs	D	E	B

The outcome of this experiment can help the researcher to understand three key aspects.

1. The impact of pricing on the sales of a product
2. The impact of in-store promotions on the sales
3. The sales–effect interrelations between in-store promotions and pricing.

Summary

An experiment refers to the process of manipulating one or more variables and measuring their effect on one or more variables, while controlling for external variables. The variable, which is manipulated, is called the independent variable and the variable whose behavior is to be measured in the experiment is called the dependent variable. The aim of experimentation is to establish and measure the causal relationship between the variables under study. A well-executed experiment can depict the causal relationship between variables by controlling extraneous variables.

To conduct experiments successfully, researchers have to consider certain aspects. They include treatment or independent variable, experimental groups and control groups, selection and measurement of the dependent variable, and control of extraneous variables.

Validity is the extent to which a research process is accurate and reflects actual market conditions. There are two types of validity considered in experimentation—internal validity and external validity.

Internal validity measures to what extent the change in the dependent variable can be explained by the independent variable. External validity measures to what extent inferences derived from experiments can be generalized to the real environment. There are several threats that affect internal and external validity. They include history, maturation, testing, instrumentation, selection bias, statistical regression, and mortality. Experiments are conducted either in laboratory environments or field environments. Experimental designs are classified into four key categories—pre-experimental designs, true experimental designs, quasi-experimental designs, and statistical experimental designs. Prominent pre-experimental designs used by researchers are—one-shot design (after only design), one-group pretest–posttest design and static group design. Widely used true experimental design are, pretest–posttest control group design, posttest-only control group design, and Solomon four-group design. Four prominent statistical experimental designs are CRD, RBD, LSD, and FD.

Review Questions

1. Explain Solomon four-group designs.
2. Define Single Factor Design.
3. What is between group designs? Describe the same with examples.
4. Describe more than two randomized group design or multigroup design.
5. What is matched group design? How do we match in matched group design?
6. Define within subject design. State the two categories of within subject design.
7. Describe what is two conditions and multiple conditions in "within subject design"? Give suitable examples.
8. How do we control for order and sequence effects in "Within Subjects Design"?
9. Differentiate between within subject and between subjects experimental design.
10. Discuss with example when to use between subject research designs?
11. When to use within subject research design? Explain with examples

12

Data Preparation and Preliminary Analysis

12.1 Introduction

Data analysis plays an important role in transforming a lot of data into variable sets of conclusions and reports. Proper analysis helps the researcher to gain insights from the data and to arrive at informed judgments and conclusions. However, if the purpose of research is not defined properly or if research questions are irrelevant, even the best analytical techniques cannot produce good results. Data analysis may give faulty results even when research is done properly. This is because of the application or inappropriate methods to analyze data. The various steps in data preparation and preliminary data analysis are shown in Figure 12.1.

12.2 Validating and Editing

Validation is the preliminary step in data preparation. It refers to the process of ascertaining whether the interviews conducted complied with specified norms. The essence of this process lies in detecting any fraud or failure by the interviewer to follow specified instructions.

In many questionnaires, we find there is a separate place to record the respondent's name, address, and telephone or mobile number and other demographic details.

Though no apparent analysis can be done on such data, it is the basis for what is called "validation." Validation helps to confirm whether the interview was really conducted.

Editing is the process of checking for mistakes by the interviewer or respondent in filling the questionnaire. Editing is usually done twice before the data are submitted for data entry.

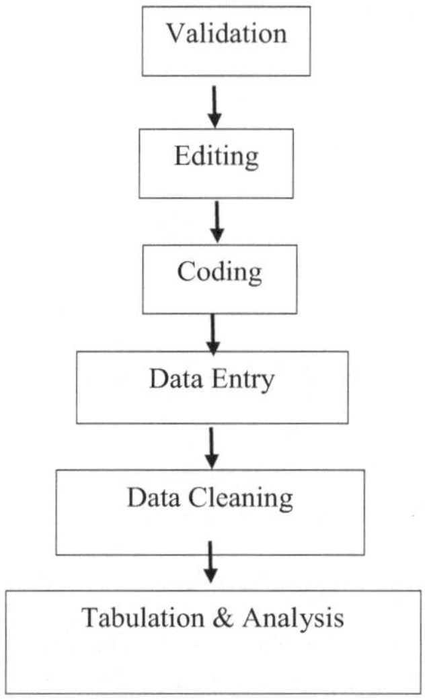

FIGURE 12.1 Data analysis stages.

DOI: 10.1201/9781315167138-12

The first editing is done by the service firm that conducted the interviews and the second editing is done by the market research firm that outsourced the interviews. Editing is a manual process that checks for problems cited below.

Finding out whether the interviewer followed the "skip pattern." The questionnaire is designed so that depending on the respondent's response, the interviewer skips to the next relevant question. This is called the "Skip Pattern."

Sometimes, it might happen that interviewers skip questions when they should not and vice versa.

In the sample questionnaire shown in Figure 12.2, the interviewer should skip to question 7 if the response to the first question is either "*a*." or "*e*." or "*f*".

Responses to open-ended questions are vital for researchers and their clients. Eliciting the correct responses to open-ended questions shows the interviewer's competence. Hence, interviewers are instructed to probe initial responses and are asked not to distort the actual wordings or interpret the response of an open-ended question.

For example, different possible responses for the second question in Figure 12.2 are shown below.

Q: Why do you eat chocolates?

Respondent 1: Because I like chocolates.

Respondent 2: I like chocolates. I like their taste and softness. (Here, interviewer probed further into the response.)

Respondent 3: I like chocolates because they give me energy.

In the first response, the interviewer failed to extract the correct response. The objective of probing is to extract the reason behind eating chocolates. In the second response, the interviewer might have asked further questions such as "Do you like ABC brand of chocolates?" On a positive reply, the interviewer might have probed more by asking. "What do you like about it?" This is the correct way to elicit responses for open-ended questions. The interviewer can even go further and probe how a specific product characteristic is attached to the individual's subconscious. Though editing is time consuming, it has to be done with care and with patience because it is important for data processing.

12.2.1 Treatment of Unsatisfactory Responses

During editing, the researcher may find some illegible, incomplete, inconsistent, or ambiguous responses. These are called unsatisfactory responses. These responses are commonly handled by assigning missing values, returning to the field, or discarding unsatisfactory respondents.

12.2.1.1 Assigning Missing Values

Though revisiting the respondent is logical, it is not always possible to revisit the field every time the researcher receives an unsatisfactory response from the participants.

In this situation, a researcher may resort to assigning missing values to unsatisfactory responses. This method can be used when the number of unsatisfactory responses is proportionately small or variables with unsatisfactory responses are not the key variables.

12.2.1.2 Returning to the Field

Sometimes the interviewer has to re-contact respondents, if the responses provided by them are unsatisfactory. This is feasible especially for industrial or business surveys, where the sample size is small and respondents are easily traceable. The responses, however, may be different from those originally given.

12.2.1.3 Discarding Unsatisfactory Responses

In this approach, unsatisfactory responses from participants are totally discarded. This method is well suited when the proportion of unsatisfactory responses is very small compared to the sample size. With respect to important, demographic, and user characteristics, respondents with unsatisfactory responses

Date…

Respondent's Mobile Number……

Respondent's Age……..

1. How many chocolates do you eat in a typical week?
    a. Less than 5
    b. Between 5 and10
    c. Between 11 and 20
    d. More than 20
    e. Don't know
    f. None

    (Interviewer-If response is "a," "e," or "f," go to Question 7.

2. Why do you eat chocolates?

    Respondent's answer…

3. Which brand of chocolates do you prefer most?
    a. Cadbury's
    b. Nutrine
    c. Nestle
    d. Amul
    e. Others (specify)…….

4. When do you like to eat chocolates?

    Response…..

5. Do you prefer chocolates to sweets? (Y/N)

6. Do you have any negative associations with chocolates?

    ……..

7. What is your age group?

a. Under 10

b. Between 10 and 20

c. Between 21 and 30

d. Above 30

e. Refused to answer, no answer, or don't know

May I know your name? My office calls about 10% of the people I visit to verify if I have conducted the interviews.

Gave name……

Refused to give name ------

Thank you for your time. Have a good day.

FIGURE 12.2 Questionnaire sample.

do not differ from other respondents. In each questionnaire, unsatisfactory responses for each respondent are proportionately more. Responses on key variables are missing.

To reiterate, editing has to be done with patience and care because it is an important step in the Questionnaire Processing.

12.3 Coding

It is the process of assigning numbers or other symbols to answers in order to group the responses into limited categories. For example, instead of using the word "landlord" or "tenant" in response to a question that asks for identification of one's residential status, one can use the codes "LLD" or "TNT". This variable can also be coded as 1 for landlord and 2 for tenant, which is then known as numeric coding.

This type of categorization and coding sacrifices some detail but is necessary for efficient data analysis. It helps researchers to pack several replies into a few categories that contain critical information required for analysis.

12.3.1 Categorization Rules

While categorizing replies obtained from a questionnaire, a researcher should follow four rules. The categories are appropriate, exhaustive, mutually exclusive, and derived from one classification principle.

12.3.1.1 Appropriate

Categorization should help validate the hypotheses of the research study. If a hypothesis aims to establish a relationship between key variables, then appropriate categories should be designed to facilitate comparison between those variables. Categorization provides for better screening of data for testing and establishing links among key variables. For example, if specific income is critical for a testing relationship, then wider income classifications may not yield the best results upon analysis.

12.3.1.2 Exhaustive

An adequate list of alternatives should be provided to tap the full range of information from respondents when multiple-choice questions are used. The absence of any response from the set of response options given will prove detrimental as specific response will be under-represented in the analysis. For example, a questionnaire designed to capture the Annual Income of respondents should list all possible alternatives that a respondent may fall into.

12.3.1.3 Mutually Exclusive

Complying with this rule requires that a specific alternative is placed in one and only one cell of a category set. For example, in a survey, the classification may be Professional, Self-employed, Government Service, Agriculture, Unemployed, etc.

Some self-employed respondents may consider themselves professionals and these respondents will fit into more than one category. A researcher should avoid having categories that are not mutually exclusive.

12.3.1.4 Single Dimension

This means every class in the category set is defined in terms of one concept. If more than one dimension is used, it may not be mutually exclusive unless there is a combination of dimensions like Engineering Student, Medical Student, Management Student, Law Student, Science and Commerce Students, and the like in the response options.

12.3.2 Code Book

To make data entry less erroneous and more efficient, a rule book called a "code book" or "coding scheme" is used, which guides research staff. A code book gives coding rules for each variable that appears in a survey. It is also the basic source for locating the positions of variables in the data file during the analysis process. Most code books generally contain the question number, variable number, location of the variables code on the input medium, descriptors for the response options, and variable name. A code book for the questionnaire in Figure 12.2 is shown in Figure 12.3.

Question number	Variable Number	Code Description	Variable Name
1	1	Number of chocolates 1=Less than 5 2=Between 5 and 10 3=Between 10 and 20 4=Above 20 5=Don't know 9=Missing	No-of-Chocolates
2	2	Reason(s) 0=Not mentioned 1=Mentioned	Reason
2	3	Taste Soft Size Low price Sweet smell Others	Taste Soft Size Cost Smell Others
3	4	Brand 1=Cadbury's 2=Nutrine 3=Nestle 4=Amul 5=Others 6=Missing	Brand
4	5	Occasion 0=Not Mentioned 1= Mentioned	Occasion
4	6	Festival	Festival

FIGURE 12.3 A Code Book for the Questionnaire.

(Continued)

Question number	Variable Number	Code Description	Variable Name
		Marriage Friends Time Pass Happy Occasion Bought new goods Gift Others	Marriage Friends Time pass Happy-Occ Bought-New Gift Others
5	7	Preference 1=Yes 0=No	Pref
6	8	Negative Association 0=Mentioned 1=Not Mentioned	Association
6	9	Tooth decay Worms Expiry date Heart disease Others	Tooth Decay Worms Expiry-Date Heart Disease Others
7	10	Age group 1=below 10 2=between 10 and 20 3=between 21 and30 4=above 30 9=missing	Age
		Name	Name

FIGURE 12.3 (CONTINUED) A Code Book for the Questionnaire.

12.3.3 Coding Close-Ended Questions

It is easy to assign codes for responses that would be generated by close-ended questions. This is because the number of answers is fixed. Assigning appropriate codes in the initial stages of research makes it possible to precode a questionnaire. This avoids the tiresome, intermediate step of framing the coding sheet prior to data entry. Coding makes it easier for data to be accessed directly from the questionnaire. The interviewer assigns appropriate numerical responses to each item or question in the questionnaire. This code is later transferred to an input medium for analysis.

12.3.4 Coding Open-Ended Questions

Questionnaire data obtained from close-ended questions are relatively easy to code as there are a definite number of predetermined responses. But, a researcher cannot always use close-ended questions. As it is impossible, to prepare an exhaustive list of responses for a question aimed at probing a person's perception or attitude to a particular product or issue. Thus, use of open-ended questions becomes inevitable in research.

However coding the data collected from open-ended questions is much more difficult as the responses are unlimited and varied.

In the questionnaire shown in Figure 12.2, Questions 2, 4, and 6 are open-ended questions. After preliminary evaluation and coding response, categories for the second question are shown in Figure 12.3. The response categories also include the "other" category to satisfy the coding rule of exhaustiveness.

12.3.5 Content Analysis for Open-Ended Questions

A qualitative method known as content analysis can be used to analyze the text provided in the response category of open-ended questions. The purpose is two fold:

1. Content analysis systematically and objectively derives categories of responses that represent homogeneous thoughts or opinions. This facilitates interpretation of large volumes and lengthy and detailed responses.
2. Content analysis identifies responses particularly relevant to the survey. This form of content analysis is known as open coding or context-sensitive scheme coding. It requires the researcher to name categories through a detailed examination of data.

Thus, rather than a predetermined framework of possible responses, the researcher works using actual responses provided by respondents to generate the categories used to summarize data.

This involves an iterative interpretation process of first reading the responses and then re-reading them to establish meaningful categories. And finally, re-reading select responses to refine the number and meaning of categories in manner, which is most representative of the respondents' text.

Each response is then coded into as many categories as necessary, to capture the "full picture" of the respondent's thoughts or opinions. To reduce potential coding errors, out-of-context responses are not coded.

Let's look at the example questionnaire and to content analysis for the second open question in Figure 12.2 Q. "Why do you eat chocolates?" (Sample responses are as follows) (a) I can afford to buy them, (b) Its shape and size are nice, (c) No other confectionery can match the test of chocolates, (d) It's very sweet, (e) I enjoy the taste, and (f) I love the taste and softness.

The first step in analysis requires that the categories selected should reflect the objectives for which the data have been collected. The research question is concerned with the reason behind the respondent's interest in eating chocolates. The categories selected are keywords.

The first pass through the data produced a few general categories as shown in Figure 12.4. These categories should contain one dimension of reason and be mutually exclusive. The use of "other" makes

Categories

- Taste -------
- Low price -------
- Soft -------
- Size -------
- Sweet smell -------
- Others -------

FIGURE 12.4 Example of coding for an open-ended question.

the category set exhaustive so that any dimension that cannot be captured in the listed categories can be assigned to the "other" category.

In general, a second evaluation of responses and categories is made so that some subcategorize can be found, which remain undiscovered in the first evaluation.

Q: "Why do you like chocolates?" (Tick as many of the following as applicable) (Presume answers will be given.) Figure 12.4

12.3.6 Coding "Don't Knows"

Although researchers include the option of "Don't know" (DK) in the possible answers to a question to ensure exhaustiveness, at times it poses problems while analyzing. This is particularly so if a considerable number of respondents choose the DK option. Respondents may choose this response either because they really don't have an answer or because they don't want to answer the question because of personal reasons.

Though the DK option is inserted in the questionnaire to assess the actual number of respondents, who don't know the answer, the number of evasive respondents choosing that option often negates this purpose.

There are two kinds of DK responses: the "legitimate DK" and the "disguised DK."

Responses of the first kind are acceptable. Respondents give such answers when they are unaware of the answer, may be due to recalling problems or memory decay.

The second type of response is mainly because of poor preparation of the questionnaire or the questioning process. At times, the respondent may be reluctant to answer the question or may feel that the question is inconsequential.

Researchers and the interviewers in the field play a major role in decreasing the proportion of "Disguised DK."

A carefully designed questionnaire can decrease the number of "Disguised DK" responses. The rest can be handled by interviewers in the field.

An interviewer must identify in advance possible questions that entail key variables for which DK responses would make things difficult.

The researcher can use various probing techniques to get definite answers or find out why the respondent has selected a DK response.

There is always a possibility that a considerable number of DK responses might be generated for some questions, despite efforts, to check the occurrence of such responses.

In such cases, the researcher can either ignore that response or allocate the frequency to all other responses in the ratio, which they occur. For example, in Table 12.1, 21% of the respondents below 10 years select the DK response. Here, the researcher can either ignore the last column or allocate the DK responses to other two response (<5 and >20) proportionally.

12.3.6.1 Handling DK Responses

Q: How many chocolates do you eat in a typical week?
Ans: Table 12.1

TABLE 12.1

Handling DK Responses

Age	Less than 5	Above 20	"Don't know" Responses
Below 10	29%	56%	21%
10–20	44%	24%	55%
21–30	21%	15%	17%
Above 30	6%	5%	7%
Total *n*	312 (100%)	142 (100%)	46 (100%)

12.4 Data Entry

Data entry helps in converting information gathered by secondary or primary methods to a medium that facilitates viewing and manipulation. The different means available for data entry are given below.

12.4.1 Optical Scanning

Optical scanners are data-processing devices that can "Read" responses on questionnaires. These instruments examine darkened small circles and ellipses. Optical scanners help process marked answers in a questionnaire and store answers in a data file. This technology is generally used for routine data collection. It reduces the number of times the data are handled, thereby reducing the number of errors possible. A common application of optical scanning is scanning of answer sheets to evaluate examination papers of competitive exams, which have a huge number of participants.

12.4.2 Barcode Reader

It can be used to simplify the interviewer's role as a data recorder. Instead of writing respondent's answers by hand or typing them, the interviewer can pass a barcode wand over the appropriate codes. This technique, however, requires codes for all possible answers.

12.4.3 Voice Recognition

It provides interesting alternatives for the telephone interviewer. This preprogrammed system, upon getting a voice response, automatically branches to the logically next question in the questionnaire. Currently, systems are just recording voice responses. These are rapidly developing to translate voice data into data files.

12.5 Data Cleaning

Data cleaning includes consistency checks and treatment of missing responses. Compared to the preliminary consistency checks during editing, checking at this stage is more thorough and extensive, as it uses computers. This is done in by Error-checking routines and marginal reports.

Error-checking routines are computer programs that check for various conditions that could lead to potential errors. For example, if a particular field on the data records should only have a code in the range of 1–4, then a logical statement can be programmed to check for the occurrence or an invalid code in that field.

Technological advancements have also made it possible to generate reports that specify the number of times a particular condition was not met and the list of data records on which the condition was not met. Another approach to error checking is the marginal report or one-way frequency table. The rows of this report are fields of the data record. The columns depict the frequency with which each possible value was encountered in each field. The report assists in determining the use of inappropriate codes and detecting whether skip patterns were properly followed. If all the numbers are consistent and comply with the coding, there is no need for further cleaning. However, in case logic errors are detected, necessary corrections can be made on the computer data file. This is the final error check in the process, after which the computer data file is deemed "clear" and ready for tabulation and statistical analysis.

12.6 Tabulation of Survey Results

Once data are cleansed-off all errors and stored in a database, they should be tabulated to facilitate further analysis.

A researcher can tabulate data by frequency tabulation and cross tabulation.

12.6.1 One-Way Frequency Tabulation

The most commonly used tabulation technique is the one-way frequency table, as summarized in Table 12.2. The one-way frequency table depicts the number of respondents who gave each possible answer to each question. Table 12.2 lists that 225 respondents (45%) like Cadbury's chocolate, 155 respondents (31%) prefer Nestle chocolates, and 83 respondents (16.6%) prefer Amul chocolates.

For every question, the responses are tabulated in the manner to get the first summary report of the survey.

Table 12.2 summarizes a computer-generated one-way frequency table for Question 3 in the questionnaire show in Figure 12.2, when a survey is administered for 500 respondents. In addition to frequencies, one-way frequency tables indicate the percentage of those responding to a question that gave each possible response.

Table 12.2 One-Way Frequency Table

Q.3: Which brand of chocolates do you prefer most? Table 12.2

12.6.2 Cross Tabulation

Frequency tables and percentage distribution averages provide a glimpse into the survey responses: response data can be further organized in a variety of ways. For example, each question can be categorized on gender basis, like how male, female respondents answered the sample question. These are known as Cross Tabulations. This simple yet powerful tool is the most often used tool in the next stage, i.e., analysis. Many researches would not need to go any further than cross tabulation in doing analysis. The idea is to look at responses to one question in relation to responses to other questions. Here, data are organized into groups, categories, or classes to facilitate comparisons. Table 12.3 summarizes a simple cross tabulation. This cross tabulation table shows frequencies and percentages of respondents according to their preferences and their consumption.

TABLE 12.2

One-Way Frequency Table

	Total
Brand	**500 (100%)**
Cadbury's	225 (45%)
Nestle	155 (31%)
Amul	83 (16.6%)
Nutrine	31 (6.2%)
Don't know/other	6 (1.2%)

TABLE 12.3

Frequencies and Percentages of respondents according to their preferences and their consumption

	Ages of Respondents				
Brand	**Less than 10**	**10–20**	**21–30**	**More than 30**	**Total**
Cadbury's	93	73	39	20	225 (45%)
Nestle	69	44	22	20	155 (31%)
Amul	39	20	12	12	83 (16.6%)
Nutrine	11	9	6	5	31 (6.2%)
Others	3	3	0	0	6 (1.2%)
Total	215 (43%)	149 (43%)	79 (43%)	57 (43%)	500 (100%)

Table 12.3 Simple Cross Tabulation

The most common way of designing cross tables is to create a table, where the columns represent various demographic factors such as age and lifestyle characteristics like working people, retired personnel, etc. These are indicators of state of mind. The behavior of these indicators is captured in the rows. This approach permits easy comparison of the relationship between the state of mind and behavior.

The question might be directed at probing how people in different age groups differ with regard to the particular factor under examination.

An example of this type of table is shown in Table 12.4. Here, we look demographic factor age as the data given in the columns. Behavior toward different brands is the factor under consideration. Behavior toward each brand is captured in the rows, i.e., the number of chocolates of a particular brand say Nestle, by respondents in the age group 10–20 is 44.

Table 12.4 Cross Tabulation

Cross tables can be produced on almost all parameters for a given survey data. A careful exercise should be undertaken before any cross table is prepared to ensure that the cross tables are true in delivering information that is synchronous with research objectives.

Apart from cross and frequency tables, there are many different ways of representing survey data. Graphical representation of data includes line charts, pie charts, bar charts, etc.

TABLE 12.4

Demographic Factor Age and Behavior toward Different Brands

		Age of Respondents					
Brand	**Number of chocolates**	**Less than 10**	**10–20**	**21–30**	**More than 30**	**Subtotal**	**Total**
Cadbury's	<5	42	34	20	11	107	225 (45%)
	5–10	31	21	12	5	69	
	10–20	15	12	5	2	34	
	More than 20	5	6	2	2	15	
Nestle	<5	36	26	13	12	87	155 (31%)
	5–10	21	12	6	5	44	
	10–20	8	4	2	2	16	
	More than 20	4	2	1	1	8	
Amul	<5	21	11	6	5	43	83 (16.6%)
	5–10	12	5	3	4	24	
	10–20	4	3	2	2	11	
	More than 20	2	1	1	1	5	
Nutrine	<5	5	4	2	3	14	31 (6.2%)
	5–10	3	3	2	1	9	
	10–20	2	1	1	1	5	
	More than 20	1	1	1	0	3	
Others	<5	2	1	0	0	3	6 (1.2%)
	5–10	1	2	0	0	3	
	10–20	0	0	0	0	0	
	More than 20	0	0	0	0	0	
Total		**215(43%)**	**149(29.8%)**	**79(15.8%)**	**57(11.4%)**		**500(100%)**

12.7 Data Mining

Data mining involves discovering knowledge by analyzing data from various perspectives and refining it into useful information.

Uses of Data Mining

1. It is a powerful technology having great potential to help companies increase revenue and cuts costs based on information derived from huge databases.
2. They are used to identify valid novel, useful, and understandable patterns in data.
3. They can counter business questions that were traditionally too time consuming to resolve.
4. They search databases for hidden patterns and predictive information that experts may miss.
5. They attempt to discover patterns and trends in data and infer rules from these patterns.

For example, consider a simple database query. "How many units of 100 g Rin Detergent was sold in the month of August in Pune District of Maharashtra India?"

On the other hand, data mining may discover that Clinic Plus + Shampoo is often purchased together with Rin Detergent, although the products appear unrelated. With the patterns discovered from data mining, a manager can support, review, and examine decisions. Table 12.5 lists the various steps in the evolution of data mining and the typical questions that can be answered.

12.7.1 Data Mining in Research

Data mining concept is being used by various companies in retail, finance, logistics, and civil aviation industries. These companies use data mining techniques to make sense of huge historic data available with them to improve their operations and marketing strategies. Data mining uses various pattern recognition and statistical and mathematical techniques to crunch through huge volumes of data and helps analysts identify important facts, relationships, trends, patterns, exceptions, and discrepancies that might escape the researcher's attention.

In businesses, data mining is used to discover patterns and establish relationships in the data to help managers formulate better business strategies. Data mining can help reveal sales trends, develop better marketing campaigns, and precisely estimate customer loyalty.

12.7.2 Uses of Data Mining

Data mining can be used for the following.

12.7.2.1 Market Segmentation

Identify the characteristics of customers of each product line and product category of a company.

12.7.2.2 Customer Defection

Identify customers who are most likely to shift loyalties to competitors.

12.7.2.3 Fraud Detection

Identify fraudulent transactions and those that have loopholes for committing fraud.

12.7.2.4 Direct Marketing

Identify prospects for mailer promotions.

TABLE 12.5

Steps in the Evolution of Data Mining and the Typical Questions That Can Be Answered

Evolutionary Step	Business Question	Enabling Technologies	Product Providers	Characteristics
Data Collection (1960s)	"What was my total revenue in the last five years?"	Computers, tapes, disks	IBM,CDC	Retrospective static data delivery
Data Access (1980s)	"What were unit sales in New England last March?"	Relational database (RDBMS), Structured Query Language (SQL), ODBC	Oracle, Sybase, Informix, IBM, Microsoft	Retrospective, dynamic data delivery at record level
Data Warehousing & Decision Support (1990s)	"What were unit sales in New England last March Drill down to Boston."	On-line analytic processing (OLAP) multidimensional databases, data warehouses	Pilot, Camshare, Arbor, Cognos, Micro-Strategy	Retrospective, dynamic data delivery at multiple levels
Data Mining (Emerging Today)	"What likely to happen to Boston unit sales next month? Why?"	Advanced algorithms, multiprocessor computers, massive databases	Pilot, Lockheed, IBM, SGL numerous startups (nascent industry)	Prospective proactive information delivery

12.7.2.5 Interactive Marketing

Predict the tastes and preferences of visitors to a website.

12.7.2.6 Market Basket Analysis

Determine product categories that are purchased together.

For example, Tea Powder and Sugar, Bread and Jam, Soaps and Detergents, and Shampoos and Hair Oils, etc.

12.7.2.7 Trend Analysis

To reveal the difference between typical customers this month and the last month.

12.7.3 Applications of Data Mining

12.7.3.1 Banking

The Bank of America has used data mining to sculpt detailed demographic views of the banking habits and financial assets of select groups of customers. Querying their data warehouse averages at 30 seconds. The system draws data from the entire bank and its 30 business units, making it a truly enterprise-wide database able to serve 1,200 users, making over 2,500 complex queries daily.

12.7.3.2 Finance

Gilman Securities uses data mining to differentiate how the financial markets react to the volatility of different business sectors. For example, finding the relationships between rate of changes between the Japanese Yen and the Government bond market.

12.7.3.3 Retail

One of the larger retailing operations in America, the Army and Air Force. Exchange Service, i.e., known to military personnel as "the PX," has used automated data mining to predict how much a particular woman will spend annually, given her age, her dependents, and her annual wage level to target advertising and sales to reach the appropriate customer base.

12.7.3.4 Insurance

Winterthur Insurance has more than 1 million customers in Spain, given the higher cost of underwriting new customers compared to working with current ones, reducing churn is an` ongoing challenge. Winterthur must predict which customers may leave and why. After implementing data mining applications, Winterthur was able to focus more easily on reducing customer churn and retain profitable customers.

12.7.4 Process of Data Mining

Data mining is a five-step process, which includes sampling, exploring, modifying, modeling, and assessing as described below.

12.7.4.1 Sampling

The first step in data mining is to decide whether one should tune the system to work on sample data or analyze the entire data existing in the databases. This decision becomes vital when the processing power of systems available with the organization is less.

If the data are very huge and the processing power is limited or speed is more important than complete analysis, it is better to draw samples than processing the entire data.

However, if the data are not very large, and the processing power is high or if it is important to understand patterns for every record in the database, a researcher should not go for sampling.

12.7.4.2 Exploring

This stage starts with data preparation, which may involve cleaning data, selecting subsets of records and in case of datasets with large numbers of variables, i.e., "fields," performing preliminary feature selection operations to make the number of variables manageable, i.e., depending on the statistical methods considered.

12.7.4.3 Modifying

This stage pertains to data modification if errors are detected in the exploration stage. This phase is a host to clustering, fractal-based transformation, application of fuzzy logic, data reduction programs such as factor analysis, correspondence analysis, and clustering. This stage helps categorize newly discovered key variables separately.

12.7.4.4 Modeling

Different modeling techniques used in data mining consist of neural networks, decision tree models, sequence-based models, classification and estimation models, and generic-based models. Any other than these can be used in the construction of the model once the data are prepared.

12.7.4.5 Assessing

This final step helps evaluate the performance of the designed model. One way to test the model is to run it for known data. For example, if you know which segment of the given market is risky, you can check to see whether the model selected this segment or not.

Summary

Preparation of data in a presentable form is essential for good analysis. To make the data collected presentable, a researcher subjects it through various processes such as validation and editing, coding, data entry, and data cleaning. Each process screens the data in its own specific way before forwarding it to the next screening stage. The very essence of validating lies in detecting fraud or failure by the interviewer to follow specified instructions.

After validation, data move forward for editing. Editing is the process where the editor checks for mistakes on the part of the interviewer or the respondent in filling the questionnaire.

Coding is the process of assigning numbers or other symbols to answers to group the response into distinct categories.

For convenience, researchers maintain a codebook that spells out guidelines for coding each of the variables that appear in the questionnaire.

After proper coding, these data are fed into a computer where the process of data cleaning takes place.

After data cleaning, the data are ready for analysis. Tabulation can be done in various ways, prominent among which are one-way frequency tabulation and cross tabulation.

An emerging concept in data analysis is data mining. As such, it has a wide variety of applications in management research because it predicts trends and behaviors and discovers hidden patterns in data that can help companies cut costs and improve profitability.

Review Questions

1. Explain preliminary data preparation techniques.
2. Discuss various types of survey tabulations.
3. Explain Data Mining and its applications.
4. Explain coding of close-ended and open-ended questions.
5. Explain Content analysis for open-ended questions.
6. Explain different means available for data entry.
7. Explain Data Mining in Research.
8. Explain the process of Data Mining.

13

Concepts of Hypothesis Testing

13.1 Introduction to Hypothesis

We know that research begins with a problem or a felt need or difficulty. To find a solution to the difficulty is the purpose of research. It is desirable that the researcher should propose a set of suggested solutions or explanations of the difficulty that the research proposes to solve. Such tentative solutions formulated as a proposition are called hypotheses. The suggested solutions formulated as hypotheses may or may not be the real solutions to the problem. Whether they are or not is the task of research to test and establish.

Relating to the characteristics of the population, managers make many assumptions under research to help them make various tactical and strategic decisions. Most of the times these decisions are based on the inferences drawn from a sample, and hence the manager is uncertain about the validity of the sample in representing the actual population. The manager is in a fix when the sample shows a marked difference from the actual value. He or she has to decide whether the difference observed is due to chance, i.e., sampling error, or it is statistically significant to make a change. The following illustrations bring out the dilemmas faced by managers.

A marketing manager believed that sales promotion campaigns would improve his product's short-term sales substantially. To test this assumption or hypothesis, a marketing research study was conducted. This revealed that consumers are indifferent to the sales promotion offers of the company. Will this make marketing manager believe that his or her assumption is incorrect and prompt him or her to change his or her promotional strategy?

An economist has assumed that India's GDP will grow by 9% in a particular year. However, a sample survey has revealed that Indian economy has registered a growth rate of just 5.5% in that year. Will this survey result make the economist conclude that his or her theoretical assumption is wrong?

These types of questions can be evaluated using the statistical techniques called hypothesis testing procedures.

We shall study a class of problems where the decision made by a decision maker depends primarily on the strength of the evidence thrown up by a random sample drawn from a population.

For example, the purchase manager of a machine tool-making company has to decide whether to buy castings from a new supplier or not. The new supplier claims that his castings have higher hardness than those of the competitors. Our hypothesis for this example could be that the mean hardness of castings supplied by the new supplier is less than or equal to 20 (say), where 20 is the mean hardness of castings, supplied by existing suppliers.

If the claim is true, then it would be in the interest of the company to switch from the existing suppliers to the new supplier because of the higher hardness, all other conditions being similar. However, if the claim is not true, the purchase manager should continue to buy from the existing suppliers. He needs a tool that allows him to test such a claim.

Testing of hypothesis provides such a tool to the decision maker. If the purchase manager were to use this tool, he would ask the new supplier to deliver a small number of castings. The sample of castings will be evaluated and based on the strength of the evidence produced by the sample, the purchase manager will accept or reject the claim of the new supplier and accordingly make his decision. The claim made by the new supplier is a hypothesis that needs to be tested and a statistical procedure which allows us to perform such a test is called testing of hypothesis.

DOI: 10.1201/9781315167138-13

Hypothesis testing enables the researcher to determine the validity of his hypothesis concerned with a particular issue. Hypothesis testing enables the researcher to decide whether data from a sample will provide support to a particular hypothesis, based on which it can be generalized to the overall population.

13.2 Meaning of Hypothesis

A hypothesis is a tentative solution or explanation or a guess or assumption or a proposition or a statement to the problem facing the researcher, adopted on a cursory observation of known and available data, as a basis of investigation, whose validity is to be tested or verified.

In conducting research, the important consideration after the formulation of a research problem is the construction of hypothesis. As you know, any scientific inquiry starts with the statement of a solvable problem, when the problem has been stated, a tentative solution in the form of testable proposition is offered by the researcher. Hypothesis is often considered a tentative and testable statement of the possible relationship between two or more events or variables under investigation.

To be useful in any study, the hypothesis needs to be stated in such a way that it might be subjected to empirical testing. The researcher is responsible to suggest or find some way to check how the hypothesis stands against empirical data.

A hypothesis, or more specifically a statistical hypothesis, is some statement about a population parameter or about a population distribution. If the population is large, there is no way of analyzing the population or of testing the hypothesis directly. Instead, the hypothesis is tested on the basis of the outcome of a random sample.

13.3 Characteristics of Hypothesis

Hypothesis should be conceptually clear; testable; related to the existing body or theory and impact; have logical unity and comprehensiveness; capable of verification; operationizable; consistent with known facts and theories and might be even expected to predict or anticipate previously unknown data; able to explain the data in simpler terms, stated in the simplest possible terms depending on the complexity of the concepts involved in the research problem; stated in a way that it can be tested for its being probably true or probably false in order to arrive at conclusions in the form of empirical or operational statements.

A good hypothesis has several basic characteristics. We discuss some of them as follows:

1. **Providing Direction:** Hypotheses provide direction to research and prevent review of irrelevant literature and collection of useless or excessive data. For example, in a research problem, "Study habits and achievement of Children in Villages," and the researcher may frame the hypothesis as, "Children in Villages put in more study hours, and achieve more in the examination." The researcher will collect data about the number of hours being put in by children for study and their achievement in the examination.
2. **Hypothesis Should Be Testable:** Hypotheses should be stated in such a way as to indicate an expected difference or an expected relationship between the measures used in the research. For example, there is no relationship between attendances to personal contact programs in a distance education course and achievement in an examination. Such propositions can be tested by means of an empirical data.
3. **Hypothesis Should Be Brief and Clear:** Hypothesis should make problems easier for the reader to understand and also for the researcher to test. The statement should be a concise statement of the relationship expected.

13.4 Formulation of Hypothesis

After testing the hypothesis, through various statistical tests, a researcher can accept or reject the hypothesis. If the hypothesis is accepted then the researcher can replicate the results; if hypothesis is rejected then the researcher can refine or modify the results.

13.5 Forms of Hypothesis

13.5.1 Declarative Hypothesis

When a researcher makes a positive statement about the outcome of the study, we get a declarative hypothesis. For example, the hypothesis "The performance of the non-kwashiorkor healthy children on problem-solving tasks is significantly higher than the kwashiorkor children." is stated in the declarative form. Here, the researcher makes an attempt to predict the future outcome. This prediction is based on the theoretical formulation of what should happen in a particular situation if the explanations of the behavior, i.e., performance on problem-solving tasks, which the researcher has given in his or her theory are correct.

13.5.2 Null Hypothesis and Alternate Hypothesis

A null hypothesis is a non-directional hypothesis that proposes no difference or no relationship. The usual form of such hypothesis is: "There is no significant difference between the performance of two groups of students, one from the school of Mathematics and the second from the School of Statistics." Since a null hypothesis can be statistically tested, it is also known as "Statistical Hypothesis" or "Testing Hypothesis" The notation used for this is H_0.

Sometimes the null hypothesis is rejected only when the probability of its having occurred by a mere chance is 1 out of 100 or 0.01 out of 1. If the null hypothesis is found false, then alternate would be true. The alternate hypothesis, denoted by H_1 or H_A, is the opposite of H_0, which must be true when H_0 is false.

13.5.3 Hypothesis in Question Form

In the question-form hypothesis, instead of stating what outcome is expected, a question is asked as to what the outcome will be. For example, if you are interested to find out whether instructions through video programs have any positive effect on the learning of the students of the Master's Program in Mathematics & Statistics.

The declarative form of the hypothesis will be: "Will Instruction through video programs affect the learning of student of Master's Program in Mathematics & Statistics?" This statement shows that instruction through video programs is not related to learning.

It is easier to state a hypothesis in question form because it appears to be quite useful to write down all the questions that one wants to answer in a particular research study.

On the other hand, a researcher faces difficulties in predicting the outcome of the study and stating the hypothesis in declarative form. But, it is worth noting that the question form is less powerful than the declarative or null form as a tool for obtaining valid information.

13.6 Problems in Formulation of Good Hypothesis

Major possible difficulties a researcher could face during the formulation of a hypothesis are the absence of knowledge of a theoretical framework; if detailed theoretical evidences are not available or if the investigator is not aware of the availability of those theoretical evidences, when the investigator is not aware of the scientific research techniques, he or she will not be able to frame a good research hypothesis. The hypothesis should be formulated in a positive and substantive form before data are collected. The hypotheses are generally derived from the earlier research findings, existing theories, personal observations, experiences, etc.

While formulating a hypothesis a researcher should consider expected relationship or differences between the variables, operational definition of variable, hypotheses are formulated following the review of literature, and the literature leads a researcher to expect a certain relationship. Hypotheses are the statements that are assumed to be true for the purpose of testing validity. A hypothesis can be put in the form of an If … then statement; if A is true, then B should follow.

For example, verbal development theory of amnesia states that childhood amnesia is caused by the development of language. To test this theory, a researcher can make a hypothesis like this: If the lack of verbal ability is responsible for childhood amnesia, then the children should not be able to verbally recall events usually words that they did not know at the time of events.

13.7 Types of Hypothesis

13.7.1 Explanatory Hypothesis

The purpose of this hypothesis is to explain a certain fact. All hypotheses are in a way explanatory for a hypothesis is advanced only when we try to explain the observed fact. A large number of hypotheses are advanced to explain the individual facts in life; for example, a theft, a murder, or an accident.

13.7.2 Descriptive Hypothesis

Sometimes a researcher comes across a complex phenomenon. He or she does not understand the relations among the observed facts. To account for these facts descriptive hypothesis is useful. A hypothesis is descriptive when it is based on the points of resemblance of something. It describes the cause-and-effect relationship of a phenomenon. For example, the current unemployment rate of a state exceeds 20% of the work force. Similarly, the consumers of local-made products constitute a significant market segment.

13.7.3 Analogical Hypothesis

When we formulate a hypothesis on the basis of similarities or analogy, it is called analogical hypothesis. For example, senior citizens invest more on long-term investments.

13.7.4 Working Hypothesis

Sometimes, by existing hypotheses, certain facts cannot be explained adequately, and no new hypothesis comes up. Thus, the investigation is held up. In this situation, a researcher formulates a hypothesis, which enables to continue investigation. Such a hypothesis, though inadequate and formulated for the purpose of further investigation only, is called a working hypothesis. It is simply accepted as a starting point in the process of investigation.

13.7.5 Null Hypothesis

Null hypothesis is symbolized as H_0. Null hypothesis is a useful tool in testing the significance of difference. In its simplest form, this hypothesis asserts that there is no true difference between two population means, and the difference found between sample means is accidental and unimportant, which arises out of fluctuation of sampling and by chance.

If the difference between the sample means is found significant the researcher can reject the null hypothesis. It indicates that the differences are statistically significant and acceptance of null hypothesis indicates that the differences are due to chance.

Under this type, the hypothesis is stated negatively. It is null because it may be nullified if the evidence of a random sample is unfavorable to the hypothesis. If the calculated value of the test is less than the permissible value, null hypothesis is accepted, otherwise it is rejected. The rejection of a null hypothesis implies that the difference could not have arisen due to chance or sampling fluctuations.

13.7.6 Alternative Hypothesis

Alternative hypothesis is symbolized as H_1 or H_A, which is the hypothesis that specifies those are the values that the researcher believes to hold true, and the researcher hopes that sample data will lead to acceptance of this hypothesis as true. Alternative hypothesis represents all other possibilities and it indicates the nature of relationship.

13.7.7 Statistical Hypothesis

Statistical hypotheses are the statements derived from a sample. These are quantitative in nature and are numerically measurable. For example, the market share of product *X* is 60%, the average life of a tube light is 2,500 hours, etc.

13.8 Errors in Hypothesis Testing

Hypotheses are assumptions that may be prove to be either correct or incorrect. It is possible to arrive at an incorrect conclusion about a hypothesis if faulty sampling procedure is adopted, data collection method is inaccurate, study design selected is faulty, inappropriate statistical methods are used, conclusions drawn are incorrect, etc.

Two common errors exist when testing a hypothesis:

1. **Type I Error:** Rejection of a null hypothesis, when it is true.
2. **Type II Error:** Acceptance of a null hypothesis, when it is false.

13.9 Importance of Hypothesis Formulation

Hypothesis is the basic function of the scientific research. If simple, brief, and clear scientific hypothesis has been formulated, there will be no problem for the investigator to proceed in the research field. Its utility or importance for a research may be studied as under. Formulation of hypothesis links between theory and investigation, which lead to discovery of addition to knowledge.

13.10 Stages of Hypothesis Testing

There are four stages:

1. The first stage is feeling of a problem. The observation and analysis of the researcher reveals certain facts. These facts pose a problem.
2. The second stage is formulation of a hypothesis or hypotheses. A tentative supposition or guess is made to explain the facts, which call for an explanation. At this stage, some past experience is necessary to pick up the significant aspects of the observed facts. Without previous knowledge, the investigation becomes difficult, if not impossible.
3. The third stage is deductive development of hypothesis using deductive reasoning. The researcher uses the hypothesis as a premise and draws a conclusion from it.

4. The fourth stage is the verification or testing of hypothesis. This consists in finding whether the conclusion drawn at the third stage is really true. Verification consists in finding whether the hypothesis agrees with the facts. If the hypothesis stands the test of verification, it is accepted as an explanation of the problem. But if the hypothesis does not stand the test of verification, the researcher has to search for further solutions.

For example, suppose you have started from your home for college on your Two Wheeler. A little while later the engine of your Two Wheeler suddenly stops. What can be the reason? Why has it stopped?

From your past experience, you start guessing that such problems generally arise due to either petrol or spark plug. Then start deducing that the cause could be that the petrol knob is not on; that there is no petrol in the tank; that the spark plug has to be cleaned.

Then start verifying them one after another to solve the problem.

1. First see whether the petrol knob is on. If it is not, switch it on and start the vehicle.
2. If it is already on, then see whether there is petrol or not by opening the lid of the petrol tank. If the tank is empty, go to the nearby petrol pump to fill the tank with petrol.
3. If there is petrol in the tank, this is not the reason, then you verify the spark plug. You clean the plug and fit it. The vehicle starts. That means the problem is with the spark plug. You have identified it. So you got the answer. That means your problem is solved.

13.11 Hypothesis Testing Procedure

Testing of a hypothesis is done by using statistical methods. Testing is used to accept or reject an assumption or hypothesis about a random variable using a sample from the distribution. The assumption is the null hypothesis (H_0), and it is tested against some alternative hypothesis (H_1). Statistical tests of hypothesis are applied to sample data.

The procedure involved in testing a hypothesis constitutes the following steps:

1. Formulation of the hypothesis.
2. Selection of an appropriate statistical test to be used for the data such as *t*-test, *Z*-test, F-test etc.
3. Selection of the level of significance 1%, 5% etc.
4. Calculation of the standard error of the sample statistics and standardization of the sample statistic.
5. Determination of the critical value.
6. Comparison of the values of the sample statistic with the critical value and identify, whether the value falls within the accepted region or rejection region.
7. Finally, drawing the inference of accepting or rejecting the null hypothesis and, hence, deducing the research conclusion.

13.11.1 Hypothesis Formulation

The first step in hypothesis testing is to state the research question in the form of a null hypothesis (H_0) and the alternate hypothesis (H_1 or H_A). H_0 assigns a value to the population parameter. It usually represents as "There is not difference between the sample statistic and the hypothesized population parameter."

Thus, the null hypothesis assumes that the difference, if any, in the observed data is attributed to random error, i.e., the occurrence of the event by chance and unusual.

In general, a null hypothesis consists of a claim that a researcher hopes to reject. On the other hand, the alternate hypothesis (H_1) is an opposite statement or claim from that which appears in the null hypothesis (H_0). The alternate hypothesis is a hypothesis that requires empirical evidence to accept it.

The researcher tends to believe that the claim in the null hypothesis is true until he or she finds some statistical evidence that supports the alternative hypothesis.

In general, a null hypothesis consists of value or claim, relating to a population parameter such as mean, standard deviation, a proportion or percentage, and state the equality($=, \leq, \geq$). Here, we use "=" sign as the company will not tolerate a difference in the quality standards on the both the sides, i.e., above and below the levels of the specified quality standards.

The alternate hypothesis would be that the mean quality standard of the sample differs from the required quality standards.

Following are the considerations while stating the Hypothesis:

1. The null and alternate hypotheses need to be formulated before the sample is drawn. This is because, if the hypotheses are developed after the sample is drawn and data are collected, they may be biased. Thus, in order to avoid such an error, the researcher has to develop the hypotheses before the sample is drawn and data are collected.
2. The hypotheses need to be specific and devoid of any ambiguity.
3. The hypotheses need to be formulated in such a way that they are fit for testing and take less time and effort for testing.
4. Generally, a hypothesis that is chosen as the null hypothesis will be the one intended to be disproved, and hypothesis that is chosen as the alternate hypothesis will be the one intended to be proved.
5. Concluding that we are accepting the null hypothesis doesn't imply that the null hypothesis is true. It only means that there is no significant statistical evidence to reject the null hypothesis. However, we use the term "accept the null hypothesis" for our convenience.

13.11.2 Selection of an Appropriate Statistical Test to be Used

After formulating the hypothesis, the next step in hypothesis testing is selecting the appropriate statistical test. The following are the key factors that will influence the selection of the appropriate statistical test: the type of research questions formulated; the number of samples involved in the hypothesis testing; and the scale of measurement used in the hypothesis testing.

13.11.2.1 Type of Research Questions Formulated

The type of research questions that are framed by the researcher is one of the key factors in deciding the statistical test to be used. The research question based on the Means and Proportions will use *Z*-test or *t*-test. For example, a research problem that requires the Comparison of the Mean Life in Hours of Two Different Brand of shoes calls for a *Z*-test or a *t*-test; for research questions, which are Framed Based on the Frequency Distribution will use chi-square test.

13.11.2.2 Number of Samples

The number of samples involved in the hypothesis test also influences the selection of the statistical test. For example, a Chemicals Company which is interested in testing the Concentration Level of an Acid and Base in the sample uses hypothesis testing about a Single Mean. In such cases, the researcher uses *Z*-test, *t*-test, or chi-square test; A Marketing Analyst, who is interested in studying the Preference of Two Different Customer Segments with regard to a company's product uses Tests of Difference about Means and Proportions; If there are more than two samples involved in the problem then we use tests such as ANOVA and Chi-square.

13.11.2.3 Measurement Scales Used

While selecting a statistical test, the type of measurement scale used is another factor that needs to be considered. Research problems that contain Interval Scales are tested using *Z*-test and *t*-test; Problems containing Ordinal and Nominal Scales are tested using chi-square test.

TABLE 13.1

Types of Statistical Tests and its Characteristics

Hypothesis Testing	Number of Samples	Measurement Scale	Type of Statistical Test	Requirement
Hypotheses about frequency distributions	1	Nominal	χ^2	
	2 or more	Nominal	χ^2	
	1 (large sample)	Interval or ratio	*Z*-test	$n \geq 30$, when σ is known
Hypothesis about means	1 (small sample)	Interval or ratio	*t*-test	$n \geq 30$, when σ is not known
	2 (large sample)	Interval or ratio	*Z*-test	$n \geq 30$, when σ is known
	2 (small sample)	Interval or ratio	*t*-test	$n < 30$, when σ is not known
	2 (small sample)	Interval or ratio	One-way ANOVA	
	1 (large sample)	Interval or ratio	*Z*-test	$n \geq 30$, when σ is known
Hypothesis about Proportions	1 (small sample)	Interval or ratio	*t*-test	$n < 30$, when σ is not known
	2 (large sample)	Nominal	*Z*-test	$n \geq 30$, when σ is known
	2 (small sample)	Interval or ratio	*t*-test	$n < 30$, when σ is not known
Variance	2 or more samples	Interval or ratio	*F*-test (or ANOVA test)	

There are other factors such as sample size and whether the population standard deviation is known or unknown. For example, if the sample size is large, i.e., >30, and the data are normally distributed then we use *Z*-test; If the population standard deviation is not known and the sample is small, we use *t*-test to test the hypothesis. Table 13.1 provides the criteria for the selection of statistical test and the recommended statistical test to be used.

13.11.3 Selection of the Level of Significance

The next step in testing hypotheses is to set up a suitable significance level to test the validity of H_0 as against H_1. The confidence with which a null hypothesis is adopted or rejected depends on the adopted significance level (Figure 13.1).

Conventionally, the significance level is expressed as a percentage, such as 5% or 1%. In the former case, it would mean that there is 5% probability of rejecting a null hypothesis, even if it is true. This means that there are 5 out of 100 chances that the investigator would reject a true hypothesis (see Figure 13.2).

The level of significance is a measure of degree of risk that a researcher might reject the null hypothesis when the null hypothesis is true. The commonly used level of significance by researchers is 5%. A 5% level of significance implies that there is 5% probability that we may wrongly conclude that there is a difference between the sample statistic and the hypothesized population parameter, when there is no difference between them.

In other words, if we take 100 samples from a population, we are identifying that 5 or <5 samples display a difference between the sample statistic and the hypothesized population parameter, when there exists no difference in the population. There is a risk that samples drawn by the researcher might be one of those five samples. However, such risk is minimal.

We can also deduce that the level of significance indicates the percentage of sample means that are outside a specific cut-off point, assuming the hypothesized population parameter is true.

There is no proper mechanism to decide the level of significance. The level of significance is set by the researcher based on various factors, such as cost involved for each type of errors, i.e., Type I and II errors. The level of significance also indicates the risk of rejecting the null hypothesis, when it is true, which is otherwise known as Type I error. In all tests of hypothesis, Type I error is assumed to be more serious than Type II error and so the probability of Type I error needs to be explicitly controlled. This is done through specifying a significance level at which a test is conducted. The significance level, therefore, sets a limit to the probability of Type I error and test procedures are designed, so as to get the lowest probability of Type II error subject to the significance level.

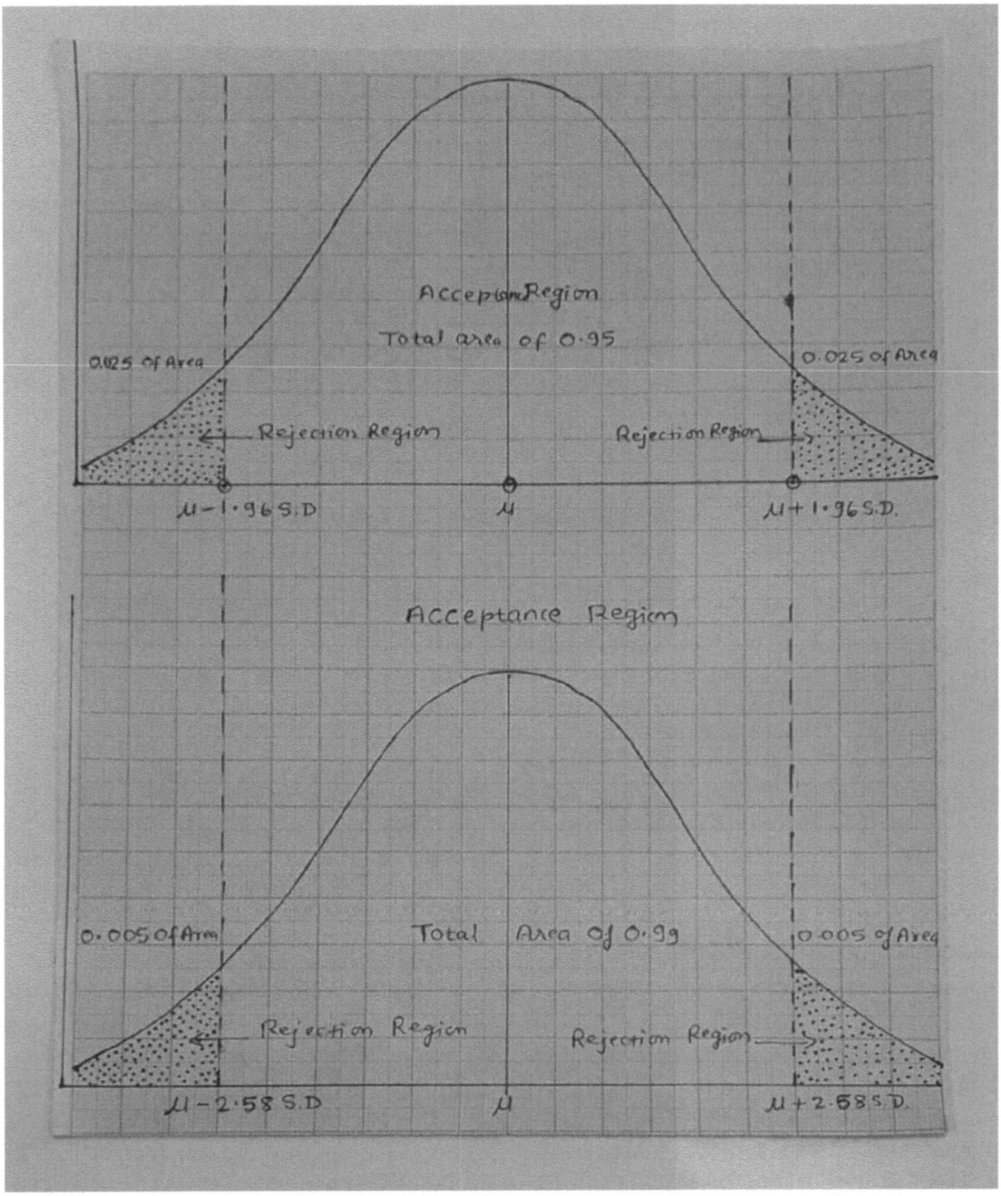

FIGURE 13.1 Acceptance or rejection of null hypothesis two tailed at 5% and 1%.

The probability of Type I error is usually represented by the symbol α, read as alpha, and the probability of Type II error is represented by β, read as beta.

Suppose, we have set up our hypotheses as follows:

$$H_0 : \mu = 40 \; H_1 : \mu \neq 40.$$

We would perhaps use the sample mean $\bar{x}$ to draw inferences about the population mean μ. Also, since we are biased toward H_0, we would be compelled to reject H_0 only when the sample evidence is strongly against it. For example, we might decide to reject H_0 only when $x > 42$ or $x < 38$ and in all other cases, i.e., when x is between 38 and 42 and so is close to 40, we might conclude that the sample evidence is not strong enough for us to be able to reject H_0.

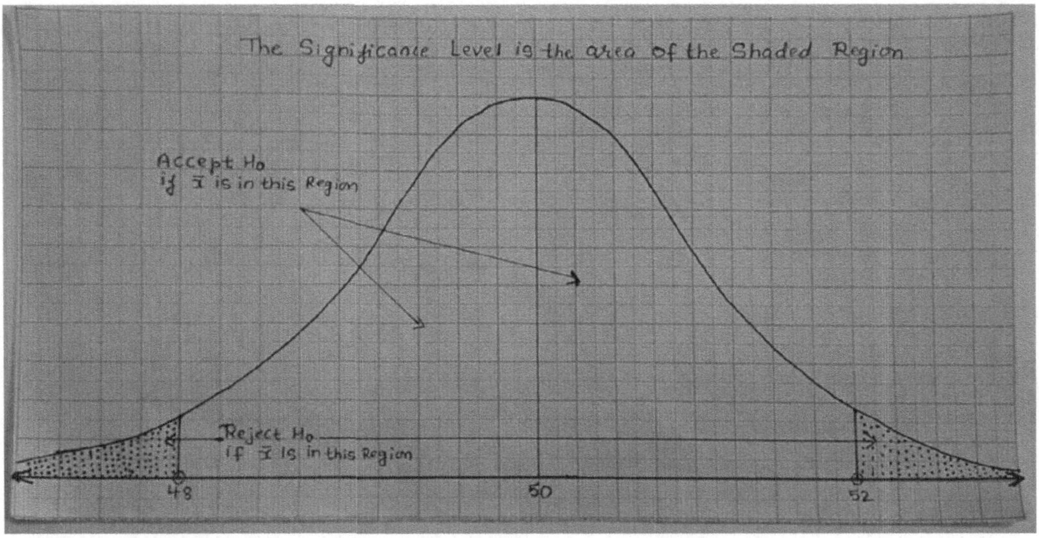

FIGURE 13.2 Two-tailed test with 5% level of significance.

Now, suppose the H_0 is in reality true, i.e., the true value μ is 40. In that case, if the population distribution is normal or if the sample size is sufficiently large ($n > 30$), the distribution of z will be normal as shown in Figure 13.2. Remember that our criterion for rejecting H_0 states that if $x < 38$ or $x > 42$, we shall reject H_0.

Referring to Figure 13.2, we find that the shaded area under both tails of the distribution represents the probability of rejecting H_0, when H_0 is true, which is the same as the probability of Type I error.

All tests of hypotheses hinge upon this concept of the significance level and it is possible that a null hypothesis H_0 is rejected at $\alpha = 0.05$, whereas the same evidence is not strong enough to reject the null hypothesis at $\alpha = 0.01$. In other words, the inference drawn can be sensitive to the significance level used.

Disadvantages of testing of hypothesis:

1. The financial or the economic costs of consequences are not considered explicitly.
2. In practice, the significance level is supposed to be arrived at, after considering the cost consequences.
3. It is very difficult to specify the ideal value of a in a specific situation; we can only give a guideline that the higher the difference in costs between Type I and Type II errors, the greater is the importance of Type I error compared to Type II error. Consequently, the risk or probability of Type I error should be lower, i.e., the value of should be lower. In practice, most tests are conducted at $\alpha = 0.01$, $\alpha = 0.05$ or $\alpha = 0.1$ by convention as well as by convenience.

13.11.3.1 The p-Value of a Test

A test of hypothesis is designed for a significance level and at the end of the test we conclude that we reject the null hypothesis at 1% significance level and so on. The significance level is somewhat arbitrarily fixed and the mere fact that a hypothesis is rejected or cannot be rejected does not reveal the full strength of the sample evidence. An alternative, and in some ways, a better way of expressing the conclusion of a test, is to state the p-value or the probability value of the test. The p-value of a test expresses the probability of observing a sample statistic as extreme as the one observed, if the null hypothesis is true.

The p-value is the probability of obtaining a test statistic, equal to or more extreme (in the direction of supporting H_1 than the actual value obtained), when null hypothesis is true. Nowadays use of p-value is becoming more and more popular because most of the statistical software provide p-value rather than critical value and p-value provides more information compared to critical value as far as rejection or do not rejection of H_0. Moving in this direction, we note that in scientific applications one is not only interested simply in rejecting or not rejecting the null hypothesis but he/she is also interested to assess how strong the data have the evidence to reject H_0.

For example, general procedure of testing a hypothesis where we tested the null hypothesis $H_0 : \mu \leq 50$ g against $H_1 : \mu > 50$ g.

To test the null hypothesis, we calculated the value of test statistic as 2.78 (say) and the critical value (z_α) at $\alpha=0.01$ was $z_\alpha=2.33$ (From Statistical Table).

Since calculated value of test statistic (= 2.78) is greater than critical (tabulated) value (= 2.33). Therefore, we reject the null hypothesis at 1% level of significance. Now, if we reject the null hypothesis at this level (1%) surely we have to reject it at higher level because at $\alpha=0.05$, $z_\alpha=1.645$ and at $\alpha=0.10$, $z_\alpha=1.28$. However, the calculated value of test statistic is much higher than 1.645 and 1.28, therefore, the question arises "Could the null hypothesis also be rejected at values of α smaller than 0.01?" The answer is "yes" and we can compute the smallest level of significance (α) at which a null hypothesis can be rejected. This smallest level of significance (α) is known as "p-value." The p-value is the smallest value of level of significance (α) at which a null hypothesis can be rejected using the obtained value of the test statistic and can be defined as: the p-value is the probability of obtaining a test statistic equal to or more extreme (in the direction of supporting H_1) than the actual value obtained when null hypothesis is true.

Procedure of taking the decision about the null hypothesis on the basis of p-value: To take the decision about the null hypothesis based on p-value, the p-value is compared with level of significance (α), and if p-value is equal or less than α, then we reject the null hypothesis, and if the p-value is greater than α, we do not reject the null hypothesis.

13.11.3.2 Type I and Type II Errors

As the hypothesis tests are based on probability, a researcher may not arrive at a conclusion with certainty. Thus, there always exists a chance of a researcher committing an error. Since, we are basing our conclusion on the evidence produced by a sample. And, variations from one sample to another can never be eliminated until the sample is as large as the population itself.

It is possible that the conclusion drawn is incorrect, which leads to an error. As summarized in Table 13.2, there can be two types of errors and for convenience, each of these errors has been given a name. Table 13.2 provides various errors a researcher may commit while testing a hypothesis. When null hypothesis H_0 is true and the researcher decides to accept the null hypothesis, then there exists no error.

However, there are two types of errors that a researcher will commonly make while testing a hypothesis. They are Type I and Type II errors.

Type I error refers to a situation where the researcher rejects the null hypothesis when it is true. Probability of Type I error is represented by α (alpha), i.e., the level of significance.

In other words, Type I error occurs when the researcher believes that there is a difference between the sample statistic and the hypothesized population parameter; Type II error occurs when there exists no difference between the sample statistic and the hypothesized population parameter.

Type II error refers to a situation where the researcher accepts the null hypothesis when it is false. Probability of Type II error is denoted by β (beta).

A researcher has to decide on the levels of these errors optimally as there exists an inverse relationship between Type I and Type II errors, i.e., any reduction in one error may increase the probability of committing another error. Thus, the researcher has to decide upon the levels optimally depending on the need.

TABLE 13.2

Types of Errors in Hypothesis Testing

		Decision Based on Sample	
States of Population	H_0 is True (H_1 is False)	Accept H_0 (Reject H_1)	Reject H_0 (Accept H_1)
	H_0 is False (H_1 is True)	Correct (no error)	Wrong (Type I error α)
		Wrong (Type II error β)	Correct

If we wrongly reject H_0, when in reality H_0 is True, the error is called a Type I error. Similarly, when we wrongly accept H_0 when H_0 is False, the error is called a Type II error. Both these errors are bad and should be reduced to a minimum. However, they can be completely eliminated only when the full population is examined, in which case there would be no practical utility of the testing procedure.

This implies that if the testing procedure is designed as to reduce the probability of occurrence of Type I error, simultaneously the probability of Type II error would go up and vice versa.

In all testing of hypothesis procedures, it is implicitly assumed that Type I error is much more severe than Type II error and so needs to be controlled.

13.11.4 Calculation of the Sample Statistics

For calculating the sample statistic we need to compute the standard error of the sample statistic or sample mean, and using the standard error, we need to standardize the sample statistic. In case when the sample size is large and population standard deviation is known, we identify that Z-test is appropriate.

When the population standard deviation is known, the standard error is calculated using the following formula:

$$\sigma_{\bar{x}} = \frac{\sigma}{\sqrt{n}} \tag{13.1}$$

Where, $\sigma_{\bar{x}} \rightarrow$ Standard Error of the Mean, $\sigma \rightarrow$ Population Standard Deviation, $n \rightarrow$ Size of the sample

Then, using the value so obtained, we will standardize the value of sample statistics. As we are using Z-test, the equation given below can be used to standardize the sample mean

$$Z = \frac{\bar{x} - \mu_0}{\sigma_{\bar{x}}} \tag{13.2}$$

Where, $\bar{x} \rightarrow$ Sample Mean, $\mu_0 \rightarrow$ Hypothesized Population Parameter, $\sigma_{\bar{x}} \rightarrow$ Standard Error of the Mean

13.11.5 Determination of the Critical Values

The next step is, to determine the critical values associated with the standardized value. Critical values are determined so as to evaluate whether the standardized value will fall into the acceptance region or rejection region. Critical values demarcate the acceptance and rejection regions. Before understanding how to determine the critical values, we need to know about two key concepts: two-tailed and one-tailed tests.

13.11.5.1 Two-Tailed Tests

A two-tailed test is a hypothesis test where the null hypothesis is rejected when the value of sample statistic falls above or below the hypothesized population parameter. A two-tailed test of hypothesis will reject the null hypothesis if the sample statistic is significantly higher or lower than the population parameter.

Two-tailed tests are administered when the hypotheses are represented as

$$H_0 : \mu = \mu_0 \quad H_1 : \mu \neq \mu_0$$

If we set a 5% level of significance, then the acceptance region will have 95% probability and rejection region will have 5% probability, which is split on the both sides of the acceptance region with areas of 0.025 probability. Figure 13.6 shows this scenario.

In other words, in a two-tailed test, the acceptance region falls between two rejection regions. Thus, in a two-tailed test of hypothesis, the rejection region is located on both the tails and the size of the rejection region is 0.025, whereas the central acceptance region is 0.95 as shown in Figure 13.3.

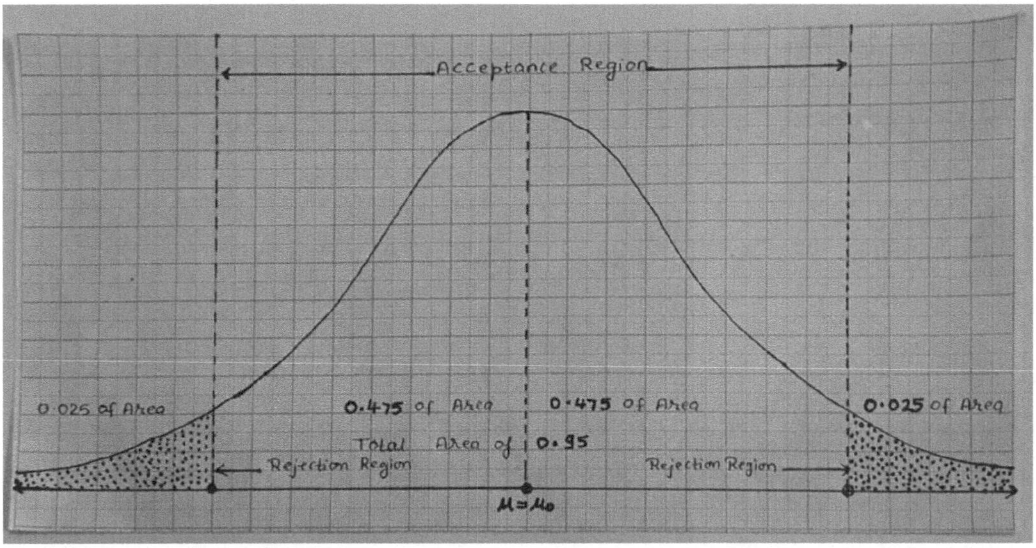

FIGURE 13.3 Two-tailed test with 5% level of significance.

If the sample mean falls within $\mu \pm 1.96\ SD$, i.e., in the acceptance region, the hypothesis is accepted. If, on the other hand, it falls beyond $\mu \pm 1.96\ SD$, then the hypothesis is rejected, as it will fall in the rejection region.

13.11.5.2 One-Tailed Test

A one-tailed test is a test where the null hypothesis is rejected when the value of the sample statistic is less than or higher than the hypothesized population parameter. Such hypothesis is represented by notation: $H_0 : \mu \leq \mu_0 \quad H_1 : \mu > \mu_0$ or $H_0 : \mu \geq \mu_0 \quad H_1 : \mu < \mu_0$. The one-tailed test can either be a left-tailed or a right-tailed test.

13.11.5.3 Left-Tailed Test

The left-tailed test will reject the null hypothesis if the sample mean value is lesser than hypothesized population parameter. If we take a significance level of 5%, then the acceptance region will be 95% and the rejection region will be 5% to the left of the distribution curve. Figure 13.4 shows this scenario.

13.11.5.4 Right-Tailed Test

The right-tailed test will reject the null hypothesis if the sample mean value is greater than the hypothesized population parameter. If we set the level of significance of 5%, then the acceptance region will have a probability of 95% and then rejection region will have a probability of 5%. Figure 13.5 shows this scenario.

The selection of two-tailed or one-tailed test depends on the formulation of hypotheses. Two-tailed test is used when the researcher is concerned with testing the hypothesis as to whether there is any difference between the sample mean and the hypothesized population parameter. One-tailed test is used when the researcher is concerned with testing the hypothesis as to whether the sample mean is greater than the hypothesized population parameter, i.e., this is more specifically a right-tailed test or whether the sample mean is lesser than the hypothesized population parameter, i.e., this is more specifically a left-tailed test.

The critical values depend on the significance level, the type of hypothesis test, and statistical test used.

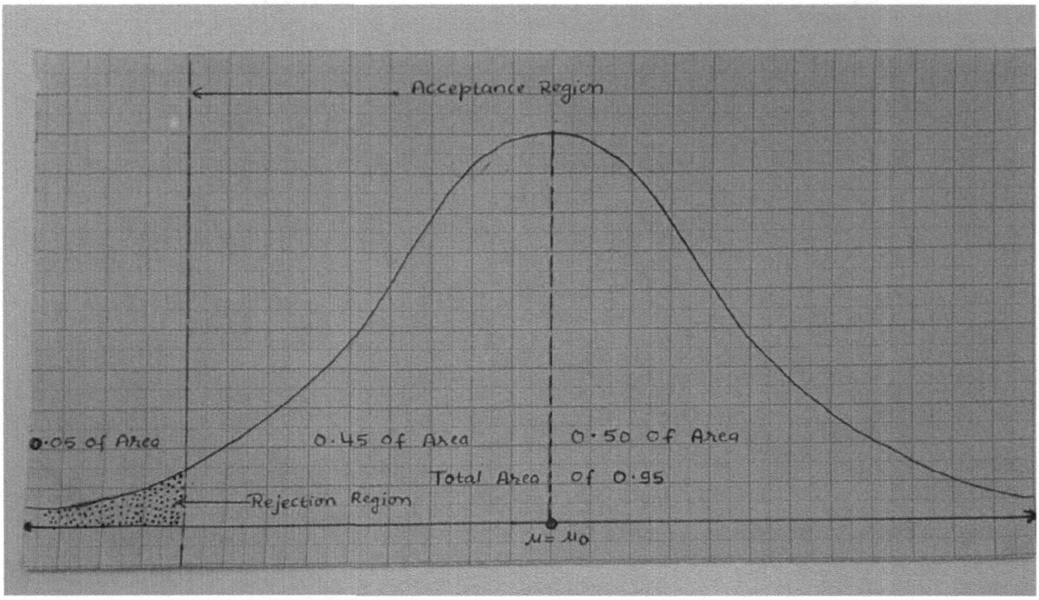

FIGURE 13.4 Left-tailed test with 5% level of significance.

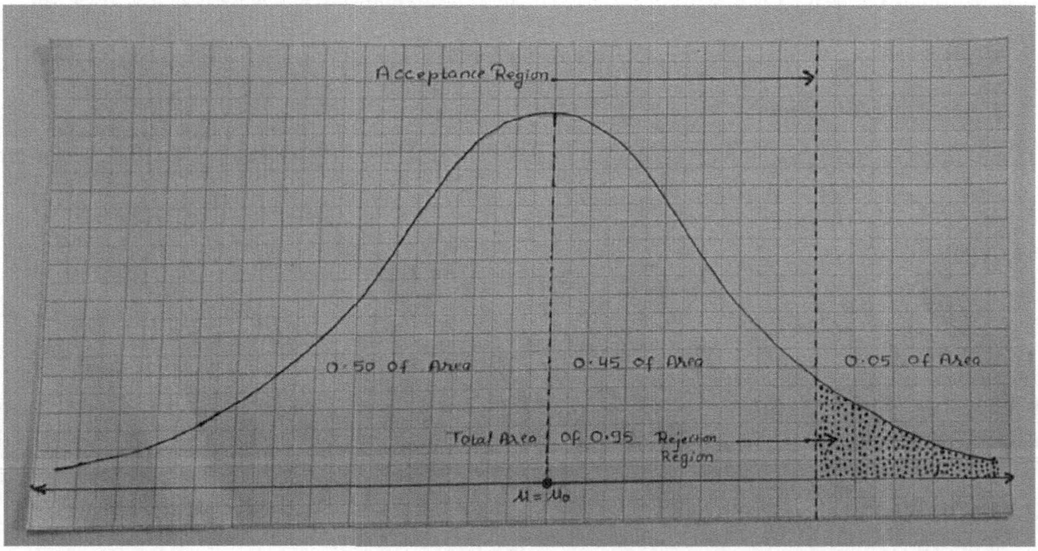

FIGURE 13.5 Right-tailed test with 5% level of significance.

For example, the hypothesis requires a left-tailed test with 2% level of significance and the *Z*-test is being used. Two percent significance level means that the accepted region covers 98% of the area of normal distribution curve and the rejection region consists of 2% of the area at the left tail of the normal distribution curve. Being a one-tailed test, acceptance region, i.e., 98%, is split into two parts: 50% on the right side and 48% on the left side. The critical value can be obtained by looking the corresponding *Z*-value in the Normal Distribution Table in Appendix, for 48% on the area under the curve, i.e., test being left-tailed. Thus, the corresponding *Z*-value is 2.05.

For example, let us take an example of the two-tailed hypothesis. Suppose a researcher is interested in knowing whether there is gender difference in Exam Result. You can formulate the following hypotheses.

Exam Results of Female Students = Exam Results of Male Students (Null hypothesis)

Exam Results of Female Students ± Exam Results of Male Students (Alternative hypothesis) or, in other words, Exam Results of Female Students may be lower or higher than that of male students.

In contrast to the two-tailed hypothesis,

1. In one-tailed hypothesis the rejection region will be located only on one tail (see Figure 13.6). In this case, the size of the rejection region will be 0.05 if one is testing the hypothesis at 5% probability level or level of significance.
2. If the sample mean falls above $\mu + 1.645SD$ (Case A: Figure 13.6) or below $\mu - 1.645SD$ (see Case B of Figure 13.6), then the hypothesis is rejected, as it will fall in the rejection region.

13.11.6 Comparison of the Values of the Sample Statistic with the Critical Value

As the boundaries are defined, the next step is to compare the standardized value with the critical value. And check whether it falls within the acceptance region.

In a two-tailed test, we reject the null hypothesis if the standardized sample mean or observed Z-value is greater than the upper critical value or lesser than the lower critical value.

In the right-tailed test, we reject the null hypothesis if the observed Z-value is greater than the upper critical value.

In the left-tailed test, we reject the null hypothesis if the observed Z-value is lesser than the lower critical value.

$$\text{From the equation } Z = \frac{\bar{x} - \mu_0}{\sigma_{\bar{x}}}$$

The calculated Z-value is –1. The test is a two-tailed test. The value is within the acceptance region of ±1.96, i.e., $-1.96 < -1 < +1.96$.

13.11.7 Finally Draw the Inference and Deduce the Research Conclusion

We reject the null hypothesis if the value of the standardized sample statistic falls in the rejection region. And accept or not reject the null hypothesis if the standardized sample statistic falls within the accepted region. As the standardized value –1 falls within the acceptance region, company can accept or cannot reject the null hypothesis. Hence, we conclude that there is no difference between the mean quality standard of the sample and the hypothesized mean quality standard.

13.12 Uses of Hypothesis

If a clear scientific hypothesis has been formulated, half of the research work is already done. The advantages or utility of having a hypothesis is summarized as follows:

1. It is a starting point for many a research work.
2. It helps in deciding the direction in which to proceed.
3. It helps in selecting and collecting pertinent facts.
4. It is an aid to explanation. It works as a basis for future knowledge.
5. It helps in drawing specific conclusions and in testing theories.

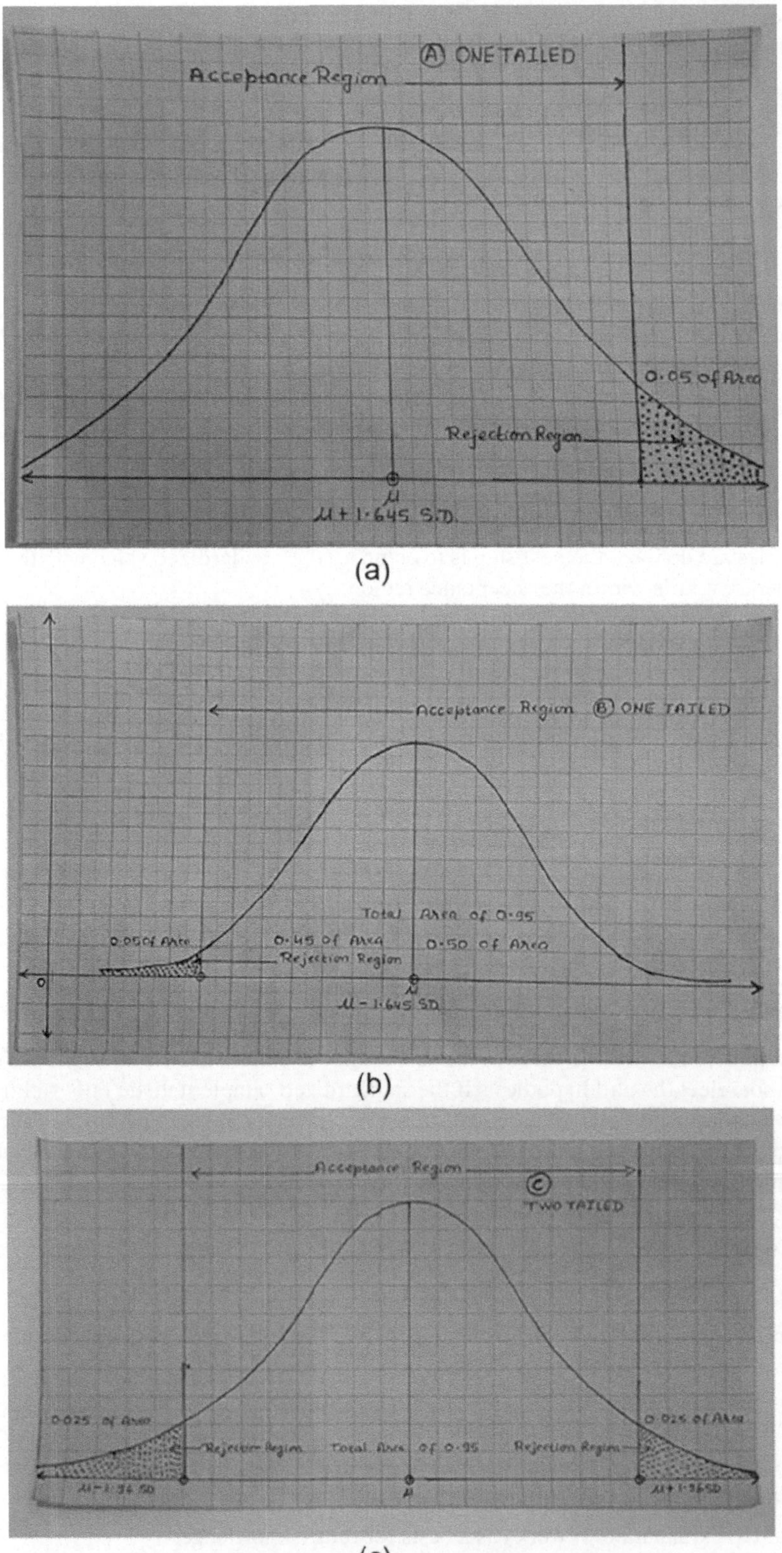

(c)

FIGURE 13.6 One-tailed and two-tailed tests of hypothesis. (a) and (b) are one-tailed, whereas (c) is two-tailed.

Summary

Hypothesis testing is one of the key statistical technique that are used in decision making. The key purpose of hypothesis testing is to analyze the difference between the value of the sample statistic and the hypothesized population parameter. Hypothesis testing enables a researcher to decide whether sample data will provide support to a particular hypothesis based on which it can be generalized to the overall population. A hypothesis is a speculative statement that is subjected to verification through a research study. A statistical hypothesis is a statement about a population parameter or about a population distribution.

The hypotheses are of various types such as explanatory hypothesis, descriptive hypothesis, analogical hypothesis, working hypothesis, null hypothesis, alternative hypothesis and statistical hypothesis. There are four stages in a hypothesis: (a) feeling a problem; (b) formulating hypothesis; (c) deductive development of hypothesis; and (d) verification or testing of hypothesis. Verification can be done either directly or indirectly or through logical methods. Testing is done using statistical methods.

The steps involved in the hypothesis testing procedure are as follows: formulate the hypothesis, select the statistical test to be used, select the significance level, calculate the standard error of the sample statistics and standardize the sample statistic, determine the critical value, compare the value of sample statistic with the critical value, identify whether the value falls within the accepted region, and deduce the research conclusion.

The null and the alternative hypotheses are set up such that one of them, and only one of them, is always true. In the absence of a strong evidence to the contrary, the decision maker would be willing to accept the null hypothesis.

The next step is to decide on the appropriate statistical test. Then we need to decide upon the level of significance to be fixed. While fixing the level of significance we need to consider the Type I and Type II errors. Of the two errors, Type I error is considered to be more serious than the other one and so is subject to explicit control.

The next step is to calculate the sample statistic. The next step is to determine the critical values. The critical value demarcates the acceptance region and the rejection region. We need to compare the standardized sample statistic with the critical value. We reject the null hypothesis if the value of standardized sample statistic falls in the rejection region and accept or not reject the null hypothesis if the standardized sample statistic falls within the accepted region.

Questions

1. Define hypothesis and explain its characteristics.
2. List out different types of hypotheses.
3. What is meant by null hypothesis and alternate hypothesis?
4. What are the characteristics of good hypothesis testing?
5. What are the stages in a hypothesis?
6. Explain the formulation of hypothesis.
7. Explain the problems in Formulation of Good Hypothesis.
8. What are the methods used to prove or reject a hypothesis?
9. What is meant by hypothesis? Explain the criteria for a workable hypothesis.
10. What are the different stages in a hypothesis? How do you verify/test a hypothesis?

11. Write short notes on:
    a. Formulation of hypothesis
    b. Null hypothesis
    c. Alternative hypothesis
    d. Level of Significance
12. Describe procedure of hypothesis testing.
13. Give the uses of hypothesis testing.

14

Hypothesis Testing: Tests of Differences

14.1 Introduction

In testing of hypothesis and estimation of parameters, we generally assume that the random variable follows a distribution such as Normal, Binomial, Poisson distribution, etc., but in many real world situations, like in business and other areas, the data are collected in the form of counts. In some cases, the collected data are classified into different categories or groups according to one or more attributes. Such type of data are known as categorical data. For example, the number of people of a colony can be classified into different categories according to age, sex, income, job, etc., or the books of library can be classified according to their subjects such as books of Science, Commerce, Art, Management, Engineering, Pharmacy, Medical Science, Law, Education, etc. Now the question arises, "How do we tackle the inference problems arising out of categorical data?" The chi-square (χ^2) test is usually used in such situations.

The hypothesis testing procedures can be classified into following two key types: tests of association and test of differences.

14.1.1 Tests of Association

Test of association are used in situations where researchers evaluate the statistical relationship between the variables. Tests of differences are concerned with making judgments with regard to the differences between populations.

14.2 χ^2 Test and Cross-Tabulation

Test of association are used in situations where the researcher has to evaluate whether there is any association between the variables under study. For example, researchers face situations where they want to know whether there is any association between brand preference and income levels or whether the inflation in an economy and the stock market index are related. Prominent tests of association are the χ^2 test, correlation analysis, and regression analysis.

14.2.1 Contingency Table

A contingency table is an arrangement of data into a two-way classification. One of the classifications is entered in rows and the other in columns. Cross-tabulations are one of the widely used data analysis techniques in research. Cross-tabulation shows the relationship between two or more variables. Cross-tabulation involves merging of frequency distributions of two or more variables in a single table. These tables will help the researcher in understanding the impact of one variable on the other variable. An example of cross-tabulation is summarized in Table 14.1.

In Table 14.1, channel viewership is segregated according to the age group. Thus, cross-tabulation proves to be the simplest method to summarize the survey results. From the above example of cross-tabulation, a researcher can find answers such "how many viewers in the age group 15–25 view Channel A" or "what is the proportion of viewership of Channel B in the sample."

DOI: 10.1201/9781315167138-14

TABLE 14.1

Channel Viewership Distribution According to Age-Group

Channel → Age Group ↓	A	B	C	Total
15–25 years	50	40	50	140
25–45 years	60	70	50	180
45 years and above	50	60	70	180
Total	**160**	**170**	**170**	**500**

To evaluate the statistical significance of association among the variables involved in the cross-tabulation, researchers use statistical technique called the χ^2 test. The χ^2 test is usually used by the researchers in two ways: test of independence and test of goodness of fit.

14.2.2 Test of Independence

The test of independence is used to evaluate whether there is any association between two variables. The goodness-of-fit test is used to identify whether there is any significant difference between the observed frequencies and the expected frequencies.

Before understanding these two tests in detail, let us examine some general aspects that are to be considered while performing the χ^2 test.

1. The χ^2 test can be performed on the actual numbers but not on percentages and proportions. If the data are in percentage or proportion form, then they need to be converted into absolute numbers before performing the χ^2 test.
2. The expected frequency of a cell should be more than five. If the cell contains a value less than five, then some of the rows or columns are to be combined so that new frequencies will have values greater than five.
3. The χ^2 test works only when the sample size is large enough. Usually the sample size needs to be more than 50.
4. Observations drawn need to be random and independent.

14.3 χ^2 Test—Goodness of Fit

Generally, the parametric tests are based on the assumption of a normal population. The suitability of a normal distribution or some other distribution may itself be verified by means of a goodness-of-fit test. The χ^2 test for goodness of fit was given by Karl Pearson in 1900. It is the oldest nonparametric test.

With the help of this test, we test whether the random variable under study follows a specified distribution such as Binomial, Poisson, Normal, or any other distribution, when the data are in categorical form.

Here, we compare the actual or observed frequencies in each category with theoretically expected frequencies that would have occurred, if the data followed a specified or assumed or hypothesized probability distribution.

This test is known as "Goodness-of-Fit Test" because we test how well an observed frequency distribution, i.e., distribution from which the sample is drawn, fit to the theoretical distribution such as Normal, Uniform, Binomial, etc.

14.3.1 Assumptions

This test works under the following assumptions:

1. The sample observations are random and independent.
2. The sample size is large.

3. The observations may be classified into nonoverlapping categories.
4. The expected frequency of each class is greater than five.
5. Sum of observed frequencies is equal to sum of expected frequencies, i.e.,

$$\sum O_i = \sum E_i$$

As usual, the first step in testing of hypothesis is to set up null and alternative.

Hypotheses, so our null and alternative hypotheses are setup, below in Step 1.

Step 1: Generally, we are interested to test whether data follow a specified or assumed or hypothesized distribution $F_0\ (x)$ or a sample has come from a specified distribution or not. So, here we consider only two-tailed case. Thus, we can take the null and alternative hypotheses as

H_0 : Data follow a specified distribution

H_1 : Data do not follow a specified distribution

In symbolical form,

$$H_0 : F(x) = F_0\ (x) \text{ for all values of } x$$

$$H_0 : F(x) \neq F_0\ (x) \text{ for at least one value of } x$$

Step 2: After setting null and alternative hypotheses, our next set up is to draw a random sample. So, let a random sample of size N be drawn from a population with unknown distribution function *F*(*x*) and the data are categorized into *k* groups or classes.

Here, we take the notation for sample size is "*N*" instead of "*n*," since in this test, generally we deal with frequencies of the observations and to represent the sum of frequencies, we take notation "*N*."

Also, let $O_1, O_2, \ldots, O_k \rightarrow$ are the observed frequencies and

$E_1, E_2, \ldots, E_k \rightarrow$ are the corresponding expected frequencies.

If the parameters of assumed distribution is (are) unknown, then in this step we estimate the value of each parameter of the assumed distribution with the help of sample data, by calculating Sample Mean, Variance, Proportion, etc., as may be the case.

Step 3: After that, we find the probability of each category or group in which an observation falls with the help of the assumed probability distribution.

Step 4: If $p_i\ (i = 1, 2, \ldots, k)$ is the probability that an observation falls in i^{th} category, then we find the expected frequency by the formula given below:

$$E_i = Np_i; \text{ all } i = 1, 2, \ldots, k$$

Step 5: Test statistic, since this test compares observed frequencies with the corresponding expected frequencies. Therefore, we are interested in the magnitudes of the differences between the observed and expected frequencies. Specifically, we wish to know whether the differences are small enough to be attributed to chance; they are large due to some other factors; with the help of observed and expected frequencies we may compute a test statistic that reflects the magnitudes of differences between these two quantities when H_0 is true.

The test statistic is given by $\chi^2 = \sum_{i=1}^{k} \frac{(O_i - E_i)^2}{E_i} \sim \chi^2_{(k-1)}$ Under H_0

where $k \rightarrow$ represents the number of classes.

If any expected frequency is less than 5, then it is pooled or combined with the preceding or succeeding class, and *k* represents the number of classes that remain after the combining classes. The χ^2-statistic follows approximately chi-square distribution with $(k-1)$ degrees of freedom.

If the parameters of the distribution to be fitted is (are) unknown, that is, not specified in null hypothesis then test statistic χ^2 follows approximately chi-square distribution with $(k-r-1)$ degrees of freedom, that is,

$$\chi^2 = \sum_{i=1}^{k} \frac{(O_i - E_i)^2}{E_i} \sim \chi^2_{(k-r-1)} \text{ Under } H_0$$

Where $r \rightarrow$ is the number of unknown parameters, which are estimated from the sample.

Step 6: Obtain critical value of test statistic. At a given level of significance under the condition that null hypothesis is true. The table given in the Appendix provides critical values of the test statistic χ^2 for various degrees of freedom and different level of significance.

Step 7: Take the decision about the null hypothesis

To take the decision about the null hypothesis, the test statistic, calculated in Step 5, is compared with chi-square critical (tabulated) value, observed in Step 6. For a given level of significance (α) under the condition that the null hypothesis is true. If calculated value of test statistic is greater than or equal to critical value, with $(k - r - 1)$ degrees of freedom at α level of significance. Then, we reject the null hypothesis H_0 at α level of significance. Otherwise we do not reject H_0.

14.3.2 Numerical

The following data are collected during a test to determine consumer preference among five leading brands of Bath Soaps (Table 14.2):

Sol:

Here, we want to test that the preference of customers over five brands is uniform. So our claim is "the preference of customers over five brands is uniform" and its complement is "the preference of customers over five brands is not uniform."

So we can take claim as the null hypothesis and complement as the alternative hypothesis. Thus,

H_0 : The preference of customers over the five brands of bath soap is uniform

H_1 : The preference of customers over the five brands of bath soap is not uniform

In other words, we can say that

H_0 : The probability distribution is uniform

H_1 : The probability distribution is not uniform

Since the data are given in the categorical form and we are interested to fit a distribution, so we can go for χ^2 goodness-of-fit test.

If $X \rightarrow$ denotes preference of customers over the five brands of bath soap that follows uniform distribution, then the probability mass function of uniform distribution is given by

$$P[X = x] = \frac{1}{N}; x = 0,1,2,\ldots,N$$

The uniform distribution has a parameter N, which is given, so for testing the null hypothesis, the test statistic is given by

$$\chi^2 = \sum_{i=1}^{k} \frac{(O_i - E_i)^2}{E_i} \sim \chi^2_{(k-1)} \text{ Under } H_0$$

Where,

O_i and $E_i \rightarrow$ are the observed and expected frequencies of ith brand of Bath Soap, respectively.

TABLE 14.2

Consumer Preference among Leading Brands of Bath Soaps

Brand Preferred	A	B	C	D	E	**Total**
Number of Customers	195	201	205	197	202	**1000**

Here, we want to test the null hypothesis that the preference of the customers is uniform, i.e., follows uniform distribution.

Uniform distribution is one in which all outcomes have equal or uniform probability. Therefore, the probability that the customers prefer one of any brand is same. Thus,

$$p_1 = p_2 = p_3 = p_4 = p_5 = p = \frac{1}{5}$$

The theoretical or expected number of customers or frequency for each brand is obtained by multiplying the appropriate probability by total number of customers, that is, sample size *N*. Therefore (Table 14.3),

$$E_1 = E_2 = E_3 = E_4 = E_5 = Np = 1{,}000 \times \frac{1}{5} = 200$$

Calculation for $\frac{(O_i - E_i)^2}{E_i}$

From the above calculations, we have

$$\chi^2 = \sum_{i=1}^{k} \frac{(O_i - E_i)^2}{E_i} = 0.32$$

Note: Degrees of freedom (df) refers to the number of values that are free to vary after restriction has been placed on the data. For instance, if you have four numbers with the restriction that their sum has to be 50, then three of these numbers can be anything, they are free to vary, but the fourth number definitely is restricted. For example, the first three numbers could be 15, 20, and 5, adding up to 40; then the fourth number has to be 10 in order that they sum to 50. The degrees of freedom for these values are then three.

The critical value of chi-square with $k - 1 = 5 - 1 = 4$ degrees of freedom at 5%

Level of significance is 9.49.

Since the calculated value of test statistic (=0.32) is smaller than critical value (=9.49)

$$\because 0.32 < 9.49$$

$$\therefore \chi^2_{\text{Calculated}} < \chi^2_{\text{Tabulated, at } \alpha=5\%}$$

TABLE 14.3

Calculation for χ^2

Soap Brand	Observed Frequency (O_i)	Expected Frequency (E_i)	$(O_i - E_i)$	$(O_i - E_i)^2$	$\frac{(O_i - E_i)^2}{E_i}$
A	195	200	–5	25	0.125
B	201	200	1	1	0.005
C	205	200	5	25	0.125
D	197	200	–3	9	0.045
E	202	200	2	4	0.02
Total	1000	1000			$\sum_{i=1}^{5} \frac{(O_i - E_i)^2}{E_i} = 0.32$

So, we do not reject the null hypothesis, i.e., we support the claim at 5% level of significance.

Thus, we conclude that the sample fails to provide us sufficient evidence against the claim, so we may assume that the preference of customers over the five brands of bath soap is uniform.

14.4 χ^2 Test—Test of Independence

The χ^2 (pronounced as Ki-square) test is used with discrete data in the form of frequencies. There are many situations where we need to test the independency of two characteristics or attributes of categorical data.

14.4.1 Assumptions

This test works under the following assumptions:

1. The sample observations are random and independent.
2. The observations may be classified into nonoverlapping categories.
3. The observed and expected frequencies of each class are greater than five.
4. Sum of observed frequencies is equal to sum of expected frequencies, i.e., $\sum O_i = \sum E_i$.
5. Each observation in the sample may be classified according to two characteristics, so that each observation belongs to one and only one level of each characteristic.

The χ^2 test for independence of attributes can be used in the situation in which the data are classified according to two attributes or characteristics. It is a test of independence and is used to estimate the likelihood that some factor other than chance accounts for the observed relationship.

Since the null hypothesis states that there is no relationship between the variables under study, the χ^2 test merely evaluates the probability that the observed relationship results from chance. The formula for χ^2 is

$$\chi^2 = \sum\left[\frac{(f_0 - f_e)^2}{f_e}\right] \quad (14.1)$$

Where,

$f_0 \rightarrow$ Frequency of the occurrence of observed or experimentally determined facts, i.e., observed frequency of cell *ij*.

$f_e \rightarrow$ Expected frequency of occurrence i.e. the expected frequency of cell *ij*.

To test the significance of χ^2, we enter from the table of the Appendix with the computed value of χ^2 for the appropriate number of degrees of freedom.

The number of degrees of freedom $df = (r-1)(c-1)$,

Where,

$r \rightarrow$ is the number or rows and

$c \rightarrow$ is the number of columns in which the data are tabulated.

14.4.2 Numerical

Consider the following data as summarized in Table 14.4 of 500 subjects who have been categorized into three age groups—elder, middle-aged, and younger—on the basis of age and their preference of four Colours: black, white, orange and violet.

TABLE 14.4

The Chi-Square Test of Independence

Colour → **Age Group ↓**	**Black**	**White**	**Orange**	**Violet**	**Total**
Elder	40(38.42)	50(45.22)	35(39.10)	45(47.26)	170
Middle	35(36.16)	42(42.56)	44(36.80)	39(44.48)	160
Younger	38(38.42)	41(45.22)	36(39.10)	55(47.26)	170
Total	**113**	**133**	**115**	**139**	**500**

14.4.3 The χ^2 Test of Independence in Contingency Table

Across the first row of the table, we find that out of 170 subjects in the older age group, 40 have given their preference for black color, 50 for white, 35 for orange, and 45 for violet.

Reading down the first column, we find that out of 113 subjects giving preference for black color, 40 belong to the older age-group, 35 to middle, and 38 to younger age group.

The other columns and rows are interpreted in the same way. The hypothesis to be tested is the null hypothesis, that is, age and color preferences are essentially unrelated or independent.

To compute χ^2 we must calculate an independent value, i.e., expected frequency for each cell in the contingency table.

Independent values are represented by the figures in parentheses within the different cells. They give the number of subjects whom we should expect to fall in a particular age group, showing their preference for a particular color in the absence of any real association.

The calculations of expected frequencies (f_e) and χ^2 are shown as under:

14.4.3.1 Calculation of Expected Frequencies (f_e)

Calculation of expected values enables the researcher to decide whether null hypothesis is true or false.

Expected frequencies are the values that are expected when the null hypothesis is true. In Table 14.4, we can see that out of the total sample of 500 respondents, the figures for preference of four colors—black, white, orange, and violet—are 113, 133, 115, and 139.

In other words, the figures for preference of four colors—black, white, orange, and violet—are in proportion of 113:133:115:139.

If there is no influence of age group on the preference of four colors, i.e., if the null hypothesis is true, then for each age group the preference figures should be in a similar proportion.

From Table 14.4, we find the number of respondents in the elder age group is 170.

If the age group variable does not have any relationship with the preference figure, then the proportion of preference figures for all the four colors in the elder age group will be in line with the overall proportion of the sample, i.e., 113:133:115:139.

On the other hand, if the null hypothesis is not true, then the proportion of preference figures for elder age group may not be in line with the overall proportion of the sample.

The expected frequency value for each category (age group) can be calculated using the

$$E_{ij} = \frac{n_i \times n_j}{n} \tag{14.2}$$

Where

E_{ij} → Expected frequency of a cell, corresponding to a particular age group and a preference group,

n_i → Represents the row total,

n_j → Represents the column total,

n → The total sample size.

TABLE 14.5

Observed and Expected Frequencies

Color Preference → Age Group ↓	Black	White	Orange	Violet	Total
Elder	$\frac{(113\times170)}{500}=(38.42)$	$\frac{(133\times170)}{500}=(45.22)$	$\frac{(115\times170)}{500}=(39.10)$	$\frac{(139\times170)}{500}=(47.26)$	170
Middle	$\frac{(113\times160)}{500}=(36.16)$	$\frac{(133\times160)}{500}=(42.56)$	$\frac{(115\times160)}{500}=(36.80)$	$\frac{(139\times160)}{500}=(44.48)$	160
Younger	$\frac{(113\times170)}{500}=(38.42)$	$\frac{(133\times170)}{500}=(45.22)$	$\frac{(115\times170)}{500}=(39.10)$	$\frac{(139\times170)}{500}=(47.26)$	170
Total	113	133	115	139	500

Table 14.5 provides the observed frequencies and corresponding expected frequencies in brackets for all the three age groups.

14.4.3.2 Computation of the χ^2 Value, Using Formula (1)

The next step in the test of independence is the calculation of the test statistic. That is calculation of chi square, which is denoted by χ^2. It is calculated using Equation (14.1)

$$\chi^2 = \sum\left[\frac{(f_0 - f_e)^2}{f_e}\right]$$

The χ^2 test statistic is calculated as follows:

$$\chi^2 = \frac{(40-38.42)^2}{38.42} + \frac{(50-45.22)^2}{45.22} + \frac{(35-39.10)^2}{39.10} + \frac{(45-47.26)^2}{47.26} + \frac{(35-36.16)^2}{36.16} + \frac{(42-42.56)^2}{42.56}$$

$$+\frac{(44-36.80)^2}{36.80} + \frac{(39-44.48)^2}{44.48} + \frac{(38-38.42)^2}{38.42} + \frac{(41-45.22)^2}{45.22} + \frac{(36-39.10)^2}{39.10} + \frac{(55-47.26)^2}{47.26}$$

$$\chi^2 = 5.856$$

14.4.3.3 Decide on the Degrees of Freedom

The degrees of freedom is calculated using the following formula:

$$\text{Degrees of freedom}(v) = (r-1)(c-1) \qquad (14.3)$$

where $v \rightarrow$ Degrees of Freedom, $r \rightarrow$ Number of Rows, $c \rightarrow$ Number of Columns

$$\text{Degrees of Freedom}(v) = (r-1)(c-1) = (3-1)(4-1) = 8$$

The χ^2 critical values for 8 d.f. as given in Table (See Appendix) are 15.507 and 20.090, respectively, for 0.05 and 0.01 levels of significance and, the obtained value, 5.856, is less than the table value even at 0.05 level of significance.

$$\because 5.856 < 15.507$$

$$\therefore \chi^2_{\text{Calculated}} < \chi^2_{\text{Tabulated at 5\% level of significance, with 8 d.f.}}$$

This indicates that there is no relationship between the age and the color preference and thus the hypothesis that age and color preference are essentially independent may be accepted at 0.05 level of significance.

Note:

In the case of 2×2 contingency table, with $(r-1)(c-1)=1$ d.f., there is no need of computing the expected frequencies (independence values) for each cell. The following formula is used.

$$\chi^2 = \frac{N\left(|AD-BC|\right)^2}{(A+B)(C+D)(A+C)(B+D)} \tag{14.4}$$

In the above formula, A, B, C, and D are the frequencies in the first, second, third, and fourth cells, respectively, and the vertical lines in $|AD-BC|$ mean that the difference is to be taken as positive.

14.4.4 Numerical

To illustrate the use of Formula (14.4), let us determine whether Item 4 of an achievement test differentiates between male and female achievers. The responses to items are given in the 2 × 2 contingency Table 14.6 given below:

Sol:

Using Formula (14.4)

$$\chi^2 = \frac{N\left(|AD-BC|\right)^2}{(A+B)(C+D)(A+C)(B+D)}$$

$$\chi^2 = \frac{280\left(|120\times 95-30\times 35|\right)^2}{(120+30)(35+95)(120+35)(30+95)}$$

$$\chi^2 = \frac{280\left(|11{,}400-1{,}050|\right)^2}{(150)(130)(155)(125)} = \frac{29{,}994{,}300{,}000}{3{,}77{,}812{,}500} = 79.3893786526$$

Since the computed χ^2 value of 79.3893786526 exceeds the critical χ^2 value of 6.635 to be significant at 0.01 level, i.e.,

$$\because 79.3893786526 > 6.635$$

$$\therefore \chi^2_{\text{Calculated}} > \chi^2_{\text{Tabulated at 1\% level of significance}}$$

Hence, we reject the hypothesis that Item 4 of the test does not discriminate significantly between male and female achievers.

TABLE 14.6

The Chi-Square Test in 2×2 Fold Contingency Table

	Passed Item 4	Failed Item 4	Total
Male Achiever	$(A)120$	$(B)30$	$(A+B)150$
Female Achiever	$(C)35$	$(D)95$	$(C+D)130$
Total	$(A+C)155$	$(B+D)125$	280

In other words, it may be concluded that Item 4 of the achievement test discriminates significantly between the two groups, namely male and female achieving students.

Note:

Further, when entries in 2×2 table are less than ten, Yate's correction for continuity is applied to Formula (14.4). The corrected formula reads:

$$\chi_C^2 = \frac{N\left(|AD - BC| - \frac{N}{2}\right)^2}{(A+B)(C+D)(A+C)(B+D)} \quad (14.5)$$

The following example illustrates the use of Formula (14.5).

14.4.5 Numerical

Twelve mathematics and Fifteen statistics MSc final year students were asked to express their attitude toward computer education. Both the groups of subjects were administered the attitude scale and were classified as having either positive or negative attitude toward computer education. Test whether there is any significant difference in the attitude of male and female counselors towards population education. The distribution of the sample is summarized in Table 14.7.

Sol: Given :

$N = 27, A = 6, B = 6, C = 10, D = 5$

$$\chi_C^2 = \frac{N\left(|AD - BC| - \frac{N}{2}\right)^2}{(A+B)(C+D)(A+C)(B+D)}$$

$$\chi_C^2 = \frac{27\left(|6 \times 5 - 6 \times 10| - \frac{27}{2}\right)^2}{(6+6)(10+5)(6+10)(6+5)}$$

$$\chi_C^2 = \frac{27(|30-60| - 13.5)^2}{(12)(15)(16)(11)} = \frac{7350.75}{31,680} = 0.23203125$$

Since the calculated value of χ^2, 0.23203125, is less than the table value of 3.842 to be significant at 0.05 level of significance, i.e.,

$$\because 0.23203125 < 3.842$$

TABLE 14.7

Distribution of Male and Female Subjects in Terms of Their Positive or Negative Attitude Toward Population Education

	Positive Attitude	Negative Attitude	Total
Mathematics Students	$(A)6$	(B) 6	$(A+B)$ 12
Statistics Students	$(C)10$	$(D)5$	$(C+D)15$
Total	$(A+C)16$	$(B+D)11$	27

$$\therefore \chi^2_{\text{Calculated}} < \chi^2_{\text{Tabulated at 5\% level of significance}}$$

Hence, it may be inferred that there is no true difference in the attitude of male and female counselors toward population education.

14.4.6 Strength of Association

The test of independence will only enable the researcher to identify, whether there is association between the two variables. However, this test will not describe the strength or magnitude of the association. The strength of association can be evaluated using the following key techniques: phi-coefficient and coefficient of contingency.

14.4.7 Phi-Coefficient ϕ

In order to measure the strength, the phi-coefficient denoted by ϕ is used. The phi-coefficient measures the strength of association between two variables:

$$\phi = \sqrt{\frac{\chi^2}{n}}$$

Where,

$\chi^2 \rightarrow$ calculated value of χ^2

$n \rightarrow$ sample size.

However, this measure is suitable only for 2×2 tables, i.e., with two rows and two columns. For 3×3 tables, i.e., with three rows and three columns, the phi-coefficient cannot be used to measure the strength of association between the variables.

14.4.8 Coefficient of Contingency (C)

Another measure, to test the strength of association, is the coefficient of contingency. This measure can be calculated for tables of any size. The coefficient of contingency can be calculated using the following formula

$$C = \sqrt{\frac{\chi^2}{\chi^2 + n}}$$

Where,

$\chi^2 \rightarrow$ calculated value of χ^2

$n \rightarrow$ sample size.

The coefficient varies from 0 to 1. While the value 0 indicates there is no association between the variables, 1 indicates the maximum strength.

14.5 Hypothesis Testing about a Single Mean

Hypothesis testing about a single mean is used when the researcher has to deal with problems that involved testing a single population mean against a hypothesized standard (H_0).

Examples of such research problems are as follows:

1. A researcher wants to find whether the quality of the sample lot deviates from a specified standard.
2. Whether the profits of the company will increase by 10% this year.
3. Whether the performance of a bike model will exceed the industry benchmark.
4. In such cases, hypothesis testing about a single mean is performed. Prominent statistical tests used in such situations are the Z-Test and the t-Test. Researchers also use the χ^2 test in these situations.

While conducting the hypothesis testing about a population mean, there are two cases that will arise:

1. Hypothesis testing when the population standard deviation is known, and
2. Hypothesis testing when the population standard deviation is not known.

14.5.1 When Population Standard Deviation is Known

When the population standard deviation is known, the sample size is not the criterion for the selection of the hypothesis test. Hence, the researchers use the Z-Test to test the hypothesis, for both large samples and small sample tests. The test can be two-tailed or one-tailed depending on the research problem.

14.5.1.1 Numerical (Two-Tailed Test)

SSS Private Limited is a leading LED TV manufacturer. It has decided to fix the price of a commercial LED TV at Rs. 21,100. From past experience in fixing prices, the company has determined that standard deviation is Rs. 3,300. Any decrease in the price will make the product unprofitable and any increase in price will make the price uncompetitive, i.e., industry average. To evaluate whether the price set by the company is optimal, it has undertaken a survey among select 60 customers. The survey revealed that mean of the price preferences of the sampled customers is Rs. 20,200. At 5% significance level, the company wants to test the hypothesis that Rs. 21,100 is an optimal price.

Sol: Given :

Sample size $n = 60$, Sample mean $\bar{x} =$ Rs. $20{,}200$, Standard deviation $\sigma =$ Rs. $3{,}300$, Significance level $\alpha = 0.05\%$

As we know the population standard deviation and the sample size is large (>30) we can use the Z-Test for testing the hypothesis. The steps involved in testing the hypothesis about a single mean when population standard deviation is known are as follows:

Step 1: Formulate the null hypothesis and alternate hypothesis; determine the level of significance; and decide on the types of test to be performed.

Step 2: Find critical values for the test.

Step 3: Find the Standard Error for the sample mean and standardize the sample mean.

Step 4: Compare the standardized sample statistic with the critical value.

Step 5: Deduce research conclusion.

Step 1: The hypothesis can be formulated as follows:

Null hypothesis $H_0 : \mu =$ Rs. 21,100, alternate hypothesis $H_1 : \mu \neq$ Rs. 21,100

Thus, from the above hypotheses, we can infer that the two-tailed test needs to be applied to test the hypothesis. The significance level was set at 5%.

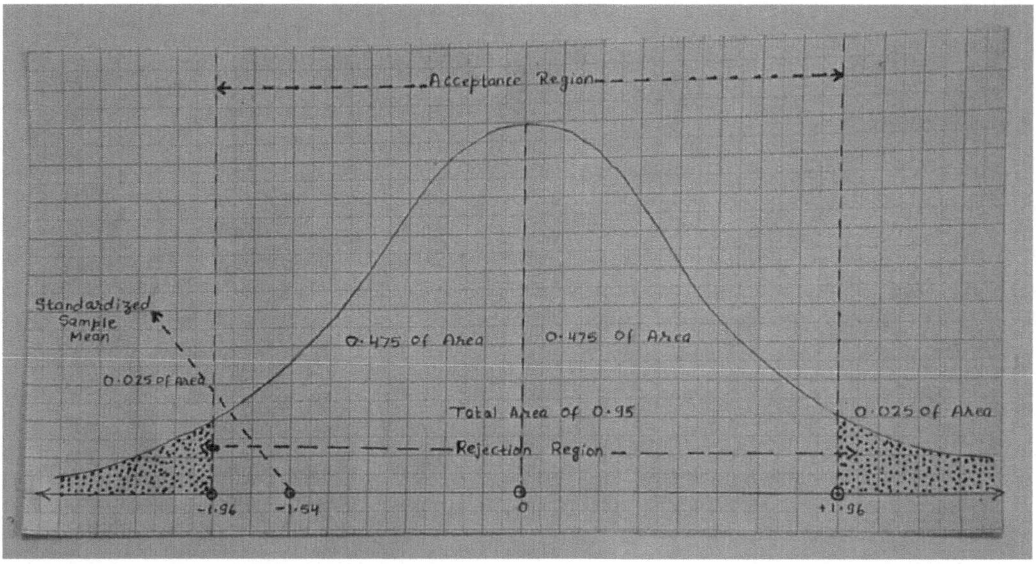

FIGURE 14.1 Two-tailed test about single mean.

Step 2: As it is a two-tailed test at 5% level of significance, acceptance region will have 95% of the area under the curve. The rejection region is divided into two regions with area 2.5% area each. The 95% region is equally split into equal areas of 0.0475. See Figure 14.1.

Thus, to determine the critical values we need to look at the corresponding z-value for 0.0475 in the table given in Appendix. The corresponding value is 1.96. Thus, the critical values are ±1.96

Step 3: The standard error of the mean is calculated as follows:

$$\sigma_{\bar{x}} = \frac{\sigma}{\sqrt{n}} \tag{14.6}$$

Where, $\sigma_{\bar{x}} \rightarrow$ Standard error of mean, $\rightarrow$ Standard deviation of the samples, i.e., Rs. 3,300, $n \rightarrow$ Sample size, i.e., 60.

Substituting the above values in Equation (14.6), we get

$$\sigma_{\bar{x}} = \frac{\sigma}{\sqrt{n}} = \frac{3,300}{\sqrt{60}} = 426.02$$

Using the standard error obtained we now standardize the sample mean. The sample mean is standardized using the following formula:

$$z = \frac{\bar{x} - \mu_0}{\sigma_{\bar{x}}} \tag{14.7}$$

Where, $\bar{x} \rightarrow$ Sample mean, $\mu_0 \rightarrow$ Hypothesized population parameter, $\sigma_{\bar{x}} \rightarrow$ Standard error of the mean.

Here,

$$z = \frac{\bar{x} - \mu_0}{\sigma_{\bar{x}}} = \frac{20,200 - 21,100}{426.02} = -2.11$$

Step 4: The standardized mean or observed z-value is compared with the critical values to see whether it will fall in acceptance region. In this case, the value −1.54 falls in the accepted region, $-1.96 < -2.11 > 1.96$

Step 5: As the value falls in the acceptance region, the company fails to reject the null hypothesis that the price set for the product is optimal, and it can devise marketing strategy for the product accordingly.

14.5.1.2 Numerical (One-Tailed Test)

Commworks Inc.; a BPO company is planning to purchase LED TV for its newly constructed office complex. SSS has advertised that its unit price Rs. 21,100 for the industrial LED TV is considerably lower than the industry average. However, Commworks is skeptical about the claim. So, it has undertaken a survey with a sample of 60 customers. The survey has revealed that mean price is Rs. 20,300. From the past experiences, Commworks has determined that the standard deviation is Rs. 3,300. Based on the survey at 1% level of significance can Commworks conclude that the average price of the product is less than the hypothesized value of Rs. 21,100?

Sol: Given :

Sample size $n = 60$, Population standard deviation $\sigma = 3{,}300$, Significance level=0.01%.

First, we need to formulate the hypothesis.

Null hypothesis $H_0 : \mu \geq 21{,}100$, Alternate hypothesis $H_1 : \mu < 21{,}100$

From the hypotheses, we can understand that the test is a one-tailed test, i.e., right-tailed test. The significance level is fixed at 1%.

The next step is to determine the critical values. As it is a one-tailed test, the accepted region consists of 99% of the area under the distribution curve and the rejection region will have a 1% of the area at the right tail of the curve (see Figure 14.2).

Observe that the accepted region consists of 50% on the left tail and 49% on the right tail. As it is a right-tailed test, we need to look for corresponding *Z*-value for 49% area under the curve. See table value given in Appendix, which is 2.33. Thus, 2.33 is the critical value. The next step is to calculate the standard error of the sample mean. The standard error of the sample mean can be calculated using the Equation (14.6).

$$\sigma_{\bar{x}} = \frac{\sigma}{\sqrt{n}} = \frac{3{,}300}{\sqrt{60}} = 426.02$$

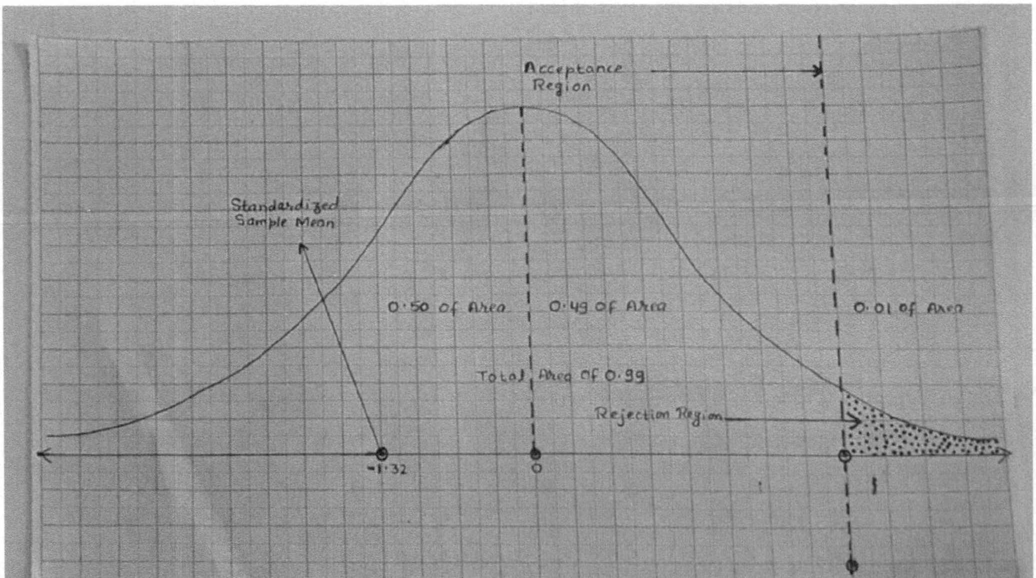

FIGURE 14.2 One-tailed test about single mean.

Using standard error we need to standardize the sample mean. This can be done using Equation (14.7):

$$z = \frac{\bar{x} - \mu_0}{\sigma_{\bar{x}}} = \frac{20,300 - 21,100}{426.02} = -1.87$$

As it is a one-tailed test, we reject the null hypothesis, when the standardized sample mean or observed Z-value falls above the critical value, and accept or do not reject the null hypothesis, when the standardized sample mean falls below the critical value. The standardized sample mean in this case is $-1.87 < 2.33$. We fail to reject the null hypothesis that the price quoted by SSS is a competitive price.

14.5.2 When Population Standard Deviation is Not Known

Another situation that researchers face in hypothesis testing related to a single mean when the population standard deviation is not known and the sample standard deviation is known.

When the population standard deviation is not known, the size of the sample is considered while selecting the statistical test to be used. Thus, for research problems involving large sample (>30), and for a known population standard deviation, the Z-Test is used.

If the sample size is <30 and the population standard deviation is not known and we need to test the hypothesis based on the sample standard deviation, we should use the t-distribution test. The procedure followed in the t-test is similar to Z-test.

Let us discuss the various steps involved in hypothesis testing about single mean when the standard deviation is not known.

14.5.2.1 Numerical

SSS Company assume that it has conducted the market survey on sample of 28 customers and the mean sample price is determined as Rs. 20,600. The population standard deviation is unknown and the sample standard deviation was identified as Rs. 3,200. At 1% level of significance, can SSS Company conclude that the average price is less than the hypothesized Rs. 21,100?

Sol: Given:

Sample size $n = 28$, Sample mean $\bar{x} =$ Rs. 20,600, Sample standard deviation $s_x =$ Rs. 3,200, Level of Significance $\alpha = 0.01$.

First, we need to formulate the hypotheses. As SSS Company wants to evaluate that quotation with unit price at Rs. 21,100 is competitively priced, the hypotheses can be formulated as follows:

$$H_0 : \mu \geq \text{Rs. } 21,100, H_1 : \mu < \text{Rs. } 21,100$$

This next step is to decide on the statistical test to be used and determine the critical values. As the sample size is <30 and the population standard deviation is not known we use *t*-test to test the hypothesis.

We also observed from the hypothesis that it is a right-tailed test. The level of significance is set at 1%. Thus, the acceptance region will be 99% and the rejection region will be 1%. Though the *t*-distribution table is similar to the Z-distribution table, there is a difference in the way distribution table has been derived. While the Z-distribution table considers the confidence level as the basis, *t*-distribution table considers the "level of significance" as the basis.

Another aspect in the t-distribution table is the "degrees of freedom." The number of degrees of freedom in *t*-test is measured as $v = n - 1$

where $v \rightarrow$ Degree of Freedom, $n \rightarrow$ Sample size

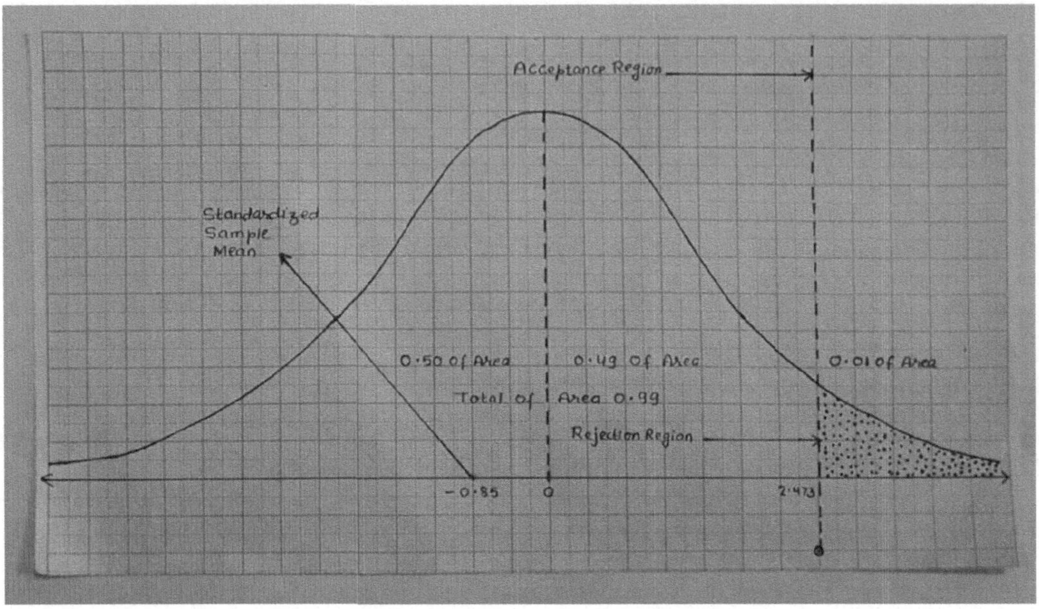

FIGURE 14.3 Hypothesis testing about single mean (small samples).

Thus, in the present case the degrees of freedom is $(28-1=27)$. The level of significance is 0.01%.

So, the critical value can be obtained by looking at the corresponding t-value in table given in Appendix, at 27 degrees of freedom under 0.01 column of one-tailed test. The critical value is 2.473 (see Figure 14.3).

The next step is to calculate the standard error of the mean. As the population standard deviation is now known, we use the sample standard deviation as an estimate to the population standard deviation. Thus, $\sigma = S_x$.

As we are calculating the standard error of the mean using the sample standard deviation, the standard error of the mean will also be an estimate. Thus, the standard error of the mean is calculated as follows:

$$s_{\overline{X}} = \frac{s_x}{\sqrt{n}} \tag{14.8}$$

$$s_{\overline{X}} = \frac{3,200}{\sqrt{28}} = 604.74$$

Then, we need to calculate the standardized sample mean or observed t-value. The standardized sample mean can be obtained using the following formula:

$$t = \frac{\overline{x} - \mu_0}{s_{\overline{x}}} \tag{14.9}$$

where $\overline{x} \rightarrow$ Sample Mean,

$\mu_0 \rightarrow$ Hypothesized population parameter,

$s_{\overline{x}} \rightarrow$ Standard error of mean.

Here,

$$t = \frac{\overline{x} - \mu_0}{s_{\overline{x}}} = \frac{20,600 - 21,100}{604.74} = -0.82$$

As it is a right-tailed test, we reject the null hypothesis when the standardized sample mean or observed t-value is greater than the critical value and fail to reject the null hypothesis, when the standardized sample mean or observed t-value is lesser than the critical value.

As the observed t-value –0.82 is <2.473, we fail to reject the null hypothesis that the price quoted by SSS is competitive.

14.6 Hypothesis Testing for Differences between Means

Till now, we discussed the hypothesis testing about a single mean. However, in many cases, researchers may be required to compare two different populations.

For example, a health department may test whether there is any difference between mortality rates in two different states. A HR manager may want to test whether female are more productive than male. In these cases the researchers can use hypothesis testing about two means. A researcher faces three situations while performing the hypothesis testing for differences between means. They undertake hypothesis testing when faced with a problem situation involving large, small, or dependent samples.

14.6.1 Test for Difference between Means: Large Samples

The steps involved in tests of differences between two means, when the problem involves large samples (i.e., >30), are similar as discussed in earlier methods. They include

Step 1: Formulate the hypotheses.
Step 2: Select the appropriate statistical test.
Step 3: Calculate the sample error and standardize the sample statistic.
Step 4: Determine the critical value.
Step 5: Compare the standardized sample statistic with the critical value.
Step 6: Deduce research conclusion.

We already know that when samples are large and data are normally distributed z-test is used to test the hypothesis.

14.6.1.1 Numerical (Two-Tailed Test)

A research wants to find out whether the Post-Graduations Final Year scores of students belonging to two different groups, namely, Mathematics & Statistics, are same in a particular university. Hence, he or she undertakes a survey and the results are summarized in Table 14.8. He or she wants to test the hypothesis with 2% level of significance that there is no difference between the Post-Graduations Final Year scores of students belonging to the two different groups using the following data.

Sol: Step 1 Formulate the hypothesis

As two samples are involved in this problem, the formulation of the hypothesis will be different from single sample tests. As the researcher feels that there is no difference between the Post-Graduations Final Year scores of the students belonging to two different groups, the hypothesis can be stated as:

Null hypothesis $H_0 : \mu_1 = \mu_2$, i.e., there is no difference between Post-Graduation Final Year scores.

TABLE 14.8

IQ Scores of Group A and Group B

	Group A	Group B
Sample Size	50	50
Mean PG FY Score	63	66
Sample Standard Deviation	7	8

Alternate hypothesis $H_1 : \mu_1 \neq \mu_2$, i.e., there is a difference between Post-Graduation Final Year scores.

Step 2: Select the appropriate statistical test

From the hypothesis, we can observe that researcher wants to test the hypothesis that there is no difference between two means. So, it is a two-tailed test. The sample sizes of both the samples are >30 and the data are normally distributed. Hence, we can use Z-Test for testing the hypothesis.

Step 3: Calculate the sample error and standardize the sample statistic

Then we need to calculate the sample error of two populations. As population standard deviations of the two groups are not given, we use the sample standard deviations of both groups as an estimate of population standard deviation of two groups to calculate the sample error. Thus,

$$\sigma_1 = s_1 = 7 \text{ and } \sigma_2 = s_2 = 8$$

As we are using the estimates of the population standard deviation, the sample error will also be an estimate. Thus, estimated sample error can be calculated as follows:

$$s_{\bar{X}_1-\bar{X}_2} = \sqrt{\frac{s_1^2}{n_1} + \frac{s_2^2}{n_2}} \tag{14.10}$$

Where

$s_1 \rightarrow$ Sample standard deviation of sample 1, i.e., Group A in this case

$s_2 \rightarrow$ Sample standard deviation of sample 2, i.e., Group B in this case

$n_1 \rightarrow$ Size of sample 1, i.e., Group A in this case

$n_2 \rightarrow$ Size of sample 2, i.e., Group B in this case

Substituting values of Table 14.11 in Equation (14.10)

$$s_{\bar{X}_1-\bar{X}_2} = \sqrt{\frac{s_1^2}{n_1} + \frac{s_2^2}{n_2}} = \sqrt{\frac{(8)^2}{n_1} + \frac{(7)^2}{n_2}} = \sqrt{\left(\frac{64}{50}\right) + \left(\frac{64}{50}\right)} = 1.50$$

Using the sample error, we standardize the difference between two sample means, i.e., $\bar{X}_1 - \bar{X}_2$. This can be done using following formula:

$$z = \frac{\left[\left(\bar{X}_1 - \bar{X}_2\right) - \left(\mu_1 - \mu_2\right)_0\right]}{s_{\bar{X}_1-\bar{X}_2}} \tag{14.11}$$

Where

$\bar{X}_1 - \bar{X}_2 \rightarrow$ Difference between sample means

$\left(\mu_1 - \mu_2\right)_0 \rightarrow$ Hypothesized difference between the population means

$s_{\bar{X}_1-\bar{X}_2} \rightarrow$ Estimated sample error of difference between two means

The hypothesized difference between the population means is 0 in this case.

Therefore,

$$z = \frac{\left[\left(\bar{X}_1 - \bar{X}_2\right) - \left(\mu_1 - \mu_2\right)_0\right]}{s_{\bar{X}_1-\bar{X}_2}} = \frac{\left[\left(63 - 66\right) - 0\right]}{1.50} = -2$$

Step 4: Then, we need to determine the critical values to assess whether standardized difference between two means falls in accepted or rejection region. The level of significance is 0.02 and it is a two-tailed test. Hence, the rejection region 0.02 of the area is split into two regions of 0.01 each on the right tail of the normal distribution curve and the left tail of the normal distribution curve.

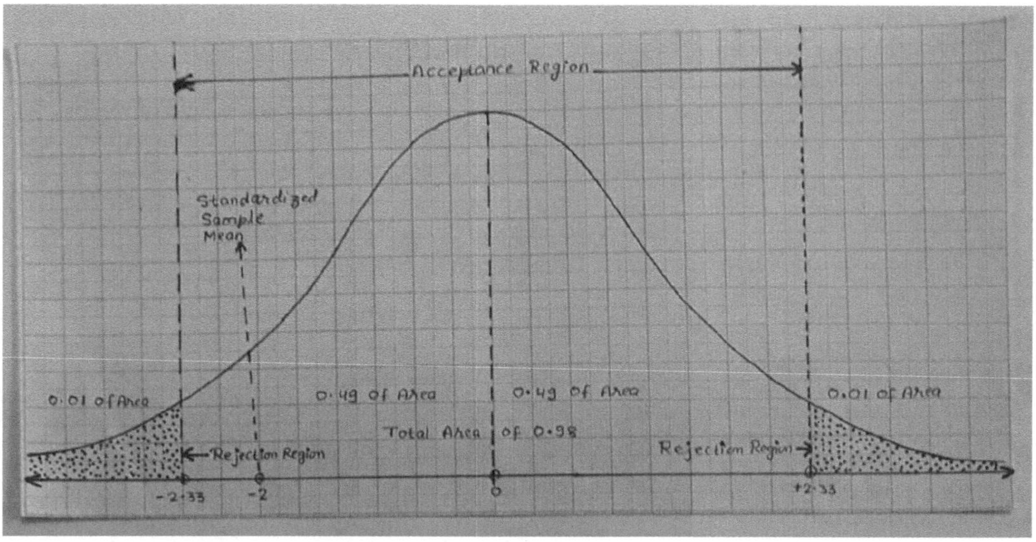

FIGURE 14.4 Tests for difference between means (large samples).

The acceptance region 0.98 is located in between, split into two equal parts, 0.49 each. The critical values can be obtained by looking at the corresponding value of Z for the 0.49 of the area. See table given in Appendix. The value is 2.33. These components are depicted in Figure 14.4.

We reject the hypothesis if the standardized difference between two means calculated Z-value fall below or above the critical value.

Step 5: Compare the standardized sample statistic with the critical value. We now compare the standardized difference between two means or calculated Z-value with the critical value.

Step 6: Deduce research conclusion: $-2.33 < -2 < 2.33$; thus, the standardized difference between two means falls within the accepted region. This implies, we accept the null hypothesis that there is no difference between the mean Post-Graduate Final Year scores between students of groups A and B.

Hypothesis testing for differences between two means for one-tailed test

For example, suppose, if the researcher wants to know whether the Quiz Score of group A students is lower than mean Quiz Score of group B, the hypothesis can be stated as follows:

Null hypothesis $H_0 : \mu_1 = \mu_2$, i.e., there is no difference between Quiz Score. Alternate hypothesis $H_1 : \mu_1 < \mu_2$, i.e., mean Quiz Score of group "A" students is less than mean Quiz Score of group "B" students. This is a one-tailed test and more specifically it is a left-tailed test.

Similarly, if the researcher wants to know whether the mean Quiz Score of group "A" is greater than mean Quiz Score of group "B."

The hypothesis can be stated as follows:

Null hypothesis $H_0 : \mu_1 = \mu_2$, i.e., there is no difference between Quiz Score.

Alternate hypothesis $H_1 : \mu_1 > \mu_2$, i.e., mean Quiz Score of group "A" students is greater than the mean Quiz Score of group B students. This is an example for right-tailed test.

TABLE 14.9

Sales Revenue Information for Region A and Region B

	Region A	Region B
Sample Size	14	17
Mean Weekly Sales (in Rs.)	6,070	5,995
Sample Standard Deviation	810	375

We need to take care while choosing between one-tailed and two-tailed test. If the researcher wants to know whether the two sample means are equal, then we need to choose the two-tailed test.

Whereas, if the researcher wants to test whether a mean of one sample is greater than the mean of another sample, then we need to select the one-tailed test and more specifically a right-tailed test.

If the researcher wants to test whether a mean of one sample is less than the mean of other sample, then we need to choose the one-tailed test, more specifically a left-tailed test.

14.6.2 Tests for Differences between Means: Small Samples

When sample sizes of two samples are <30, procedure to test the hypothesis will differ on two aspects.

1. One aspect will be that the *t*-Test is used instead of the Z-Test.
2. Another aspect is that determining standard error of the difference between means of two samples will be different from the procedure followed for large sample tests.

14.6.2.1 Numerical

A marketing manager wants to evaluate the monthly sales revenue generated by one of its newly launched product in two different regions—A and B. He studied the sales revenue pattern for a month in selected super markets in both the regions. The findings of the study are summarized in Table 14.9.

Based on this survey can the company conclude that the sales will be more in region A compared to region B?

Sol: A manager should do hypothesis testing before arriving at any conclusion regarding the above problem. The procedure is similar to the procedure we followed in large samples.

Step 1: Formulate the Hypothesis

Null hypothesis $H_0 : \mu_1 = \mu_2$, i.e., there is no difference between sales in two regions.

Alternate hypothesis $H_0 : \mu_1 > \mu_2$, i.e., average sales in region A are higher compared to average sales in region B.

Step 2: As the sample size is <30 and population standard deviation is not known, we can use the *t*-Test to test the hypothesis.

Step 3: Calculate the sample error and standardize the sample statistic

As stated earlier, the procedure for calculating standard error and standardizing the sample statistic differs from the earlier method. We can recall that, when population standard deviation is not known, we calculate the standard error using the following equation:

$$s_{\bar{X}} = \frac{s_x}{\sqrt{n}}$$

However, that formula is not appropriate for small samples tests. Thus, we use the following procedure to calculate the standard error.

We need to assume that the unknown population variances are equal, i.e., $\sigma_1^2 = \sigma_2^2$

Instead of using s_1^2 and s_2^2 in the calculation of standard error, we use weighted average of both these values; here the weights represent the degrees of freedom of each sample. The estimate so obtained is called the Pooled Estimate of σ^2.

It is given by

$$s_p^2 = \frac{\left[(n_1 - 1)s_1^2 + (n_2 - 1)s_2^2\right]}{n_1 + n_2 - 2} \tag{14.12}$$

Using the pooled estimate value, we can calculate the estimated error of the difference between two sample means, i.e., when sample size is small and population standard deviation is not known. It can be calculated using the formula given below:

$$s_{\bar{X}_1-\bar{X}_2} = s_p\sqrt{\frac{1}{n_1}+\frac{1}{n_2}} \tag{14.13}$$

After calculating the standard error, we need to standardize the difference between the sample means. This can be done using the following formula:

$$t = \frac{\left[\left(\bar{X}_1-\bar{X}_2\right)-\left(\mu_1-\mu_2\right)_0\right]}{s_{\bar{X}_1-\bar{X}_2}} \tag{14.14}$$

Substituting the value provided in Table 14.9 in Equation (14.12), we get

$$s_p^2 = \frac{\left[(n_1-1)s_1^2+(n_2-1)s_2^2\right]}{n_1+n_2-2}$$

$$s_p^2 = \frac{\left[(14-1)(810)^2+(17-1)(375)^2\right]}{14+17-2} = \frac{[8,529,300+2,250,000]}{14+17-2}$$

$$s_p^2 = 3,71,700$$

$$\therefore s_p = \sqrt{3,71,700} = 609.67$$

We now calculate the estimated error of the difference between two sample means, i.e., when sample size is small and population standard deviation is not known, using Equation (14.13), we have

$$s_{\bar{X}_1-\bar{X}_2} = s_p\sqrt{\frac{1}{n_1}+\frac{1}{n_2}}$$

$$\text{Thus, } s_{\bar{X}_1-\bar{X}_2} = 609.67\sqrt{\frac{1}{14}+\frac{1}{17}} = 609.67\times 0.36 = 219.48$$

Using the standard error, we standardize the difference between the sample means by substituting the above calculated values in Equation (14.14), and we get

$$t = \frac{\left[\left(\bar{X}_1-\bar{X}_2\right)-\left(\mu_1-\mu_2\right)_0\right]}{s_{\bar{X}_1-\bar{X}_2}} = \frac{[(6,070-5,995)-0]}{219.48} = 0.3417$$

Step 4: Determine the critical values

We may recall that in the single mean test involving a single sample, we used the formula, $\upsilon = n-1$, for determining the degrees of freedom.

However, in this case two samples are involved, thus degrees of freedom can be determined using the following formula: $\upsilon = n_1+n_2-2$.

Substituting the values of n_1 and n_2 in the above equation, we get

$$\upsilon = 14+17-2 = 29$$

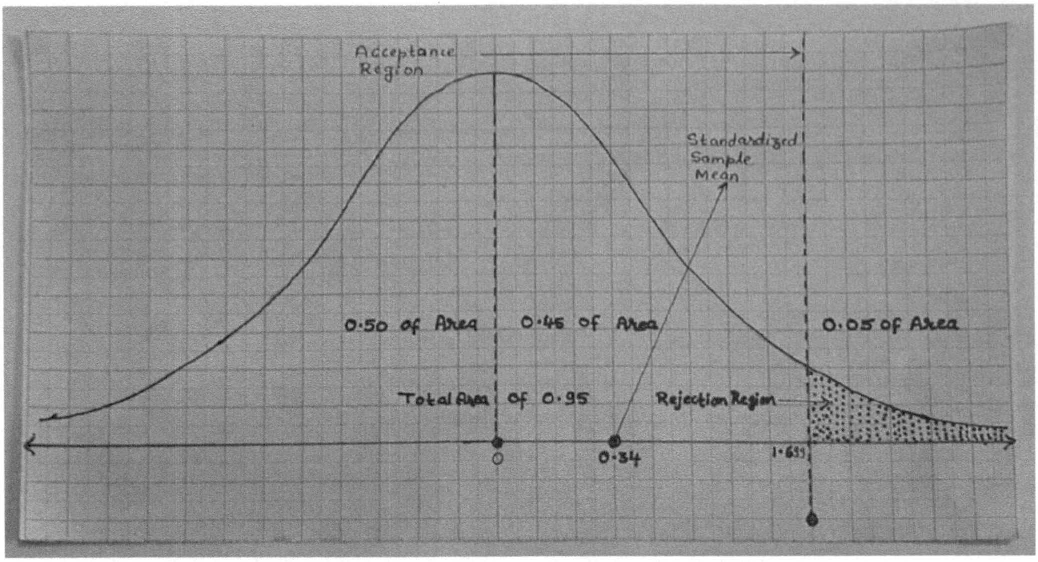

FIGURE 14.5 Test for difference between means (small samples).

The level of significance is fixed at 5%. The test is right-tailed. Hence, the accepted region consists of 0.95 of the area under the distribution curve and the rejection region is 0.05 at the right tail of the distribution curve. The value of degrees of freedom is 29. So, the critical value can be obtained by looking at the corresponding *t*-value in table given in Appendix, at 29 degrees of freedom under 0.05 column of one-tailed test. The value is 1.699.

Step 5: Compare standardized sample statistic with critical value: the standardized difference of the two means value 0.3417 is less than the critical value 1.699.

Step 6: We reject the null hypothesis if the standardized difference of two means value is greater than the critical value.

As the standardized difference of two means value is less than the critical value $(0.3417 < 1.699)$, the marketing manager fails to reject the null hypothesis that there is no difference between the sales of the product between the two regions.

Figure 14.5 gives the graphical representation of the situation.

14.6.3 Tests for Differences between Means and Paired Samples

Till now we discussed hypothesis testing for difference between two means that involve independent samples. However, in some cases, the samples involved are dependent samples. That is, the data of two different samples relate to the same population. These samples are also called paired samples. In paired samples, every data point in sample 1 is matched or related with every data point in sample 2.

For example, a pharmaceutical company may want to test the effect of a diabetes drug by measuring the sugar levels of the same sample of patients before and after the drug is administered.

TABLE 14.10

Sales Figures of Ten Customers before and after the Sales Promotion Program

Before	250	120	146	188	152	210	178	120	120	128
After	154	114	182	186	134	166	190	114	88	112

TABLE 14.11

Sales Figures of Ten Customers before and after the Sales Promotion Program

Before	250	120	146	188	152	210	178	120	120	128	
After	154	114	182	186	134	166	190	114	88	112	
D_i	96	6	−36	2	18	44	−12	6	32	16	$\sum D_i = 172$
D_i^2	9216	36	1296	4	324	1936	144	36	1024	256	$\sum D_i^2 = 14,272$

14.6.3.1 Paired Samples t-Test

It is a widely used test in such situations. Though hypothesis testing for such samples is similar to the hypothesis testing between means with independent samples, the formula for calculating the standard error and standardization of the sample statistic is different.

Let us understand how the paired sample *t*-Test can be used to test the difference between two means with dependent samples with an example.

14.6.3.2 Numerical

An advertising agency claimed that through their unique sales promotion program, sales of a company can be increased by 20% within 1 month of the launch of sales promotion program. However, the marketing manager of the company planning to use the services of the agency is not convinced of the claim. Therefore, he conducted a survey with a sample of 10 consisting of previous clients of the advertising agency. The survey provided the data of those 10 customers in terms of before and after sales figures as summarized in Table 14.10.

The marketing manager wants to test the advertising claim that there will be increase in sales by 20% at a 5% level of significance.

Sol: The samples are dependent and hence the usual *t*-Test about the differences between two means cannot be used, as it works on the assumption that the samples are independent. The procedure in this case is as follows:

1. To find the deviation between the each observation of paired sample denoted by D and
2. Use that differential variable for hypothesis testing, thus reducing the two-sample test to a one-sample test.

Assume that the difference between each observation of paired sample is denoted by D and $\bar{D}$ represents the average difference as summarized in Table 14.11.

The first step in the process is to state the hypothesis.

As the company wants to test the advertiser's claim that there will be increase in sales by 20% within a month of the launch of the sales promotion program, the null hypothesis would be that the increase in sales would be ≥20% and the alternate hypothesis would be that the increase in sales would be <20%.

In other words, the null hypothesis states that the differences in sales before and after implementing the sales promotion program would be ≥20% and the alternate hypothesis states that the difference would be <20%.

Null hypothesis $H_0 : \bar{D} \geq 20$, Alternate hypothesis $H_1 : \bar{D} < 20$

Then, we decide upon the statistical test to be used, the level of significance, and the degrees of freedom.

We can use a paired *t*-Test as the samples are not independent, and are of small size. The test statistic can be calculated using the following formula:

$$t = \frac{\bar{D} - d}{\frac{S_D}{\sqrt{n}}} \quad (14.15)$$

where $\bar{D} \rightarrow$ Mean Difference of the sample, $d \rightarrow$ Hypothesized valued difference, $S_D \rightarrow$ Standard deviation of the difference, $n \rightarrow$ Sample size, the level of significance is set at 5%.

The sample size is 10, hence the degrees of freedom is given by $\upsilon = n - 1$, and in this case, it is

$$\upsilon = n - 1 = 10 - 1 = 9$$

Then, we need to find out the mean difference of the sample and the standard deviation of the difference of sample. The mean difference of the sample can be found using the following formula:

$$\bar{D} = \frac{1}{n}\sum_{i=1}^{n} D_i \quad (14.16)$$

Where, $D_i \rightarrow$ Difference of each observation of the paired sample, $n \rightarrow$ Sample size.

Thus, substituting value from table in Equation (14.16), we get

$$\bar{D} = \frac{1}{n}\sum_{i=1}^{n} D_i = \frac{1}{10} \times 172 = 17.2$$

Then, we need to calculate the standard deviation of the difference denoted by S_D using the following formula:

$$S_D^2 = \frac{1}{n-1}\left(\sum_{i=1}^{n} D_i^2 - n\bar{D}^2\right) \quad (14.17)$$

Where $D_i \rightarrow$ Difference of each observation of the paired sample, $n \rightarrow$ Sample size, $\bar{D} \rightarrow$ Mean difference.

In this case, $n = 10$

$$\sum_{i=1}^{10} D_i^2 = 14,272 \ldots (\text{From Table 14.14})$$

By substituting the above values in Equation (14.17), we get

$$S_D^2 = \frac{1}{n-1}\left(\sum_{i=1}^{n} D_i^2 - n\bar{D}^2\right) = \frac{1}{10-1}\left[14,272 - 10(17.2)\right]$$

$$S_D^2 = \frac{1}{9}(14,272 - 172) = \frac{1}{9}(14,100) = 1566.66$$

To obtain S_D, we need to find the square root of the obtained value

$$\sqrt{S_D^2} = \sqrt{1566.66}$$

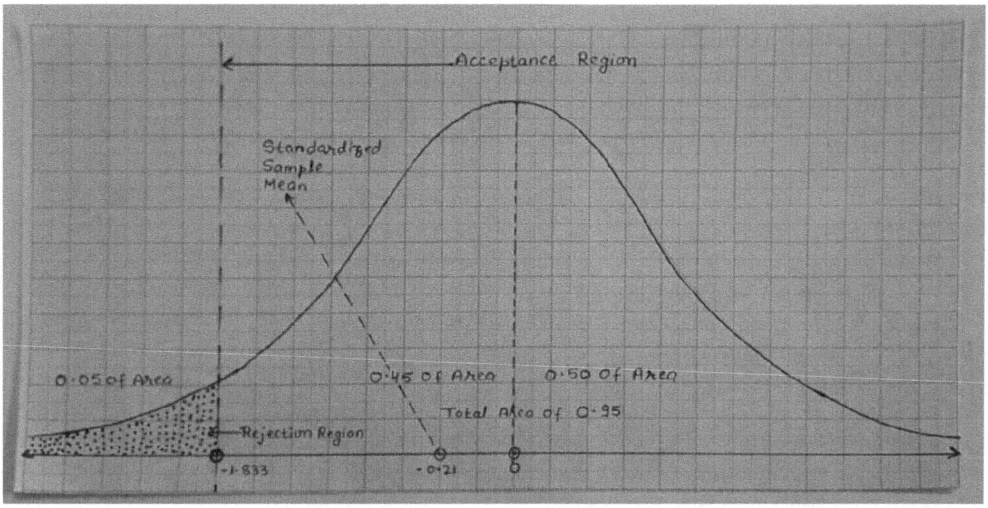

FIGURE 14.6 Testing differences between means and paired samples.

$$S_D = 39.58$$

The next step is to calculate the sample statistic using the Equation (14.15), we have

$$t = \frac{\bar{D} - d}{\frac{S_D}{\sqrt{n}}} = \frac{17.2 - 20}{\frac{39.58}{\sqrt{10}}} = \frac{-2.8}{\frac{39.58}{3.1622776602}} = \frac{-2.8}{12.5162949788} = -0.22$$

Thus, we find the critical values. As the test is one-tailed, the critical values can be determined by looking at the t-Value for 9 degrees of freedom under 0.05 column in the t-distribution table given in Appendix. The value is −1.833.

Being a one-tailed test and more specifically left-tailed test, we reject the null hypothesis, if the calculated t-value is lesser than the critical value.

However, we fail to reject the null hypothesis if the calculated t-value is greater than the critical value.

In this case, the calculated t-value is −0.22, which is greater than the observed tabular value −1.833.

Hence, we reject the null hypothesis that the increase in sales is greater than or equal to 20%. Figure 14.6 gives the graphical representation of the situation.

14.7 Analysis of Variance

Till now, we dealt with the research problems, where two means are involved. However, if the problem requires the comparison of the means of more than two populations, using Z-Tests and t-Tests becomes complex and tedious.

Moreover, when faced with more than two means the researcher should conduct these tests for every possible pair of means.

To avoid this tedious process, one can use Analysis of Variance (ANOVA), which is one of the commonly used statistical techniques to test the differences between two or more means.

As its name suggests, the ANOVA focuses on variability. It involves the calculation of several measures of variability, all of which comes down to one or another version of the measure of variability such as the Sum of Squared Deviations or Mean Sum of Squared Deviations.

The statistical technique known as "Analysis of Variance," commonly referred to by the acronym ANOVA, was developed by Professor R. A. Fisher in the 1920s.

Variation is inherent in nature, so ANOVA means examining the variation present in data or parts of data.

In other words, ANOVA means to find out the cause of variation in the data. The total variation in any set of numerical data of an experiment is due to causes such as Assignable causes and Unassignable or Chance causes.

The variation in the data due to assignable causes can be detected, measured, and controlled, whereas the variation due to chance causes is not in the control of human being and cannot be traced or find out separately.

The reason this analysis is called ANOVA rather than multigroup mean analysis or something like that is because it compares group means by analyzing comparisons of variance estimates.

ANOVA facilitates the analysis and interpretation of data from field trials and laboratory experiments in agriculture and biological research.

Today, it constitutes one of the principal research tools of biological scientists, and its use is spreading rapidly in social sciences, physical sciences, engineering, management, etc.

We compared means from two independent groups by using *t*-Test. But, if we are interested to test more than two independent groups, then *t*-Test cannot be applied.

First, we have to apply ANOVA technique.

An F-test is used, to test the means of several groups. This F-test was named "F" in honor of Professor R. A. Fisher by G. W. Snedecor.

ANOVA is helpful because it possesses an advantage over a two-sample *t*-test. The multiple two-sample *t*-test would result in an increase of chance of committing a type I error. The test of significance based on *t*-distribution is an adequate procedure only for testing the significance of the difference between two population means. In a situation, when we have more than two population to consider at a time and want to test the means of these population are same.

For example, six doses of a drug are applied to five patients each and responses or values of dependent variable or observations of these thirty patients are obtained. Now, we may be interested in finding out whether the effect of these six doses of drug on the patients differs significantly.

The answer to this problem is provided by the technique of ANOVA. Thus, ANOVA technique is used to test the homogeneity of several population means, and it is a powerful statistical tool for tests of significance in comparing more than two means.

According to Professor R. A. Fisher, ANOVA is "Separation of variance ascribable to one group of causes from the variance ascribable to other group." So, by this technique, the total variation present in the data is divided into two components of variation: one is due to assignable causes (between the group variability) and other is variation due to chance causes (within group variability).

ANOVA is one of the powerful techniques of statistical analysis and is used for testing of equality of means of several populations. It tests the variability of the means of several populations. Multivariate analysis of variance (MANOVA) is used when there is more than one response variable. The F-test in ANOVA has been used with normality assumption.

The key objective of the ANOVA are as follows:

1. To test whether there is any significant difference between the means of various samples.
2. ANOVA test uses the variability between the sample means as the basis for analysis.
3. It measures the variability in data points within the samples and
4. It also measures the variance between the sample means. These variations are compared using the F-test.

If the value of the F-test or the variability between the sample means and the variability within the samples is large, then we can deduce that there is significant difference between the means of the samples.

The process followed in the one-factor ANOVA is as follows:

Step1: Formulate the hypothesis.

TABLE 14.12

Daily Output of Three Manufacturing Systems

	Daily Output		
Day	**System 1**	**System 2**	**System 3**
1	25	20	20
2	30	20	15
3	10	10	10
4	35	30	15

Step 2: Obtain the mean of each sample.

Step 3: Find the mean of all the means, i.e., grand mean.

Step 4: Calculate the variation between samples denoted by $SS_{between}$.

Step 5: Obtain the mean square of the variation between samples denoted by $MS_{between}$ using $SS_{between}$.

Step 6: Calculate the variation within the sample denoted by SS_{within}.

Step 7: Calculate the mean square of variation within samples denoted by MS_{within} using SS_{within}.

Step 8: Calculate the total variance denoted by SS_y.

Determine F-ratio using the formula given below.

$$\text{F-ratio} = \frac{MS_{between}}{MS_{within}}$$

Step 9: The F-ratio is compared with the corresponding value in the F-distribution table. If the F-ratio is less than the table value, then we accept the null hypothesis that there is no difference between population means. However, if the F-ratio is equal to or above the Table Value, we reject the null hypothesis and accept the alternate hypothesis that there is difference between population means.

14.7.1 Explanation of Analysis of Variance through an Example

14.7.1.1 Numerical

A computer components manufacturer has developed three different manufacturing systems for producing a particular component. Table 14.12 summarizes the daily output of each of the systems for 4 days. The data were gathered by conducting a number of test runs of each of the systems. The company wants to test whether the three systems produce similar outputs. The level of significance is set at 5%.

Step 1: Formulate the hypotheses

First, we need to formulate the hypotheses. In the ANOVA test we try to determine whether there is any significant difference between sample means denoted $\mu_1, \mu_2, \ldots, \mu_n$. Thus, the null hypothesis is that there is no significant difference between the sample means. The alternate hypothesis would be that there is a significant difference between the sample means. As the problem contain three samples, the hypotheses can be stated symbolically as follows:

$$\text{Null hypothesis}: H_0 : \mu_1 = \mu_2 = \mu_3 \text{ Alternate hypothesis}: H_1 : \mu_1 \neq \mu_2 \neq \mu_3$$

Step 2: Obtain mean of each sample

The problem consists of three samples. We need to calculate the mean of each sample. Thus,

$$\bar{X}_1 \left(\text{System 1 in this case}\right) = \frac{(25 + 30 + 10 + 35)}{4} = 25$$

$$\bar{X}_2\left(\text{System 2 in this case}\right) = \frac{(20+20+10+30)}{4} = 20$$

$$\bar{X}_3\left(\text{System 3 in this case}\right) = \frac{(20+15+10+15)}{4} = 15$$

Step 3: Find the mean of all the means, i.e., Grand Mean
Mean of all sample means is given by the formula

$$\bar{\bar{X}} = \frac{\left(\bar{X}_1 + \bar{X}_1 + \bar{X}_1\right)}{3} = \frac{(25+20+15)}{3} = 20$$

Step 4: Calculate the variation within groups denoted by SS_{Within}
The next step is to calculate the variation within the groups. This can be done using the following process:

1. First, we need to calculate the deviations of each sample mean from the grand mean and square them.
2. Then, each squared deviation is multiplied by the number of items present in each sample. This can be shown in an equation as follows:

$$SS_{between} = \sum_{i=1}^{k} n_k \left(\bar{X}_k - \bar{\bar{X}}\right)^2$$

Where, $SS_{between} \rightarrow$ Variation between samples, $\bar{X}_k \rightarrow$ Mean of each sample, $\bar{\bar{X}} \rightarrow$ Mean of all sample means

$$SS_{between} = n_1\left(\bar{X}_1 - \bar{\bar{X}}\right)^2 + n_2\left(\bar{X}_2 - \bar{\bar{X}}\right)^2 + n_3\left(\bar{X}_3 - \bar{\bar{X}}\right)^2 + \cdots + n_n\left(\bar{X}_n - \bar{\bar{X}}\right)^2 \quad (14.18)$$

Substituting values in Equation (14.18), we have

$$SS_{between} = 4(25-20)^2 + 4(20-20)^2 + 4(15-20)^2$$

$$SS_{between} = 100 + 0 + 100 = 200$$

Step 5
Obtain mean square between samples denoted by $MS_{between}$ using SS_{within}
Using the $SS_{between}$ value we now determine the mean square between samples. The mean square between samples denoted by $MS_{between}$ can be calculated using the formula

$$MS_{between} = \frac{SS_{between}}{(k-1)} \quad (14.19)$$

Where,
$(k-1) \rightarrow$ Represents the degrees of freedom between the samples
There are three samples; hence, the degrees of freedom would be $3-1=2$.
Substituting values in Equation (14.19), we get

$$MS_{Between} = \frac{200}{2} = 100$$

Step 6: Calculate the variation within the samples denoted by SS_{within}

This can be done in the following manner. We need to first determine the deviation of each sample item from its corresponding sample mean and finding the square of those deviations. Later, all the squared deviations are added to obtain the SS_{within}. This can be shown in the mathematical form:

$$SS_{within} = \sum \left(X_{1i} - \bar{X}_1\right) + \sum \left(X_{2i} - \bar{X}_2\right) + \sum \left(X_{3i} - \bar{X}_3\right) + \cdots \tag{14.20}$$

$$\text{Where, } i = 1,2,3,4,5,\ldots.$$

Substituting values in Equation (14.20), we get

$$SS_{within} = [(25-25)^2 + (30-25)^2 + (10-25)^2 + (35-25)^2]$$
$$+[(20-20)^2 + (20-20)^2 + (10-20)^2 + (30-20)^2]$$
$$+[(20-15)^2 + (15-15)^2 + (10-15)^2 + (15-15)^2]$$

$$SS_{within} = [0 + 25 + 225 + 100] + [0 + 0 + 100 + 100] + [25 + 0 + 25 + 0]$$

$$SS_{within} = [350 + 200 + 50] = 600$$

Step 7: Calculate the mean square within samples denoted by MS_{within} **using** SS_{within}

The mean square within samples denoted by MS_{within} can be calculated using following formula:

$$MS_{within} = \frac{SS_{within}}{(n-k)} \tag{14.21}$$

Where,

$(n-k) \rightarrow$ Denotes degrees of freedom
$n \rightarrow$ Refers to the total number of items in all the samples
$k \rightarrow$ Refers to the number of samples As $n = 12$ and $k = 3$

$$\text{Degrees of freedom } (n-k) = 12 - 3 = 9$$

Substituting values in Equation (14.21), we get

$$MS_{within} = \frac{600}{9} = 66.66$$

Step 8: Calculate the total variance denoted by SS_y

TABLE 14.13

Analysis of Variance Table for One-Factor ANOVA Test

Sources of Variation	Sum of Square (SS)	Degrees of Freedom (d.f.)	Mean Square (MS)	F-Ratio	F-Critical Value
Between samples	200	2	100	$\frac{100}{66.6} = 1.50$	4.26 (For 2,9 d.f.)
Within samples	600	9	66.6		
Total	800	11			

Then, we need to calculate the total variance. We can calculate the total variance by determining the deviation of each sample item with the grand mean and squaring them. The sum of all the squared deviation will give the total variance. This can be shown mathematically as follows:

$$SS_y = \sum \left(X_{ij} - \overline{\overline{X}}\right)^2 \tag{14.22}$$

where $i = 1,2,3,4,\ldots$

$$SS_y = (25-20)^2 + (30-20)^2 + (10-20)^2 + (35-20)^2 + (20-20)^2 + (20-20)^2$$
$$+(10-20)^2 + (30-20)^2 + (20-20)^2 + (15-20)^2 + (10-20)^2 + (15-20)^2$$
$$= 25 + 100 + 100 + 225 + 0 + 0 + 100 + 100 + 0 + 25 + 100 + 25$$
$$= 800$$

We recheck the total variance using the equation

$$SS_y = SS_{between} + SS_{within}$$
$$SS_y = 200 + 600 = 800 \tag{14.23}$$

For our convenience, we can tabulate all the values using ANOVA as summarized Table 14.13.

Step 8: Determining F-Ratio

F-Ratio is given by

$$\text{F-Ratio} = \frac{MS_{between}}{MS_{within}}$$
$$= \frac{100}{66.6} = 1.50 \tag{14.24}$$

Step 9: The F-Ratio is compared with the corresponding value in the F-Distribution Table

The calculated F-ratio should be compared with the corresponding value in the F-Distribution table or the critical value. The critical value of F can be determined by looking at the value that meets the degrees of freedom of the numerator $(MS_{between})$ and the degrees of freedom of denominator (MS_{within}) under the relevant level of significance F-distribution table. Thus, in this case, we need to look at the value that meets the 2, 9 degrees of freedom under the 5% level of significance table (See table given in Appendix). The value is 4.26.

We accept the null hypothesis, when the calculated F value is less than the critical F value. However, we reject the null hypothesis if the calculated F value is equal to or greater than the critical F value. In the present case, the calculated F value is less than the critical F value $(1.50 < 4.26)$. Thus, we accept the null hypothesis that there is no difference between the population means.

Summary

In this chapter, we discussed the hypothesis tests of differences in detail. We first discussed about the χ^2 test and its application in the hypothesis testing. The χ^2 test is usually used by researchers in two ways: as a test of independence and as a test of goodness of fit.

We have discussed that when the data are classified into different categories or groups according to one or more attributes, such types of data are known as categorical data. The χ^2 test for goodness of fit. The χ^2 test for independence of two attributes.

Later, we discussed hypothesis testing about a single mean. While comparing the sample with a known population mean, the researcher faces two situations: hypothesis testing when the population standard deviation is known and hypothesis testing when the population standard deviation is not known.

When the population standard deviation is known, we use the z-test to test the hypothesis.

The later part of the chapter dealt with test of differences about two means. There are three cases in the problems that involve testing differences between two means testing of differences between two means for large samples, testing of differences between two means for small samples, and testing differences between two means for dependent samples. Finally, we examined the ANOVA test that is used to test the differences between more than two means.

Review Questions

1. Describe Chi-square test.
2. Describe ANOVA.
3. Give the basic assumptions in ANOVA.
4. Give the uses of ANOVA.
5. Explain the uses of *t*-test and F-test indicating in each case what exactly is sought to be tested and under what assumptions the test will be valid.
6. What is χ^2 test of goodness of fit? What precautions are necessary in using this test?
7. Illustrate with examples the usefulness of χ^2 test as a test for independence.
8. Describe the use of the χ^2 test in testing independence of attributes in a 2×2 contingency table.
9. Explain how the χ^2 distribution can be used
    i. to test the goodness of fit, and
    ii. to test the independence of the cell frequencies in a 2×2 contingency table.
10. Explain how the χ^2 distribution can be used for judging the agreement between a hypothetical and an observed distribution.
11. Discuss briefly the use of χ^2 test as a test of goodness of fit. State the conditions to be satisfied for the applicability of the test.
12. Describe the χ^2 test of significance and state the various uses to which it can be put.
13. What is ANOVA? Explain clearly the technique of ANOVA for data with one-way classification.
14. Explain, with illustrations, the ANOVA technique.
15. Discuss the fundamental principles of ANOVA with special reference to the assumptions made therein.

15

Correlation and Regression Analysis

15.1 Introduction

Researchers often face situations where they want to understand the statistical relationship between two variables. Examples of such situations are: Is there any relationship between Television Viewing and Study Habits of children? Are Multinational Companies more profitable than Domestic Companies in a particular country? Such problems can be answered using statistical techniques like correlation and regression analysis.

These techniques are also called Measures of Association. They help researchers in understanding the statistical relationship between two variables, their nature, and the magnitude of the relationships. Correlation and regression analysis is used for measuring the relationship between two variables measured on interval or ratio scales.

15.1.1 Scatter Diagrams

Scatter diagrams provide the relationship between two variables in graphical form. The variable X is plotted on the X-axis or the horizontal axis and the variable Y is plotted on the Y-axis. The diagram summarizes the nature of relationship between two variables, i.e., whether the relationship is positive or negative. The diagram also explains the magnitude of the relationship, whether it is stronger or weaker. When we can draw a straight line passing through most of the points, either on the straight line or close to the straight line, then we can conclude that the relationship is stronger. If more points are away from the straight line, or the points are scattered, then we can conclude that the relationship is weaker.

15.2 Correlation Analysis

Correlation analysis is a statistical technique used to measure the magnitude of linear relationship between two variables. Correlation analysis cannot be used in isolation to describe the relationship between variables. It can be used along with regression analysis to determine the nature of the relationship between two variables. Thus, correlation analysis can be used as a basis for further analysis. To analyze the relationship between two variables, two prominent correlation coefficients are used—Pearson's Product Moment Correlation Coefficients and Spearman's rank correlation coefficient.

15.3 Pearson's Product Moment Correlation Coefficient Correlation

Pearson's product moment correlation coefficient measures the strength of the linear relationship between two variables. This is also known as a Simple Correlation Coefficient and is denoted by "r". The "r" value ranges from −1 through 0 to +1.

If the "r" value is −1 then it indicates that there is a perfect negative relationship between the two variables. If the "r" value is +1 then it indicates that there is a perfect positive relationship between the two

DOI: 10.1201/9781315167138-15

variables. If the "r" value is 0 then it indicates that there is no relationship between the two variables. The correlation coefficient "r" can be calculated using the means of the two variables as shown below:

$$r_{xy} = r_{yx} = \frac{\sum (X_i - \bar{x})(Y_i - \bar{y})}{\sqrt{\sum (X_i - \bar{x})^2 \sum (Y_i - \bar{y})^2}} \quad (15.1)$$

Where, $r_{xy} = r_{yx} \rightarrow$ is the correlation coefficient for the variable X and Y, $\bar{x} \rightarrow$ The mean of X, $\bar{y} \rightarrow$ The mean of Y, $X_i \rightarrow$ The values of variable X, $Y_i \rightarrow$ The values of variable Y.

There is more convenient way of calculating r. That is by using the formula given below:

$$r_{xy} = r_{yx} = \frac{n\sum XY - \sum X \sum Y}{\sqrt{\left[n\sum (X^2) - \left(\sum X\right)^2\right]\left[n\sum (Y^2) - \left(\sum Y\right)^2\right]}} \quad (15.2)$$

Alternatively, the correlation coefficient can also be calculated using the variance of both the variables as shown below:

$$r_{xy} = r_{yx} = \frac{\sigma_{xy}}{\sqrt{\sigma_x^2 \sigma_y^2}} \quad (15.3)$$

Where, $\sigma_{xy} \rightarrow$ Represents the covariance of X and Y, $\sigma_x{}^2 \rightarrow$ Represents the variance of X, $\sigma_y{}^2 \rightarrow$ Represents the variance of Y.

Let us now understand the process of calculating the correlation coefficient using an example.

Numerical: SSS limited, a Fast Moving Consumer Goods (FMCG) manufacturer, has gathered data on advertising spending and corresponding impact on sales (see Table 15.1).

Now, the company wants to know whether the two variables, i.e., ad spending and sales are related (Table 15.2).

TABLE 15.1

Ad Spending and Corresponding Sales Data of SSS Products

Advertising Spending (Crores) *X*	4	12	6	8	6	8	8	6	10	10
Sales (in '000 Units) *Y*	8	10	8	8	8	6	10	6	10	8

TABLE 15.2

Calculation of Person's Product Moment Correlation

Advertising Spending (Crores) *X*	**Sales (in '000 units) *Y***	**X^2**	**Y^2**	***XY***
4	8	16	64	32
12	10	144	100	120
6	8	36	64	48
8	8	64	64	64
6	8	36	64	48
8	6	64	36	48
8	10	64	100	80
6	6	36	36	36
10	10	100	100	100
10	8	100	64	80
$\sum X = 78$	$\sum Y = 82$	$\sum X^2 = 660$	$\sum Y^2 = 692$	$\sum XY = 656$

Sol: Calculation of Person's Product Moment Correlation

Substituting the values in Equation (15.2), we get

$$r_{xy} = r_{yx} = \frac{n\sum XY - \sum X \sum Y}{\sqrt{\left[n\sum (X^2) - \left(\sum X\right)^2\right]\left[n\sum (Y^2) - \left(\sum Y\right)^2\right]}}$$

$$r_{xy} = r_{yx} = \frac{(10)\,(656) - (78)(82)}{\sqrt{\left[(10)(660) - (78)^2\right]\left[(10 \times 692) - (82)^2\right]}}$$

$$r_{xy} = r_{yx} = \frac{6,560 - 6,396}{\sqrt{[6,600 - 6,084][6,920 - 6,724]}} = \frac{164}{\sqrt{[516][196]}}$$

$$r_{xy} = r_{yx} = \frac{164}{\sqrt{101,136}} = \frac{164}{318.0188673648} = +0.5156926737 = +0.51$$

This indicates that there is a positive correlation between ad spending and sales volume, i.e., when the ad spending is high the sales volume shows an increase.

15.4 Rank Correlation Coefficient

Researchers often face situations where they have to make decisions based on data measured on ordinal scales. In such cases, Pearson's product moment correlation coefficient is not appropriate as it is suitable only for interval-scaled variables. Instead, they have to use the rank correlation coefficient, which is also known as Spearman's Rank Correlation Coefficient. The rank correlation coefficient describes the linear relationship between two ordinal scaled variables. The rank correlation coefficient is denoted by "r_s". The value of "r_s" ranges from 0 to 1.

It can be calculated using the following formula

$$r_s = 1 - \frac{6\sum_{i=1}^{N} D^2}{N\left(N^2 - 1\right)} \tag{15.4}$$

The procedure for calculating the rank correlation coefficient for a pair of variables is discussed below.

The researcher has to rank the actual data by giving rank 1 to the highest value and rank 2 to the second highest value and so on. If two or more than two values are equal, then he or she has to take the average of the ranks that are supposed to be assigned, had they been different values, and the same rank is assigned to all values.

For example, if a situation arises where there are three values competing for rank 3, then the researcher has to take the average of 3,4,5, i.e.,4 and assign rank 4 to all three variables.

After assigning ranks, the difference between the ranks of each observation has to be determined. This difference is denoted by "D". We then need to find the squares of the differences $\left(D^2\right)$ and substitute the values in the above equation.

Let us understand this procedure using an example.

Numerical: A quality-rating agency has been following a certain quality measurement system to rank various brands of a Television and its product. However, the agency is contemplating to shift to a new quality measurement system, which is easier to use. The agency wants to assess whether the new system will provide results similar to that of the existing system. The ranking of ten popular Television models using the existing and the new system are summarized in Table 15.3.

TABLE 15.3

The Ranking of Computer Models

Television Models	A	B	C	D	E	F	G	H	I	J
Existing System X	5	8	10	2	4	9	1	7	3	6
New System Y	3	1	6	8	2	9	5	7	10	4

TABLE 15.4

Calculation of Rank Correlation Coefficient

Television Models	**Existing System *X***	**New System *Y***	***D***	***D*²**
A	5	3	2	4
B	8	1	7	49
C	10	6	4	16
D	2	8	–6	36
E	4	2	2	4
F	9	9	0	0
G	1	5	–4	16
H	7	7	0	0
I	3	10	–7	49
J	6	4	2	4
				$\sum D^2 = 178$

Sol: In this problem, the data are already in the form of ranked data. Hence, we can directly calculate the difference between the ranks of two variables. The differences and square of the differences are calculated in Table 15.4.

15.4.1 Calculation of Rank Correlation Coefficient

Substituting the values in the above equation, we get

$$r_s = 1 - \frac{6\sum_{i=1}^{N} D^2}{N(N^2-1)} = 1 - \frac{6 \times 178}{10\left[(10)^2 - 1\right]} = 1 - \frac{1,068}{990} = +0.07$$

This indicates that there is a positive correlation between two variables. This means that both the systems give similar results.

15.4.2 Testing the Significance of Correlation Coefficient

Although the above calculations show that there is a positive correlation between two variables, it is not yet clear whether this result is statistically significant or it is a chance occurrence. In such cases, researchers can use significance test to check whether the correlation between the variables is statistically significant or due to chance occurrence.

Numerical: Let us continue with the example of SSS limited, on which we used Pearson's product moment correlation coefficient. The correlation coefficient has been determined as 0.51. Can we assume that it is true for the whole population? There can be two conclusions with regard to the value of the correlation coefficient. One is that there is no relationship between the two variables and that the value obtained is because of chance. Another conclusion can be that there exists a statistically significant correlation between two variables and the sample data represents that.

In such cases, we can use the test of hypothesis to analyze whether the correlation is statistically significant.

Given: Correlation coefficient$=r=0.51$, Sample size$=n=10$, $\Rightarrow$ It is a small sample.

Step I: Setup null and alternative hypotheses.

Here, we want to test that is there any relationship between the two variables, i.e., ad spending and sales volume.

Since our claim contains equality sign, we can take the claim as the null hypothesis and complement as the alternative hypothesis, that is,

Our claim is

H_0 : "There is no relationship between the two variables , i.e., ad spending and sales volume," i.e., $\mu=0$.

Where,

$\mu \rightarrow$ is the population correlation coefficient, which is considered to be zero.

This implies we are assuming that there is no relationship between the two variables X and Y, i.e., ad spending and sales volume in this particular case.

And our complement is

H_1 : "There is a relationship between the two variables , i.e., ad spending and sales volume," i.e.,

$\mu \neq 0$. Since the alternative hypothesis is two-tailed, our test is a two-tailed test.

Step II: The level of significance α. Here, $\alpha=0.05$ (= 5% level).

Step III: Define a test statistic to test the null hypothesis as

As the sample is small $(n=10\ )$, we can use the t-test to test the hypothesis. The formula for calculating the sample statistic is

$$t=\frac{r}{\sqrt{\frac{1-r^2}{n-2}}} \tag{15.5}$$

Step IV: Calculate the value of test statistic on the basis of sample observations as

Substituting these values in Equation (15.5), we get

$$t=\frac{0.51}{\sqrt{\frac{1-0.51}{10-2}}}=\frac{0.51}{\sqrt{0.06125}}=\frac{0.51}{0.2474873734}=2.060711191 \cong 2.060$$

Step V: Now, we find the critical value.

We know that the "degree of freedom" for t-test is given by $n-2$. In this case, $n=10$, hence the degrees of freedom is $n-2=10-2=8$. Now, let us assume that the level of significance is 5%. From the hypothesis, it is clear that the test is two-tailed. Hence, we need to look for tabular t-value for 8 degrees of freedom at 0.05 level of significance (see table given in Appendix). The value is 2.306.

Step VI Testing of Hypothesis: Now, to take the decision about the null hypothesis, we compare the calculated value of test statistic with the critical value. Since the calculated t-value $(=2.060)<$than the tabular t-value (Critical Value) (=2.306).

So, we accept the null hypothesis and reject the alternate hypothesis. Since null hypothesis is our claim, we support the claim.

Thus, sample does not provide us sufficient evidence against the complement. Hence, we conclude that there is no relationship between ad spending and sales volume.

15.5 Regression Analysis

Regression analysis finds out the degree or relationship between a dependent variable and a set of independent variables by fitting a statistical equation through the method of least square. Whenever we are interested in the combined influence of several independent variables upon a dependent variable our study is that of multiple regression. For example, demand may be influenced not only by price but also by growth in industrial production, extent of import prices of other goods, consumer's income, taste, preferences, etc.

Market researchers could use regression for explaining per cent variation independent variable caused by a number of independent variables and also problems involving prediction or forecasting.

Regression analysis is another statistical tool for measuring the association between two variables. It is a technique used to predict the nature and closeness of relationships between two or more variables. This method is different from correlation analysis. Regression analysis helps researchers evaluate the casual effect of one variable on another variable. Regression analysis is used to predict the variability in the dependent (or criterion) variable based on the information about one or more independent (or predictor) variables.

Regression analysis provides the answers for the question "What are the expected values of dependent variable given the data about the independent variables?"

Regression analysis that involves two variables is termed bivariate linear regression analysis. Regression analysis that involves more than two variables is termed multiple regression analysis. The bivariate linear regression analysis involves analyzing the straight line relationship between two continuous variables. The bivariate linear regression can be expressed as:

$$Y = \alpha + \beta X \tag{15.6}$$

Where,

$Y \rightarrow$ Represents the dependent variable, $X \rightarrow$ Represents the independent variable, α and $\beta \rightarrow$ are the two constants, which are known as regression coefficients, $\beta \rightarrow$ is the slope coefficient, i.e., β is the change in value of Y with the corresponding change in one unit of X, $\beta \rightarrow$ can by symbolically represented as $\frac{\Delta Y}{\Delta X}$, $\alpha \rightarrow$ Represents Y-intercept when $X=0$.

If the mean value of X and Y is given, then α value can be calculated using the following formula:

$$\alpha = \bar{Y} - \beta\bar{X} \tag{15.7}$$

Let us suppose that a marketing analyst at a shopping mall wants to forecast the inflow of visitors to the mall. From the correlation analysis he or she could be able to establish that the footfalls in the mall are positively related to the advertisement frequency in the local media. Using linear bivariate linear regression on the data, the analyst wants to forecast traffic to the mall based on advertisement frequency in the local media. Table 15.5 provides the observed data with regard to the number of ads published in leading local newspaper per week and the corresponding weekly inflow of visitors.

The data can be plotted on a Scatter Diagram with the independent variable X on the X-axis (customer inflow in this case) and the dependent variable Y on the Y-axis (number of advertisements) (see Figure 15.1).

From Figure 15.1, we can clearly establish that there is a perfect positive correlation between two variables, i.e., correlation coefficient $r = +1.0$. For forecasting, we need to establish the regression equation as in Equation (15.6). We need to first calculate the regression coefficient α and β.

As we know, β is the measure of change of Y corresponding to one unit change in X. It can be calculated as shown below:

$$\beta = \frac{(Y_i - Y_j)}{(X_i - X_j)} \tag{15.8}$$

TABLE 15.5

Data With Regard to the Ad Frequency and Weekly Inflow of Visitors

Number of Ads (*X*)	4	8	12	16	$\bar{X} = 10$
Weekly Inflow of Visitors (*Y*)	4000	6000	8000	10000	$\bar{Y} = 7{,}000$

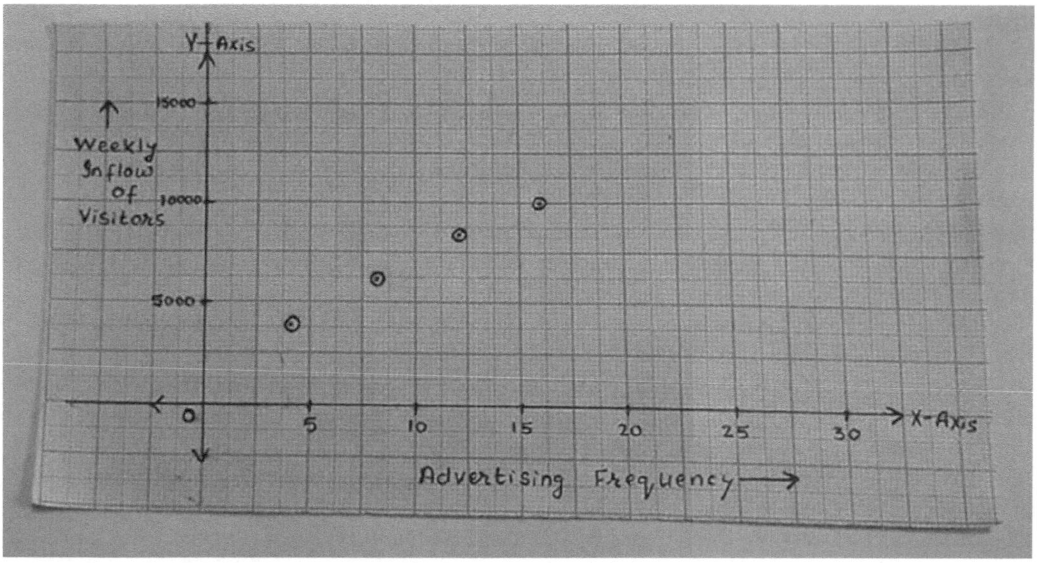

FIGURE 15.1 Scatter diagram for the data in Table 15.5.

Where, X_iY_i and $X_jY_j \rightarrow$ Are any two consecutive data points.

By substituting the values in Equation (15.8), we get

$$\beta = \frac{(Y_i - Y_j)}{(X_i - X_j)} = \frac{(6{,}000 - 4{,}000)}{(8-4)} = \frac{2{,}000}{4} = 500$$

We can calculate α using the formula:

$$\alpha = Y - \beta\overline{X} \tag{15.9}$$

From Table 15.5, we have $\overline{X} = 10$ and $\overline{Y} = 7{,}000$

Substituting the above values in Equation (15.9), we get

$$\alpha = \overline{Y} - \beta\overline{X} = 7{,}000 - 500(10) = 2{,}000$$

Now, by substituting the values of α and β in the Equation (15.6), we get the simple regression equation as

$$Y = \alpha + \beta X = 2{,}000 + 500X_i \tag{15.10}$$

Using Equation (15.10), the marketing analyst can predict the customer inflow into the mall. If the analyst wants to know how much customer inflow will there, if the number of ads in the media is increased to 20, he or she can predict the value by substituting the value in Equation (15.6):

$$Y = \alpha + \beta X = 2{,}000 + 500(20) = 2{,}000 + 10{,}000 = 12{,}000$$

15.5.1 Least Squares Method

In the previous example, we could establish a clear straight line relationship between two variables because of perfect correlation between two variables. However, such possibilities are very limited. In most cases, researchers face situations where the data points do not fit in a straight line, thus leading to a possibility of depicting the relationship between two variables using various lines. The selection of any line will lead to a situation, where the line will be closer to some points and farther from some other points.

For example, an HR manager has gathered data with regard to man-hours spent and the units produced by each employee (see Table 15.6).

The manager wants to identify a regression line that best fits the data.

Sol:

If we plot the data of Table 15.6 on a scatter diagram with X (i.e., number of man hours) on the X-axis and Y (i.e., units produced) on the Y-axis, we may find that no straight line can totally represent every data point on the scatter diagram (see Figure 15.2).

Consider a straight line drawn in Figure 15.2 that describes the data. The line does not cover all the points of the data. The researcher cannot decide whether the line is a best fit or not.

A regression line is termed as a best fit if it minimizes the average vertical distance between observed values and the estimated regression line. The vertical distances from the observed values or points to be estimated regression line is termed as error. It is denoted by e_i.

The Least Squares methods will help the researcher to minimize error, helping him or her find a line that best fits the set of data.

The equation for regression line assumed by Least Squares method is shown below:

$$Y = a + bx + e_i \tag{15.11}$$

TABLE 15.6

Number of Man-hours and the Corresponding Productivity (in Units)

Man-Hours (X)	2.6	4.8	2.4	6.2	6.8	8.4	10.6	10.2	6.2	6.8	8.4	4.4
Productivity in Units (Y)	8.2	10.2	8.6	10.4	12	14.2	18.6	28.4	12.2	10.8	22.6	12.2

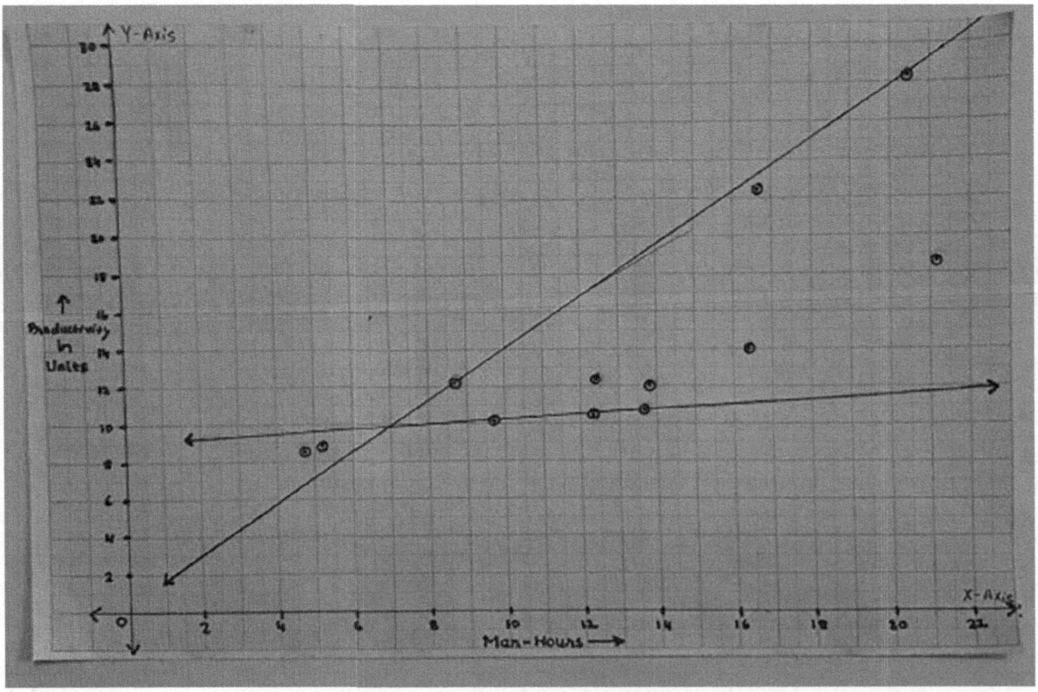

FIGURE 15.2 Scatter diagram and possible regression lines for the data in Table 15.6.

Where, $Y \rightarrow$ is the dependent variable, $X \rightarrow$ is the independent variable, $a \rightarrow$ is the Y-intercept, $b \rightarrow$ is the slope of the line.

The constant b can be calculated using the formula

$$b = \frac{n\sum(XY) - \sum X \sum Y}{n\sum(X^2) - \left(\sum X\right)^2} \tag{15.12}$$

Where, $Y \rightarrow$ is the independent variable, $X \rightarrow$ is the dependent variable "a" is calculated as:

$$a = \bar{Y} - b\bar{X} \tag{15.13}$$

Where, $\bar{Y} \rightarrow$ is the mean of values of dependent variable, $\bar{X} \rightarrow$ is the mean of values of independent variable, $e_i \rightarrow$ is the error.

It is also called residual. The criterion for the Least Squares method is given by:

$$\sum_{i=1}^{n} e_i^2$$

Where,

$e_i = Y_i - \hat{Y}_i$, $\bar{Y}_i \rightarrow$ is the mean actual value of the dependent variable,
$\hat{Y}_i \rightarrow$ is the value lying on the estimated regression line.

Let us solve the example previously discussed using the Least Squares method.

We need to determine the constants "a" and "b" to develop the regression equation. The required calculations for determining the constants are summarized in Table 15.7.

Substituting values in Equation (15.12), we get

$$b = \frac{n\sum(XY) - \sum X \sum Y}{n\sum(X^2) - \left(\sum X\right)^2}$$

TABLE 15.7

Calculations for Determining Constants "a" and "b"

Man-Hours (X)	Productivity in Units (Y)	XY	X^2
2.6	8.2	21.32	6.76
4.8	10.2	48.96	23.04
2.4	8.6	20.64	5.76
6.2	10.4	64.48	38.44
6.8	12	81.6	46.24
8.4	14.2	119.28	70.56
10.6	18.6	197.16	112.36
10.2	28.4	289.68	104.04
6.2	12.2	75.64	38.44
6.8	10.8	73.44	46.24
8.4	22.6	189.84	70.56
4.4	12.2	53.68	19.36
$\sum X = 77.8$	$\sum Y = 168.4$	$\sum XY = 1244.72$	$\sum X^2 = 581.8$

$$b = \frac{12(1244.72)-(77.8)(168.4)}{12(581.8)-(77.8)^2} = \frac{14936.64-13101.52}{6981.6-6052.84} = 1.9758818209$$

The next step is to calculate "*a*" using Equation (15.13).

To calculate the value of "*a*", we need to first determine the means of values of variables *X* and *Y*.

$$\bar{Y} = \frac{\sum Y}{n} = \frac{168.4}{12} = 14.03,\ \bar{X} = \frac{\sum X}{n} = \frac{77.8}{12} = 6.48$$

Substituting values in Equation (15.13), we get

$$a = \bar{Y} - b\bar{X} = 14.03-(1.975)(6.48) = 14.03-12.798 = 1.232$$

We now develop the estimated regression equation by substituting the values of "*a*" and "*b*" in Equation (15.11), we get

$$\hat{Y} = a + bX = 1.232 + 1.975X$$

Where, $\hat{Y} \rightarrow$ Represents the estimated value of dependent variable for a given value of *X*.

Thus, as per the estimated regression equation for every additional increase of one man-hour, the productivity will increase by 1.97 units.

15.5.1.1 Plotting a Regression Line

Using the estimated regression equation, we can determine the predicted values of *Y* (see Table 15.8).

We plot the regression line based on the predicted values of *Y* in Figure 15.3.

The regression line best fits the given set of data as it represents the data in a more effective manner. This can be cross checked by finding the error or residual values for each observation. A regression line is best fit if the sum of positive and negative values of error equals zero.

15.5.2 The Strength of Association—R^2

The above developed estimated regression equation can only explain the nature of relationship between two variables. However, if the researcher wants to know how strong or weak the relationship is, i.e., to what degree that the variation in *Y* can be explained by *X*, the coefficient of determination denoted by R^2 is used. R^2, which is measured in percentage, will explain how much of the total variation in *Y* variable is explained by *X* variable.

For example, if the R^2 value is 0.10, then there is a weak relationship between two variables, i.e., only 10% of the total variation in *Y* variable can be explained by the variation in *X* variable.

Similarly, if R^2 is 0.90, then we can conclude that there is a strong relationship between the two variables. In turn, this also implies that nearly 90% of the total variation in *Y* can be explained by the variation in *X* variable.

The R^2 value ranges from 0 to 1. Thus, if R^2 value is 1, then all the observed values of *Y* will be on the regression line, i.e., which is the same as variables having a perfect correlation. In contrast, if the R^2 value is zero, then it implies there is no relationship between the two variables, and the variation in *X* variable will not affect the variation in *Y* variable. R^2 can be calculated using the following formula:

$$R^2 = \frac{\text{Explained variance}}{\text{Total variance}}$$

$$\text{Total variance} = \text{Explained variance} - \text{Unexplained variance}$$

TABLE 15.8

Predicted Values and Errors for each Observation

Man-Hours (X)	**Productivity in Units (Y)**	XY	X^2	$\hat{Y}$	$Y-\hat{Y}$	$(Y-\hat{Y})^2$	$(\hat{Y}-\bar{Y})^2$	$(Y-\bar{Y})^2$
2.6	8.2	21.32	6.76	6.357	1.833	3.359889	58.721569	33.9889
4.8	10.2	48.96	23.04	10.712	–0.512	0.262144	11.009124	14.6689
2.4	8.6	20.64	5.76	5.972	2.628	6.906384	64.931364	29.4849
6.2	10.4	64.48	38.44	13.477	–3.077	9.467929	0.305809	13.1769
6.8	12	81.6	46.24	14.662	–2.662	7.086244	0.399424	4.1209
8.4	14.2	119.28	70.56	17.822	–3.622	13.118884	14.379264	0.0289
10.6	18.6	197.16	112.36	22.167	–3.567	12.723489	66.210769	20.8849
10.2	28.4	289.68	104.04	21.377	7.023	49.322529	53.978409	206.4969
6.2	12.2	75.64	38.44	13.477	–1.277	1.630729	0.305809	3.3489
6.8	10.8	73.44	46.24	14.662	–3.862	14.915044	0.399424	10.4329
8.4	22.6	189.84	70.56	17.322	4.778	22.829284	14.379264	73.4449
4.4	12.2	53.68	19.36	9.922	2.278	5.189284	16.875664	3.3489
$\sum X = 77.8$	$\sum Y = 168.4$	$\sum XY = 1244.72$	$\sum X^2 = 581.8$			$\sum (Y-\hat{Y})^2 = 146.811833$	$\sum (\hat{Y}-\bar{Y})^2 = 301.898074$	$\sum (Y-\bar{Y})^2 = 413.4268$

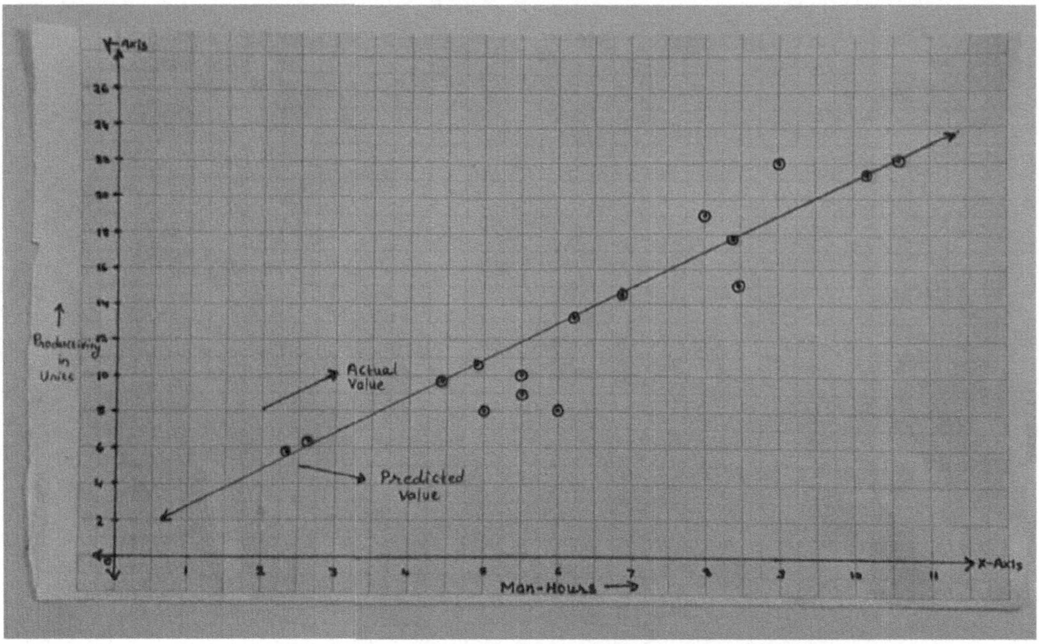

FIGURE 15.3 Least squares regression line.

$$\text{Explained variance} = \text{Total variance} - \text{Unexplained variance}$$

$$\therefore R^2 = \frac{\text{Total variance} - \text{unexplained variance}}{\text{Total variance}}$$

$$\therefore R^2 = 1 - \frac{\text{Unexplained variance}}{\text{Total variance}}$$

The unexplained variance is given by $\sum \left(Y_i - \hat{Y}\right)^2$

The total variance is given by $\sum \left(Y_i - \bar{Y}\right)^2$

$$\therefore R^2 = 1 - \frac{\sum_{i=1}^{n} \left(Y_i - \hat{Y}\right)^2}{\sum_{i=1}^{n} \left(Y_i - \bar{Y}\right)^2} = 1 - \frac{146.811833}{413.4268} = 1 - 0.355109618 = 0.64$$

This implies that of the total variation of Y, nearly 64% is explained by the variation in X. Hence, there is a strong linear relationship between the two variables.

15.5.3 Test of Statistical Significance of Regression Equation

As discussed earlier, the total variation can be decomposed into two components.

$$\text{Total variance} = \text{Explained variance} - \text{unexplained variance}$$

$$\text{where, Total variance} = \sum \left(Y_i - \bar{Y}\right)^2$$

i.e., total variation calculates the squared deviation of the actual values of Y from the mean $\bar{Y}$. It is also called Total Sum of Squares.

The explained variation is termed as Sum of Squares due to Regression denoted by SSR.

$$\text{SSR} = \sum \left(\hat{Y}_i - \bar{Y}\right)^2 = 301.898074$$

The unexplained variation is termed as Error Sums of Squares denoted by SSE.

$$\text{SSE} = \sum \left(Y_i - \hat{Y}_i\right)^2 = 146.811833$$

We may recall that SSE is the residual difference or Error between the actual values and the predicted values. In simple terms, the unexplained variation is the measure of squared deviations of the data points from the estimated regression line in the scatter diagram.

If there is a perfect correlation between two variables, the error or residual difference will be zero. We can use, the F-test or Analysis of Variance to evaluate the linear relationship between two variables. The F-test is used to test the significance of the linear relationship between two variables. The hypothesis can be stated as follows:

Null hypothesis H_0 : There is no linear relationship between two variables

Alternate hypothesis H_1 : There is a linear relationship between two variables

We assume the level of significance to be 5%. We reject the null hypothesis if the calculated F value falls above the table F value. However, we accept the null hypothesis if the calculated F value is below the table F value.

Before conducting the F-test, we develop a summary ANOVA table that is used for hypothesis test. Table 15.9 provides these details.

Where, $k \rightarrow$ represents number of variables involved in the problem. As we are dealing with two variables, the degrees of freedom for SSR would be $k-1=2-1=1$.

The sample size n is 12, hence the degrees of freedom for SSE would be

$$n-k=12-2=10$$

From Table 15.9, SSR = 301.898074 SSE = 146.811833; Table15.10 provides the actual calculations.

We obtained the calculated F value as 20.56. The table F value at 5% level of significance for 1 degrees of freedom in numerator and 10 degrees of freedom in denominator is 4.96. As the calculated F value is greater than the table F value (see table given in Appendix) we reject the null hypothesis. Hence, we conclude that there is a linear relationship between two variables.

TABLE 15.9

Analysis of Variance Table

Sources of Variation	Degrees of Freedom	Sum of Squares	Mean Square	F Value
Variation explained due to regression (explained variation)	$k-1$	SSR	$\text{MSR} = \frac{\text{SSR}}{k-1}$	$F = \frac{\text{MSR}}{\text{MSE}}$
Residual difference of error (unexplained variation)	$n-k$	SSE	$\text{MSE} = \frac{\text{SSE}}{n-k}$	
Total	$n-1$	SST		

TABLE 15.10

Analysis of Variance Table

Sources of Variation	Degrees of Freedom	Sum of Squares	Mean Square	F Value
Variation explained due to regression (explained variation)	$k-1=2-1=1$	SSR = 301.898074	$\text{MSR} = \frac{301.898074}{1} = 301.898074$	$F = \frac{301.898074}{14.6811833} = 20.56$
Residual difference of error (unexplained variation)	$n-k=12-2=10$	SSE = 146.811833	$\text{MSE} = \frac{146.811833}{10} = 14.6811833$	
Total	$n-1=12-1=11$	SST = 448.709907		

Summary

In this chapter, we discussed bivariate correlation and regression analysis. Correlation analysis is used to measure the linear relationship between two variables. There are two prominent correlation coefficients used in correlation analysis. They are Pearson's product moment correlation coefficient and the Spearman's correlation coefficient. Hypothesis testing can be applied in correlation analysis to analyze whether the correlation between the variables is statistically significant or due to chance occurrence. Z-test or *t*-test is used to test the correlation coefficient. Regression analysis is another statistical technique used to measure the association between variables. We used the Least Squares method to find a regression line that best fits the data. The strength of association between two variables can be measured using the coefficient of determination. The ANOVA test is used to test the significance of the estimated regression line.

Review Questions

1. What is a "scatter diagram"? How does it help us in studying the correlation between two variables, with respect to both its nature and extent?
2. What is Spearman's correlation coefficient? Bring out its usefulness.
3. Describe method of least squares and show how it can be used to fit a linear regression. How is linearity of regression tested?
4. Explain Pearson's Product Moment Correlation Coefficient Correlation.
5. Explain Test of Statistical Significance of Regression Equation.
6. Explain the Spearman's rank correlation coefficient.
7. Give the application of hypothesis testing in correlation analysis.
8. Explain regression analysis and the use of least squares method in finding a best-fitting regression line.
9. Give the application of the ANOVA test in regression analysis.

16

Multivariate Analysis

16.1 Introduction

In the previous chapters, we discussed how research problems that involve one or two variables could be solved, by using the Z-Test, *t*-Test, Correlation Analysis, and Regression Analysis. However, in practice, the situations that researchers face usually require them to study more than two variables. In such situations, univariate statistical analysis, i.e., study of a single variable and bivariate statistical analysis, i.e., study of the relationship between two variables, may not prove useful. Thus, research studies aimed at solving such complex situations warrant the use of multivariate techniques.

Multivariate analysis deals with the analysis of research problems that involve more than two variables. The popularity of multivariate techniques has increased with the advent of advanced computing resources, and the availability of off-the shelf statistical software packages, which made the application of multivariate techniques easier.

16.2 Multivariate Techniques

It is defined as "all statistical techniques that simultaneously analyze more than two variables on a sample of observations." In other words, multivariate analysis helps the researcher in evaluating the relationships between multiple or more than two variables simultaneously.

For example, consider the sales level of a company's product is not influenced only by demand, but also by various other variables such as company strategy, company personnel, technology, automation, e-commerce, finance availability, pricing, distribution, product features, promotion, and competitors' strategies. While analyzing sales, the manager of the company has to take these variables into consideration as well.

Multivariate techniques can be categorized as dependency techniques and Interdependency techniques. Those techniques that deal with problems involving one or more dependent variables, while the remaining variables are considered independent, are called as Dependency techniques. Dependency techniques are further classified as the measurement scales and the number of dependent variable in the problems.

Those multivariate techniques that deal with more than two variables, where the variables are not segregated as dependent variables and independent variables, are called as Interdependency techniques. These techniques aim at analyzing interrelationships between the variables.

16.3 Dependency Techniques

Dependency techniques aim at explaining or predicting one or more dependent variable based on two or more independent variables. Here, the focus is on defining a relationship between one dependent variable and many independent variables that affect it. A typical research question that calls for the employment of a dependency technique is, to what extent can the percentage of defects produced by a machine be explained by the factors?

Age of the machine, quality of raw material, improper maintenance, poor electrical connections, overrunning of machines, nonreplacement of worn parts, weather-related issues, ignoring of warning signals, untrained operator, and the experience of the operator. Here, the number of defects is a dependent

DOI: 10.1201/9781315167138-16

variable and is influenced by various independent variables. There are various dependency techniques that a researcher can use to analyze the dependency in this problem; prominent among them are Multiple regression analysis, Discriminant analysis, Multivariate analysis of variance (MANOVA), and Canonical correlation analysis (CCA).

However, the selection of an appropriate technique depends on the number of dependent variables in the problem and the measurement scale used.

If the problem consists of single dependent variable and the measurement scale used is interval or ratio scales then, we use multiple regression analysis. However, if the problem consists of single dependent variable and the measurement scale used is nonmetric, i.e., ordinal or nominal, then we use multiple discriminant analysis. If the problem consists of more than one dependent variable, then we use MANOVA and CCA.

16.3.1 Multiple Regression Analysis

Bivariate regression analysis measures the association between an independent variable and the dependent variable. However, in most real-life situations a dependent variable is not influenced by a single variable, but more than one variable.

For example, sales may not be influenced by just price but also influenced by other marketing mix variables like product features, distribution strategy, and promotional strategy. Even external factors such as competition and economic situation influence the sales. In such situations, a multivariate technique called Multiple Regression Analysis can help the researcher to evaluate the association between a single dependent variable and two or more than two independent variables.

16.3.1.1 Uses of Multiple Regression Analysis

1. To identify relationships between variables and to predict the outcomes.
2. To find answers to questions such as "How is sales volume related to the pricing and advertising expenditure?"
3. To know the strength of the relationship between the variable.

For example, it helps the researcher to determine how Pricing and Advertising expenditure affects the sales volume more strongly; to predict outcomes; to find the impact on the dependent variable, given the values of the independent variable, etc.

Thus, while bivariate regression analysis helps in finding a straight line that best fits the data in a two-dimensional space, multiple regression analysis helps in finding a plane that best fits the data in a multi-dimensional space.

In multiple regression analysis the regression equation is shown in the form of:

$$Y = a + \beta_1 X_1 + \beta_2 X_2 + \beta_3 X_3 + \beta_4 X_4 + \beta_5 X_5 + \cdots + \beta_n X_n + e$$

Where, $Y \rightarrow$ is the dependent variable, $X \rightarrow$ is the independent variable, $\beta's \rightarrow$ are the slope coefficients that represent the change in dependent variable Y when there is a change of 1 unit in X variable, $a \rightarrow$ is the Y-intercept when $X = 0$.

For example, a consumer product company that has an exclusive tie-up with a retail chain wants to identify the factors that influence the sales of one of its products in the stores of the retail chain. After conducting a multiple regression analysis on the available data it arrived at the following equation:

$$\hat{Y} = 150 + 6X_1 + X_2$$

Where,

$\hat{Y} \rightarrow$ is the estimated dependent variable, sales of the product (in units)

$X_1 \rightarrow$ is an independent variable, the in-store promotion expenditure (in rupees)

$X_2 \rightarrow$ is another independent variable, the shelf space (in square inches)

We see from the equation that both in-store promotion and shelf space have a positive effect on the sales. From the equation, we can see that for every increase of single rupee in in-store promotion expenditure, sales increase by 6 units and for every increase of 1 square inch of shelf space there is an increase of sales by 1 unit.

The β coefficients, i.e., $\beta_1, \beta_2, \beta_3$, etc., are called coefficients of partial regression.

These coefficients are different from the coefficients, which were used in the bivariate regression analysis.

In bivariate regression analysis as there is no second independent variable, the change in variable Y, i.e., dependent variable, can be explained by the change in variable X, i.e., independent variable.

However, that is not the case in multiple regression analysis. There are two or more than two independent variables that describe the variation in the dependent variables, and these independent variables are correlated.

$\beta_1 \rightarrow$ represents the change in variable Y given a change of 1 unit of an independent variable (X_1), when all other independent variable $(X_2, X_3, \ldots, X_n)$ are kept constant.

In other words,

$\beta_1 \rightarrow$ represents the isolated effect of a particular independent variable (X_1) on variable Y.

Thus, if there are two independent variable X_1 and X_2 and their regression coefficients are β_1 and β_2, then β_1 will represent the change in variable Y for every unit of change in X_1, when X_2 is kept constant and $\beta_2 \rightarrow$ will represent the change in variable Y for every unit of change in X_2 when X_1 is kept constant. These coefficients are additive in nature.

If $\beta_1 \rightarrow$ represents the change in Y variable for each unit change in X_1 variable and $\beta_2 \rightarrow$ represents the change in Y variable for each unit change in X_2 variable, then the change in Y variable would be $\beta_1 + \beta_2$. The $\beta \rightarrow$ values can be expressed as raw values or beta weights. The raw values are the values expressed in the units in which each X variable is measured. However, it is difficult to compare the regression coefficients of each variable as they are measured in different measurement units. Thus, to facilitate direct comparison, the regression coefficients are standardized. The standardized β coefficients are also called β weights.

16.3.1.2 Coefficient of Multiple Determination

Similar to the coefficient of determination used in bivariate regression analysis, the coefficient of multiple determination measures the magnitude of the association of the variables involved in multiple regression. It is denoted by R^2.

In mathematical terms, it measures the percentage of variation in variable Y explained by the independent variables.

For example, if the R^2 value is 0.70 of the total variation in Y variable is explained by variation in the independent variables.

16.3.1.3 Test of Significance

The regression equation is subjected to significance testing, to test whether the dependent variables are influenced by the independent variables. A researcher can use F-test to test the significance of the R^2 value. It tests whether the R^2 value is significant. The procedure followed in hypothesis testing for a multiple regression equation is similar to the procedure followed in bivariate regression analysis.

The hypothesis for this test would be:

$$\textit{Null hypothesis} \quad H_0 : R^2 = 0, \quad \textit{Alternate hypothesis}\ H_1 : R^2 \neq 0$$

The test statistic can be determined using the following formula:

$$F = \frac{\dfrac{\text{SSR}}{k}}{\dfrac{\text{SSE}}{(n-k-1)}}$$

Where, SSR → is the sum of squares due to regression, SSE → is the residual sum of squares, N → represents the sample size, k → represents the number of variables in the problem.

We reject the null hypothesis if the calculated F-value exceeds the tabular F-value. We accept the null hypothesis if the calculated F-value is less than the tabular F-value.

16.3.1.4 Issues in Multiple Regression Analysis

While using multiple regression analysis, the researcher has to consider the following factors:

16.3.1.5 Multicollinearity

While selecting the predictor variables, i.e., independent variables, we need to take care that they are correlated with the criterion variable, i.e., dependent variable, but not with other independent variables. In reality there exists some correlation among the independent variables. But, if the correlation between independent variables is high, this leads to difficulty in understanding the relationship between the dependent and independent variables. This is known as multicollinearity, a condition of high correlation among a set of independent variables.

Multicollinearity makes it difficult for researchers to ascertain which of the independent variables influence the dependent variable. It leads to improper estimation of β coefficients, which describe the variation of Y variable due to the variation in X variables.

It also makes it difficult for the researcher to interpret the relative effect of various independent variables on the dependent variable.

Researchers have the following ways of tackling the problem of high correlation between independent variables:

1. One way is to collect more data so that the independent variables can be explained better.
2. Another way is to remove a variable from the analysis that has high correlation with another variable. Researchers also combine two highly correlated variables into a single variable for regression analysis.

16.3.1.6 Dummy Variables

Often researchers are required to include categorical variables such as marital status and gender, i.e., variables on nominal scale, in their regression models. However, regression analysis allows for only numerical variables. In such cases, we transform the categorical variables into dummy variables, which can take the values of 0 or 1.

If the variables are dichotomous, i.e., variables having two categories, i.e., gender and marital status, one option is coded as 0 and other option is coded as 1.

For example, if the variable is gender, then male is coded as 0 and female is coded as 1.

If the variable has more than two categories, i.e., if a user has to be rated as heavy user, moderate user, and light user, then we need to keep one category aside as the reference category to prevent perfect multicollinearity.

Thus, for a variable that consists of n categories, we need to create $n-1$ dummy variables instead of n dummy variables.

For example, if a HR manager wants to study the effect of the Education Qualification of an employee on his or her performance, the variable educational qualification is put in categories such as Graduate, Post-Graduate, and Doctorate. Hence, the coding is done as summarized in Table 16.1.

The dummy variables X_1 and X_2 are considered as predicting variables. The regression equation takes the form of $Y_i = a + b_1X_1 + b_2X_2$.

We need to note that Doctorate category was considered as reference category; hence, this category is not incorporated in the regression equation. For the category Doctorate, the two variable values, i.e., X_1 and X_2, are zero. So, for the Doctorate category, the regression equation is $\hat{Y}_i = a$.

TABLE 16.1

Example for Dummy Variables

Education Qualification	Original Variable Code	X_1	X_2
Graduate	1	1	0
Post-Graduate	2	0	1
Doctorate	3	0	0

For the Graduate category $X_1 = 1$ and $X_2 = 0$.

So, the regression equation for this category is $\hat{Y}_i = a + b_1$.

Thus, the coefficient for Graduate category is the difference between predicted Y_i Graduate and the predicted value of Y_i of Doctorate (as $\hat{Y}_i$ for the Doctorate category is *a*).

Similarly, for the Post-Graduate category $X_1 = 0$ and $X_2 = 1$. So, the regression equation for this category is $\hat{Y}_i = a + b_2$

Thus, the coefficient for Post-Graduate category is the difference between the predicted Y_i Post-Graduate and the predicted value Y_i of Doctorate, i.e., as $\hat{Y}_i$ for the Doctorate category is "*a*".

16.3.2 Discriminant Analysis

Discriminant analysis is a technique used for classifying a set of observations into predefined groups based on a set of variables known as predictors of input variables.

Researchers often face situations where they need to classify the population or objects into certain groups.

For example, a financial institution may want to classify various investment options into high return, medium return, and low return investments. Also, a market research agency might want to assess the quality of various Bike models and classify them under high quality, medium quality, and low-quality categories. Discriminant analysis can be used in such situations.

By using discriminant equation, we can classify the objects into particular predefined groups, to predict the success or failure of the objects. Based on the classification of objects, we can find answers to questions such as Which investment option will provide higher returns? or Who are the potential customers?

Discriminant analysis also helps in determining the factors that aid in discriminating the objects. For example, this can be used in marketing where we can apply it to achieve an understanding of how customer preferences for different brands differ.

The general discriminant analysis equation is as follows:

$$Z = b_1X_{1i} + b_2X_{2i} + \cdots + b_nX_{ni}$$

Where, $Z \rightarrow$ is the discriminant score, $X_{ni} \rightarrow$ are the discriminating variables, i.e., independent variables, $b_1, b_2, b_3, \ldots, b_n \rightarrow$ are the discriminant coefficients or weights corresponding to each independent variable.

The discriminant score is determined for each object. Using these scores as the basis, the researcher will decide as to which group the object belongs to. This equation is also used to identify the major factors that help in discriminating the objects.

For example, a financial institution, which offers home loans to individual customers, wants to develop a model to assess the credit risk of loan applicants so that it can distinguish between defaulters and good clients among the loan applicants. For this, the company has collected data relating to past customers and their characteristics. It also collected the details about past defaulters. Based on these data, the company has arrived at a discriminant equation as given below:

$$Z = 3X_1 + 4X_2 - 7X_3 - 0.8X_4$$

Where,

$Z \rightarrow$ Represents the discriminant score to differentiate between defaults and good loans
$X_1 \rightarrow$ Represents the time period of the loan, $X_2 \rightarrow$ Represents the amount of the loan
$X_3 \rightarrow$ Represents income level, $X_4 \rightarrow$ Represents the number of years of work experience

Using the equation, the institution can now determine the credit worthiness of new loan applicants.

If the institution feels that the applicant is a potential defaulter, it can ask for more document proofs or collaterals to cover the risk.

We can observe that the variable X_3 has a higher weight compared to the other variables. This implies that the income level has a higher weightage than the other variables.

The negative sign of X_3 implies that the credit risk and the income levels are inversely related. That is, the higher the income level, the lower is the credit risk. We can also observe that the amount of loan is also given higher weightage. Thus, the company should give more importance to the income levels and the loan amount while assessing the creditworthiness of the loan applicants.

The number of discriminant equations required to carry out discriminant analysis depends on the number of categories into which the objects are to be classified.

We need to develop $n-1$ discriminant equations, where $n \rightarrow$ represents the number of categories to carry out the discriminant analysis.

For example, if the problem consists of categorizing the objects into two groups (such as eligible candidates and ineligible candidates, buyers and nonbuyers), then we need to develop a single discriminant equation. For a problem consisting of three categories, i.e., high return stocks, medium return stocks, and low return stocks, we need to develop two discriminant equations.

16.3.3 Canonical Correlation Analysis

CCA is a way of measuring the linear relationship between two multidimensional variables. CCA is an extension of multiple regression analysis. Multiple regression analysis analyzes the linear relationship between a dependent variable and multiple independent variables. However, CCA analyzes a linear relationship between multiple dependent variable and multiple independent variables.

For example, a social researcher wants to know the relationship between various work environment factors such as work culture, HR policies, compensation structure, top management, influencing various employee behavior elements such as employee productivity, attrition rate, job satisfaction, perception about the company.

The linear combination for each variable is called Canonical Variables or Canonical Variates. CCA tries to maximize the correlation between two canonical variables.

For example, $W \rightarrow$ represents the linear combination of work environment factors.

$$W = a_1X_1 + a_2X_2 + a_3X_3 + a_4X_4 + a_5X_5 + a_6X_6 + a_7X_7$$

And $V \rightarrow$ represents the linear combination of employee behavior elements.

$$V = b_1Y_1 + b_2Y_2 + b_3Y_3 + b_4Y_4 + b_5Y_5 + b_6Y_6 + b_7Y_7$$

The coefficients of each canonical variable are called Canonical Coefficients. The researcher will deduce his or her conclusions based on the relative magnitudes and the nature of canonical coefficients of each equation.

Being a complex statistical tool that requires a great investment of effort and computing resources, CCA has not gained as much popularity as statistical tools such as multiple regression.

16.3.4 Multivariate Analysis of Variance

MANOVA is another wisely used multivariate technique. MANOVA examines the relationship between several dependent variables and several independent variables. It tries to examine whether there is any difference between various dependent variables with respect to the independent variables. Table 16.2 Summarize the difference between ANOVA and MANOVA.

TABLE 16.2

Difference between ANOVA and MANOVA

Distinction Points	ANOVA	MANOVA
Deals with	ANOVA deals with problems containing one dependent variable and several independent variables	MANOVA deals with problems containing several dependent variables and several independent variables
Inter-relationship	ANOVA test ignores interrelationships between the variables. The leads to biased results	MANOVA considers this aspect by testing the mean differences between groups on two or more dependent variables simultaneously

For example, an industrial buyer wants to know whether the products from Company A, Company B, and Company C differ in terms of various parameters set by the company, such as quality, customer support, pricing, and reliability.

16.4 Interdependency Techniques

Interdependency techniques are used in situations, where no distinction is made between variables, which are independent variables and those which are dependent variables. Instead the interdependent relationships between variables are examined.

Prominent interdependency techniques are Factor analysis, Cluster analysis, Metric multidimensional scaling, and Nonmetric multidimensional scaling.

This selection of an appropriate technique depends on the measurement scale used in the problem. If the problem consists of metric data, then we use factor analysis, cluster analysis, and metric multidimensional scaling. However, if the problem consists of nonmetric data, we use nonmetric multidimensional scaling.

16.4.1 Factor Analysis

Factor analysis can be defined as a "set of methods in which the observable or manifest responses of individuals on a set of variables are represented as functions of a small number of latent variables called factors." Factor analysis is used when the research problem involves a large number of variables making the analysis and interpretation of the problem difficult. Factor analysis helps the researcher to reduce the number of variables to be analyzed, thereby making the analysis easier.

For example, consider a market researcher at a Debit Card Company who wants to evaluate the Debit Card usage and behavior of customers, using various variables. The variables include age, gender, marital status, income level, education employment status, credit history, and family background. Analysis based on a wide range of variables can be tedious and time-consuming. Using factor analysis, the researcher can reduce a large number of variables into a few dimensions called factors that summarize the available data.

In most cases, several input variables are being used to measure a part of the same underlying construct. The underlying construct is known as the factor. By grouping these input variables under a construct, we can make the data analysis easier and faster.

In the factor analysis certain variables, which are highly correlated, are combined into specific factors. These factors form the new variables and the values for these variables are obtained by adding the values of the variables that formed the factor. For example,

1. Age, gender, marital status can be combined under a factor called demographic characteristics.
2. Income level, education, employment status can be combined under a factor called socioeconomic status.
3. Credit history and family background can be combined under a factor called background status.

16.4.1.1 Benefits of Factor Analysis

Factor analysis can be used in research in the following ways:

1. Factor analysis can be used to identify the hidden dimensions or constructs, which may not be apparent from direct analysis. For example, high end projection screen LED are relatively new products in India, marketers may not have a complete idea about the attributes that a consumer looks for while selecting a model among the available options in the market. So, a sample of consumers is taken and each one is asked to rate his or her preference for a particular LED model. Similarly, they are asked to rate various other LED models. Factor analysis is performed on these data to extract the factors from the data. These factors help to provide insights into the causal links that explain the relative preferences of consumers for various LED models.
2. Factor analysis can also be used to identify relationships between variables.
3. Another major benefit of factor analysis is that it helps in data reduction. For example, if a researcher has to evaluate 60 variables, by combining them into few common factors, data can be simplified and analysis can be done using less time and effort.
4. Factor analysis can also help the researcher to cluster the products and population being analyzed. For example, several LED models can be clustered under a set of small categories based on the factors that are extracted from the data.

16.4.2 Cluster Analysis

Cluster analysis is a technique that is used in order to segment a market. The objective is to find out a group of customers in the market place that are homogeneous, i.e., they share some characteristics so that they can be classified into one group. The cluster or group so found out should be large enough so that the company can develop it profitably, as the ultimate objective of a company is to serve the customer and earn profits. The group of customers that the company hopes to serve should be large enough for a company so that it is an economically viable proposition for the company. This is also true for the customer as customer would not be willing to pay beyond a certain price for a particular product, i.e., price of course is a function of positioning of product, cost of production, etc.

Cluster analysis is used in research for various purposes. This technique is widely used in marketing in order to segment the market.

For example, let us consider the Wrist-Watch Industry. There could be many ways in which the Watch Industry could be segmented, which are as follows:

Gender (Male or Female), Technology (Digital or Analog), Design Features, Occasion of Use (Formal or Casual or Party), Price (Low or Medium or High or Jewellery), etc.

Some of the above segmentation factors are demographic, i.e., price, gender, whereas some are psychographic factors, i.e., occasion to use. This, therefore, presents a problem to the market researcher or company, as to how to identify combination of factors that can be used to segment the market place. It is not always possible to segment a market on the basis of one single factor. Thus, a combination of factors must be used to segment the market place. And, this is where Cluster Analysis technique specifically deals with how objects, i.e., people, places, products, should be assigned to groups, so that there should be similarity within the groups; and as much difference between the groups as possible.

For example, a Bond Paper manufacturing company can segment the industrial buyer market according to usage patterns, where it can identify heavy users, moderate users, and light users and devise the marketing strategy for each segment accordingly.

Cluster analysis can also be used to cluster consumers according to their buying behavior. A company can segment the market according to the stage of buying readiness of its different consumers and devise the promotional strategy accordingly.

Cluster analysis can also be used to identify new product ideas by clustering the company products into homogeneous groups, and comparing them with the offerings available in the market; this can help identify gaps in the company's product portfolio. Cluster analysis can also be used as a data reduction technique. By clustering objects into homogeneous groups a company can restrict the analysis to a few clusters rather than considering each object. The output from cluster analysis can be further analyzed by other multivariate techniques.

16.4.2.1 Procedure Followed in Cluster Analysis

1. **Defining the Problem:** We need to first define the problem and decide upon the variables based on which the objects are clustered.
2. **Selection of Similarity or Distance Measures:** The similarity measure tries to examine the proximity between the objects. Closer or similar objects are grouped together and the farther objects are ignored. The major methods to measure the similarity between objects are Euclidean distance measures, Correlation coefficients, and Association coefficients.

16.4.2.2 Selection of Clustering Approach

The next step is to select the appropriate clustering approach. The types of clustering approaches are Hierarchical clustering approach and Nonhierarchical clustering approach.

16.4.2.3 Hierarchical Clustering Approach

It consists of either a top-down approach or a bottom-up approach. In the top-down approach, all the objects are considered as a single cluster. The single cluster is then split into two clusters and so on till there is a statistical justification to do so. The top-down approach is also known as the divisive method.

In bottom-up approach, all the objects are considered as separate clusters. That is, there are as many clusters as the number of the objects present in the population given in the problem. Then objects are combined to form bigger clusters. This process is continued till all the objects are accounted for. Such an approach is known as agglomerative approach.

Prominent hierarchical clustering methods are Single linkage, Complete linkage, Average linkage, Ward's method, and Centroid method.

16.4.2.4 Nonhierarchical Clustering Approach

A cluster center is first determined and all the objects that are within the specified distance from the cluster center are included in the cluster. Then the focus moves to the remaining nonclustered objects, which are clustered using the same procedure. The process is continued till all the objects are accounted for. The prominent nonhierarchical clustering methods are Sequential threshold method, Parallel threshold method, and Optimizing portioning method.

16.4.2.5 Deciding on the Number of Clusters to be Selected

We then need to decide on the number of clusters to be chosen. There are no specific guidelines to restrict the number of clusters based on which cluster analysis is performed. There are various ways through which researchers try to determine the appropriate number of clusters. One way is to decide the number of clusters intuitively.

For example, a market researcher using cluster analysis for customer segmentation based on annual income of the consumer may know beforehand the number of clusters that need to be formed.

Another way is to get inputs from the pattern of clusters that a method generates. The researcher can use distance between the objects as the criterion. So, a researcher can set a certain distance value and he or she can limit the clustering process to the point where the values exceed that specified value.

16.4.2.6 Interpreting the Clusters

The next step is to interpret the clusters. Interpretation of clusters can be done using the centroid. The centroid is an average value of the objects in the cluster based on each of the variables making up each object's profile. The centroid helps the researcher in explaining the cluster and providing appropriate label to the cluster.

16.4.3 Multidimensional Scaling

It is defined as a technique that involves representing object's preferences and perceptions as points on a multidimensional space. Consumers usually perceive a product as similar to another product, not just because of a single attribute, but because of several attributes.

However, obtaining multiple dimensions about the issue or object is difficult for a company. This happens for the following reasons:

1. Direct interviews with the consumer may not reveal those dimensions as he or she may not be aware of the basis for his or her perception about the similarity between two products.
2. The consumer may not be interested in revealing the dimension based on which he or she arrived at this conclusion.
3. Such situations warrant the use of a statistical technique called multidimensional scaling to reveal the underlying dimensions based on which consumers perceive that two objects are similar. Multidimensional scaling is commonly used in motivational research.

Types of multidimensional scaling techniques are Metric multidimensional and Nonmetric multidimensional scaling.

While metric multidimensional scaling deals with problems involving metric data, nonmetric multidimensional scaling deals with problems involving nonmetric data.

The following process is used in multidimensional scaling:

The similarities between the objects are taken as inputs and transformed into distances. The objects are placed on the multidimensional space according to the distances between them.

For example, if a respondent has revealed that he or she considers two products Limca and Pepsi as similar products, then these objects are placed` on the multidimensional space in such a way that the distance between them is shorter than the remaining objects. Such an arrangement provides the researcher the inputs relating to the criteria that underlie the customer preferences for the products.

16.4.3.1 Applications of Multidimensional Scaling

1. **Market Segmentation:** It is the technique of trying to identify groups of consumers who exhibit commonality of perception of products and preferences, One can use MDS techniques to identify present perceptions of products by consumers, and use it modify the company's product, package, advertising, additional features so that the product offering of the company moves more and more closer to the "ideal" requirement of the consumer.
2. **Advertisement Evaluation:** The MDS technique could be used at the stage of advertisement pretesting. Once an advertisement has been developed, it could "be tested for similarity or dissimilarity" with other advertisements in the same product category. As the ultimate objective of an advertisement is to communicate, with the target consumer effectively and this is possible only if the advertisement is distinct in its message from the other competing advertisements,
3. **Product Re-positioning Studies:** If a company is interested in re-positioning its product or service, i.e., in the mind of the consumer, the first and foremost activity to be done is to assess the current perception of the product in the mind of the consumer. The classic re-positioning case is that of Cadbury chocolates, which kept on assessing its positioning platform, and successfully moved chocolates from a product perceived, as one for children, to a product that could be consumed by a person of any age, at any time of the day, and for varied occasions.
4. **New Product Development:** MDS technique shows us the various perceived perceptions of the different brands. Spaces or Gaps in the product perceptions could be used to develop new offerings for the target consumer.
5. **Test Marketing:** MDS technique can be used to identify cities that have similar demographic characteristics, and one could then identify a city that could represent a national character, and use that city for test marketing. One can thus observe that MDS is a very useful technique to help understand the market place and develop strategies for the future.

Summary

Multivariate analysis can be defined "statistical techniques, which simultaneously analyze more than two variables on a sample of observations." Multivariate techniques are used to analyze the relationships between more than two variables. Multivariate techniques can be classified under two key categories—dependency techniques and interdependency techniques. Prominent dependency techniques are multiples regression analysis, discriminant analysis, and CCA. Multiple regression analysis involves measuring the relationship between one dependent variable and two or more independent variables. Multiple regression analysis is an extension of bivariate regression analysis.

Discriminant analysis is mainly used in situations where the researcher has to classify the objects into homogeneous groups.

CCA is another dependency technique and is defined as a way of measuring the linear relationship between two multidimensional variables.

MANOVA examines the relationship between several dependent variables and several independent variables.

Interdependency techniques are a type of multivariate technique used to deal with problems that don't have dependency conditions. Prominent interdependency techniques are factor analysis, cluster analysis, and multidimensional scaling.

Factor analysis is a data reduction technique that helps the researcher to reduce the number of variables into a few factors.

Cluster analysis is a widely used multivariate technique in research. Multidimensional scaling is defined as a technique that involves representing object's preferences and perceptions as points on a multidimensional space.

Review Questions

1. Explain multivariate analysis with an example.
2. Give the characteristics of multivariate techniques.
3. Explain the nature and the classification of multivariate techniques.
4. Explain Multiple Regression Analysis.
5. Explain Discriminant Analysis.
6. Explain Multivariate Analysis of Variance.
7. Explain Factor Analysis.
8. Explain Cluster Analysis.
9. What is cluster analysis? What are its possible applications?
10. Give a few examples of marketing situations where cluster analysis can be used.
11. Discuss with the help of examples the areas where Multidimensional Scaling can be applied for marketing.
12. Brief the following dependency techniques:
    i. Multiple regression analysis
    ii. Discriminant analysis
    iii. MANOVA
    iv. CCA.
13. Brief the following interdependency techniques:
    i. Factor analysis
    ii. Cluster analysis
    iii. Multidimensional scaling.

17

Research Reports

17.1 Introduction

As we know the research reports are not only a descriptive summary of the overall findings, but also act as a guide for future research; the documentation of the report is very crucial and sensitive. The report should have a record of enough information so that it can be easily understood and followed as and when required. The last and final phase of the journey in research is writing of the report. After the collected data have been analyzed and interpreted and generalizations have been drawn, the report has to be prepared. The task of research is incomplete till the report is presented.

A research report contains many items including findings, analysis, interpretations, conclusions, and at times recommendations. These can be presented to the management either in a written form or communicated orally.

17.2 Classification of Research Reports

17.2.1 Short Reports

Short research reports usually run into 4–5 pages and are prepared for those researches, which have a well-defined problem, limited scope, and employ a clear-cut methodology.

These reports usually include a concise statement with regard to the approval for the study, followed by the objective of the study, i.e., the problem definition, the research overview, which contains in concise the main part of the research such as the methodology used, followed by conclusions based on the findings and recommendations, if any.

17.2.2 Long Reports

Long reports are more detailed than short reports. They can be further subdivided into Technical reports (TRs) and Management reports based on the objectives of the researchers and the end users. TRs are primarily meant for researchers. Management reports are meant for managers as end users, to aid their decision-making.

17.2.2.1 Technical Report

A TR is used whenever a full written report of the study is required for either evaluation or record keeping or public dissemination. For example, Ph.D. thesis. In a TR, the main emphasis on the methodology employed, objectives of the study, assumptions made, or hypotheses formulated in the course of the study; how and from what sources the data are collected and how have the data been analyzed; the detailed presentation of the findings with evidence and their limitations. A TR should focus on a specific topic logically pertaining to the research objective. The report should include a descriptive title, author name and information, date, list of keywords, informative abstract, body, acknowledgment(s), list of references, and appendices. The introduction of each TR should clearly identify its thesis and an organization plan for the same.

DOI: 10.1201/9781315167138-17

The body of each TR should consist of sources of data, research procedures, sampling design, and data collection methods; instruments used and data analysis arranged into a standard format under motivation, methods, results, and discussion.

The TR should include sufficient procedural information for other users to replicate the study. Therefore, the TR should explain the following:

1. What was done?
2. Why it was done?
3. What was discovered, and
4. What was significant in the findings?

The report should identify clearly what is the original about the work, and how it related to past knowledge. There is no minimum or maximum length requirement for a TR. However, usually they are of 10–15 pages. A good-quality TR should have the conclusions and recommendations in line with the findings. While all necessary details should be referred to, it should avoid the inclusion of nonessential information and over-simplification.

17.2.2.2 Management Report

Managers and decision-makers want information quick and straight to the point. Therefore, they show little interest in knowing the technicalities of the research. They are more interested in the ultimate findings and conclusions, which can act as a base for their decisions. As the management reports are meant for a nontechnical audience, there should be very less use of technical jargons, and wherever jargons are used, they should be explained using a footnote or in the appendices. The language of management report should be such that it is easy to understand.

Some of the other features of a good management report are short and direct statement; underlining relevant parts for better emphasis; pictures and graphs accompanying tables; graphics and animations accompanying the presentation of the report, etc.

17.2.3 Monograph

A monograph is a treatise or a long essay on a single subject. For the sake of convenience, reports may also be classified either on the basis of approach or on the basis of the nature of presentation such as Journalistic Report, Business Report, Project Report, Dissertation, Enquiry Report or Commission Report and Thesis.

17.2.3.1 Journalistic Report

Reports prepared by journalists for publication in the media may be journalistic reports. These reports have news and information value.

17.2.3.2 Business Report

A business report may be defined as report for business communication from one departmental head to another; one functional area to another; even from top to bottom in the organizational structure on any specific aspect of business activity. These are observational reports, which facilitate business decisions.

17.2.3.3 Project Report

A project report is the report on a project undertaken by an individual or a group of individuals relating to any functional area or any segment of a functional area or any aspect of business, industry, or society.

17.2.3.4 Dissertation

A dissertation is a detailed discourse or report on the subject of study generally used as documents to be submitted for the acquisition of higher research degrees from a university or an academic institution. The thesis is an example in point.

17.2.3.5 Enquiry Report (Commission Report)

An enquiry report or a commission of enquiry report is a detailed report prepared by a commission appointed for the specific purpose. Here, a detailed study is conducted on any matter of dispute or on a subject requiring greater insight. Since they contain expert opinions, these reports facilitate action.

17.3 Research Reports Components

Research reports have a set of specified components. They may be short or long, formal or informal, routine or special, public or private, daily or weekly, monthly or annual, etc. A typical research report has the following sections:

1. **Cover Page and the Title Page**: Prefatory Information contains Transmittal Letter and Authorization Statement. Introductory pages include Foreword, Preface, Acknowledgment(s), Table of Contents, Lists of Tables and Illustrations, Summary, etc.
2. **Subject Matter of Text**: It includes Headings, Quotations, Footnotes, Exhibits, etc.
3. **Introduction:** It includes Problem Statement, Research Objectives, Background, etc.
4. **Methodology:** It includes Sampling Design, Research Design, Data Collection, Data Analysis, Limitations of Research Study, etc.
5. **Results and Findings**
6. **Analysis, Interpretation, and Conclusions**
7. **Recommendations and implications**
8. **Reference Section:** It includes Appendices, Bibliography, and Glossary (if required).

17.3.1 Cover Page and the Title Page

The title should incorporate elements such as the variables taken into account in the study; the type of relationship between the variables included in the study; and the target population for whom the results can be useful.

A short informative title can be effective. The first page of the report contains the details of the topic of the research as well as the name of the organization that is being or has a product or service or program that is being researched as well as the date. The cover and the title page of a report contain the Title of the subject or project, Presented to whom, i.e., Name of the Client, On what date, For what purpose, i.e., the nature of the project in a precise and succinct manner, Written by whom, i.e., name of the organization and the researchers.

If there is any restriction on the circulation of the report, it is indicated, e.g., "For Official Use Only," in the top right corner of the cover and the title page.

Prefatory Information
It contains the Transmittal Letter and Authorization Statement.

17.3.1.1 Letter of Transmittal

The letter of transmittal is a short of authorization by the client organization, citing approval for the project. This becomes necessary when the relationship between the researcher and the client is formal.

A transmittal letter consists of a salutation of the person, who commissioned the report, the objectivity of the letter, a brief synopsis of the report, acknowledgments, and follow-up action expected of the reader.

17.3.1.2 Authorization Statement

A letter of authorization is a letter from the client to the researcher approving the project and specifying some of the details. Such letters usually accompany the research reports to federal and state governments, where detailed information about authorization factors is required. At times, a reference to the letter of authorization in the letter of transmittal is deemed enough. The letter not only helps in identifying the sponsor, but also outlines the original request.

17.3.2 Introductory pages

The introductory pages are given lower case Roman numerals (e.g., i, ii, iii). Arabic numerals (e.g., 1, 2, 3) are used from the first page of the introduction. The introductory pages contain Foreword, Preface, Acknowledgment(s) and Table of Contents, Lists of Tables and Illustrations, Summary, etc.

17.3.2.1 Foreword

The first page of the foreword is not numbered, but it is counted among the introductory pages. Usually a foreword is one page or even shorter. If a foreword is more than a page, subsequent pages of the foreword are numbered in lower case Roman numerals. The foreword is written by someone other than the author. It is written by an authority on the subject or the sponsor of the research or the book and introduces the author and the work to the reader. At the end of the foreword, the writer's name appears on the right side. On the left side, address and place of writing the foreword, and date appear. Name, address, place, and date are put in italics.

17.3.2.2 Preface

The first page of the preface is not numbered, but it is counted among the introductory pages. Subsequent pages of the preface are numbered in lower case Roman numerals. The preface is written by the author to indicate How the subject was chosen; Subject importance and need; The focus of the book's content, purpose, and audience; At the end of the preface, the author's name is given on the right side; On the left side, address and place of writing the preface, and date appear; Name, address, place, and date are put in italics.

17.3.2.3 Acknowledgment

If the acknowledgment section is short, it is treated as a part of the preface. If it is long, it is put in a separate section. The first page of the acknowledgment is not numbered, but it is counted among the introductory pages. Subsequent pages of the acknowledgment are numbered in lower case Roman numerals. At the end of the acknowledgment, only the author's name appears in italics in the right-hand corner.

17.3.2.4 Table of Contents

Great care should be taken in writing the table of contents. The contents sheet is both a summary and a guide to the various segments of the book. The table of contents should cover all the essential parts of the book and yet be brief enough to be clear and attractive. The first page is not numbered, but the subsequent pages are numbered in lower case Roman numerals.

The heading TABLE OF CONTENTS or CONTENTS in all capital letters appears at the top.

The listed on the left side includes Foreword, Preface, Acknowledgment; Numbers and titles of sections, chapters, center heads, center subheads, and side heads.

On the right side, the corresponding page numbers are given. The page numbers are aligned on the right. The section and chapter titles are put in all capital letters. The center head is put in capital and

Traditional Classification	Decimal Classification
1.	1.
A.	1.1
1.	1.1.1
a.	a.

FIGURE 17.1 The classification of the headings.

lower case letters. The center subheads and side heads are put in lower case letters, except the first letter of the first word and proper nouns. The classification of the headings can be done in the traditional or decimal system in the declining order as follows (Figure 17.1):

The headings of the text can be indented in a step form to visually highlight the classification. At the end of the headings of the text, references to appendices, bibliography, glossary, and index appear. These references are put in all capital letters from the margin.

17.3.2.5 Lists of Tables and Illustrations

Lists of tables and illustrations follow the table of contents. Each list starts on a separate page. If the items in each list are few, both the lists are put on the same page but under different headings. The headings for these lists may be in all capital letters—LIST OF TABLES, LIST OF ILLUSTRATIONS, TABLES, or ILLUSTRATIONS—and they follow the format of the heading that is used on the contents page—TABLE OF CONTENTS or CONTENTS. Only the first letter of the main words are capitalized in writing the titles of tables and illustrations. The second and subsequent lines of an item are indented. The page number appears against the first, second, or third line where the item's description ends. Tables and illustrations are numbered continuously in serial order throughout the book in Arabic numerals (e.g., 1, 2, 3) or in the decimal form (e.g., 1.1, 2.1, 2.2., 3.1). In the latter classification, the first number refers to the chapter number and the second one to the serial order of the table or illustration within the chapter.

17.3.2.6 Summary

Then the report contains an executive summary or abstract of the research and its findings. It is usually a one-page, concise overview of findings and recommendations of the research conducted. A report invariably carries an abstract or an executive summary in the initial pages as a help to the busy researcher or executive. The summary is positioned immediately before or after the contents sheet. The summary and the contents together provide an overview to the reader. The length of the summary may vary from 100 words to 1,000 words. In a short report, the preface itself becomes the summary. In a long report, the summary is given in the first chapter of the text.

This functions as a miniature report. The key findings are very concisely presented in the executive summary running into 100–200 words or a maximum of two pages. The major thrust of the executive summary should be on highlighting the objective, salient features, and analysis of the results including the recommendations. Recommendations should be given if the client wants them, else should be avoided. This is because some decision-makers do not want their thought process to be limited to the recommendations given. As the executive summary is the gist of the whole report, it is framed only after the report is completed. Conclusions should be supported later and graphics should be used if necessary.

The introduction gives an overview of the report. It highlights parts of the project such as problem definition, research objectives, background material, and the findings. It lays down the plan for the development of the project.

17.3.4 Introduction

17.3.4.1 Problem Statement

This highlights the basic problem the research will probe into. It explains the reasons why the research is being conducted and is usually followed by a set of objectives.

17.3.4.2 Research Objectives

The purpose of the report shows the aims and objectives of the research. It also shows the details of the type of the research (qualitative or quantitative) that was used by the researcher.

Research objectives form the heart of the study. They address the purpose of the project. Every research follows a set of well-planned objectives. Therefore, the general and specific objectives should be stated. These can be adjusted for sequencing without changing their basic nature. The research objectives can take the form of questions and statements. The objectives influence the choice of research methodology and the basic structure used to report the findings.

17.3.4.3 Background

This topic shows a historical background of the people/event/practice/program or organization under study. It also mentions the problem that needs to be studied and also the overall goals of the research as well as the suggested outcomes of the research. The topic also shows what questions are being answered by conducting the present research. This section may also involve the relevant literature review which was done by the researcher. Background information may include a review of the previous research or descriptions of conditions that caused the project to be authorized. It may entail preliminary results from an experience survey or secondary data from various sources. The references from secondary data, definitions, and assumptions are included in this section. Background material depending on whether it contains literature reviews or information relating to the occurrence of the problem is placed either after the research objectives or before the problem definition, respectively.

17.3.5 Methodology

The section of methodology deals with measures and procedures used for conducting the research. Basically, it contains the following details:

1. **Sample**: It represents the number of samples that are being used from the total population for the research study.
2. **Scales Used:** The details of the instruments and questionnaires that are being referred for the concerned research are mentioned in this section.
3. **Type of Data Collected**: The details of the types of data. For example, interviews, questionnaires, recordings, observations, etc., are also mentioned in the methodology section.

For short reports and management reports, it is not necessary to have a separate section on the methodology used. This can be included in the introduction section and details can be accommodated in the appendix. However, in the case of a TR, methodology needs to be explained as an independent section, and include the following:

17.3.5.1 Sampling Design

The researcher in this section defines the target population and the sampling methods put to use. This section contains other necessary information such as:

1. Type of Sampling (Probability or Nonprobability) used

2. Type of probability sampling (Simple random or Complex random) or nonprobability sampling (Quota sampling or Snowball sampling) used.
3. The factors influencing the determination of sample size and selection of the sampling elements.
4. The levels of confidence and the margin of acceptable error.

The sampling methods used should be explained and calculations should be placed in the appendix rather than in the body of the report.

17.3.5.2 Research Design

The research design has to be custom-made to the research purpose and should contain information on Nature of the research design; Design of Questionnaires; Questionnaire development and pretesting; Data that were gathered; Definition of interview and type of interviewers; Sources (Both primary and secondary) from which data were collected; Scales and instruments used; Designs of sampling, coding, and methods of data input; Strength and weaknesses, etc.

Copies of materials used and the technical details should be placed in the appendix.

17.3.5.3 Data Collection

The contents of this section depend on the research design. As the name implies, data collection pertains to the information about Time of data collection; Field conditions during data collection; The number of field workers and supervisors; The training aspects of supervisors and workers; Handling of irregularities, if any; Subject assignments to various groups; Administration of tests and questionnaires; Manipulation of variables.

In case any secondary data were used, then the relevance of that data should be given. Details of field instructions and any other necessary information should be given in the appendix.

17.3.5.4 Data Analysis

This section provides information on the different methods used to analyze the data and the justification for choosing the methods. In other words, it should justify the choice of the methods based on assumptions. It provides details on Data handling, Groundwork analysis, Rational statistical tests and analysis.

17.3.5.5 Limitations of Research Study

Limitations of the study or research in this section consist of the restrictions of limitations of the findings. It shows how and under which conditions the results can be generalized; certain researchers tend to avoid this section but this is not a sign of professionalism. There should be a tactful combination of reference and explanation of the various methodologies and their limitations or implementation problems. The limitations need not be explained in detail. Details of limitations do not be little the research. They help the reader to acknowledge its honesty and validity.

17.3.6 Results and Findings

This section deals with the analysis of the data collected. It discusses the results and findings of the research.

Most of the space in the report is devoted to this section. It presents all the relevant data but made no attempt to draw any inferences. The section attempts to bring to the fore any pattern in the industry. Charts, graphs, and tables are generally used to present quantitative data. It is better to report on finding per page and support it with quantitative data.

17.3.7 Analysis, Interpretation, and Conclusions

This section deals with the interpretation and discussion of the findings of the data analyzed. On the basis of the interpretations, the results are concluded. The conclusion section shows how the result is significant and to what extent is it helpful to the research targets and other researchers.

Conclusions should be directly related to the research objectives or hypotheses. Conclusions are inferences drawn from the findings. The researcher should always present the conclusions as he has first-hand knowledge of the research study. It is wrong to leave the inference of the conclusions on the reader.

Recommendations, on the other hand, are a few corrective actions presented by the researcher. They highlight the actions the report calls for as per the researcher. The recommendations should be in line with the results of the report and should be explicit. They may even contain plans of how future research for the same can proceed. However, recommendations ought to be given only if the client is interested.

It may happen that the client does not want any recommendations on the findings. In such a case, the report should not carry any recommendations.

17.3.8 Recommendations and Implications

The researcher recommends suggestions and implications of the study conducted.

17.3.9 Reference Section

This section follows the text. First comes the appendices section, then the bibliography and glossary. Each section is separated by a divider page on which only the words APPENDICES, BIBLIOGRAPHY, or GLOSSARY in all capital letters appear. All reference section pages are numbered in Arabic numerals in continuation with the page numbers of the text.

17.3.9.1 Appendices

The last section of the research report contains the various sources (like questionnaire, company forms, case studies, data in tabular format, testimonials), which were analyzed and used by the researcher.

Appendices are optional. They should be used to present details that were part of the research but were not necessary to the presentation of the findings or conclusion.

Appendices include raw data, calculations, graphs, copies of forms and questionnaires, complex tables, instructions to field workers, and other quantitative material that would look inappropriate in the main text. The reader can refer to them if required. However, care should be taken that they do not exist in isolation and reference to each appendix is given in the text.

17.3.9.1.1 Contents of Appendix

1. Supplementary or secondary references are put in the appendices section. But all primary reference material of immediate importance to the reader is incorporated in the text. The appendices help the author to authenticate the thesis and help the reader to check the data.
2. The material that is usually put in the appendices are Original data, Long tables, Long quotations; Supportive legal decisions, laws, and documents; Illustrative material; Extensive computations; Questionnaires and letters; Schedules or forms used in collecting data; Case studies or histories; Transcripts of interviews, etc.

17.3.9.1.2 Numbering of Appendices

The appendices can be serialized with capital letters (Appendix A, Appendix B) to differentiate from the chapter or table numbers.

17.3.9.1.3 References to Appendices

1. In the text, the reader's attention is drawn to the appendices as in the case of tables.
2. All appendices are listed in the table of contents.

17.3.9.2 Bibliography

As defined earlier bibliography is a list of citations or reference to books or periodicals on a particular topic. It is necessary to provide the details of the secondary sources used to prepare the technical or long report. Special software can help in searching, sorting, indexing, and formatting bibliographies into any required style. This software helps to cite the references from online sources and translate them into database records, which can be used for future referrals.

17.3.9.2.1 Positioning of the Bibliography

The bibliography comes after the appendices section and is separated from it by a division sheet written BIBLIOGRAPHY. It is listed as a major section in all capital letters in the table of content.

A bibliography contains the source of every reference cited in the footnote and any other relevant works that the author has consulted. It gives the reader an idea of the literature available on the subject and that has influenced or aided the author.

17.3.9.2.2 Bibliographical Information

The following information is given for each bibliographical reference:

For Books: Author(s), Title (underlined), Place of publication, Publisher, Date of publication, Number of pages, etc.

For Magazines and Newspapers: Author(s), Title of the article (within quotation marks), Title of the magazine (underlined), Volume number (Roman numerals), Serial number (Arabic numerals), Date of issue, Page numbers of the article, etc.

17.3.9.2.3 Difference between Bibliographical and Footnote Entries

The formats of bibliography and footnote differ in the following respects:

1. In a bibliography, the first line of an item begins at the left margin and the subsequent lines are indented. But in a footnote, the first line is indented and the subsequent lines of the item begin at the left margin.
2. In a bibliography, the last name of the author is given first (Dubey, Umesh), but in a footnote, the first name is given first (Umesh Dubey).
3. A bibliography is arranged within a section in the alphabetical order of the last name of the author or in the alphabetical order of the title of the work, or in the chronological order of publication. But footnotes are arranged in the sequence in which they have been referred to in the text.
4. Punctuation marks in a bibliography and in a footnote are different.
5. In a bibliography, the total number of pages of a book (205 pp.) or page numbers of the article (1–21) are given, while in a footnote only the specific page (p. 21) or pages cited (pp. 3–5) are given.

Examples of Bibliographical and Footnote Entries

....

ONE AUTHOR

Bibliography

Dubey, Umeshkumar, *Quantitative Techniques in Business, Management, and Finance: A Case-Study Approach,* CRC Press Taylor & Francis Group: Boca Raton, London, New York, 2017.

Footnote

[1]Umeshkumar Dubey, *Quantitative Techniques in Business, Management, and Finance: A Case-Study Approach,* CRC Press Taylor & Francis Group: Boca Raton, London, New York, 2017.

....

TWO AUTHORS

Bibliography

Dubey, Umeshkumar, Kothari, D.P, *Quantitative Techniques in Business, Management, and Finance: A Case-Study Approach,* CRC Press Taylor & Francis Group: Boca Raton, London, New York, 2017.

Footnote

[1]Umeshkumar Dubey and D. P. Kothari, *Quantitative Techniques in Business, Management, and Finance: A Case-Study Approach,* CRC Press Taylor & Francis Group: Boca Raton, London, New York, 2017.

....

THREE AUTHORS

Bibliography

Dubey, Umeshkumar, Kothari, D.P, Awari G.K, *Quantitative Techniques in Business, Management, and Finance: A Case-Study Approach,* CRC Press Taylor & Francis Group, Boca Raton, London, New York, 2017.

Footnote

[1]Umeshkumar Dubey and D. P. Kothari, G.K. Awari, *Quantitative Techniques in Business, Management, and Finance: A Case-Study Approach,* CRC Press Taylor & Francis Group: Boca Raton, London, New York, 2017.

....

MORE THAN THREE AUTHORS

Bibliography

Dubey, Umeshkumar, *et al., Quantitative Techniques in Business, Management, and Finance: A Case-Study Approach,* CRC Press Taylor & Frrancis Group: Boca Raton, London, New York, 2017.

Footnote

[1]Umeshkumar Dubey, *et al., Quantitative Techniques in Business, Management, and Finance: A Case-Study Approach,* CRC Press Taylor & Francis Group: Boca Raton, London, New York, 2017.

....

ARTICLE IN A JOURNAL

Bibliography

Dubey, Umesh Kumar. "Taxation of Agricultural Incomes", *Industrial Times, X, 12* (22 June 2020), 8.

Footnote

[1]Dubey, Umesh Kumar, "Taxation of Agricultural Incomes", *Industrial Times, X, 12* (22 June 2020), 8.

....

17.3.9.3 Glossary

A glossary is a short dictionary giving definitions and examples of terms and phrases, which are technical, used in a special connotation by the author, unfamiliar to the reader or foreign language in which the book is written. It is listed as a major section in all capital letters in the table of contents.

Positioning of a Glossary: The glossary appears after the bibliography. It may also appear in the introductory pages of a book after the lists of tables and illustrations.

Order of Listing: Items are listed in alphabetical and normal order. Example, **Center Heading:** It is listed under C and not under H.

17.4 Written Presentation

17.4.1 Prewriting Concerns

The effectiveness of a research report depends on how well it is presented. A report has many parts and all parts should display interconnectivity. This interconnectivity is possible only with a meticulous organization of the different parts of the report. This organization should be reflected in the initial sections of the report. A good researcher spends significant amount of time in designing this initial section wherein he or she tries to relate the purpose of the report, the audience it is meant for, the technical background, and the limitations under which the report is written.

The gap arising due to degree of difference between the subject knowledge of the writer and the reader should be taken into account. The technical knowledge of the end-users may not match that of the researcher so the report should be written in a simple manner with less technical jargons. This would enable the reader to understand the theme of the project and relate the conclusions to the specific objectives outlined in the report. In fact, all parts of the report should coherently pursue the research problem. This means that the conclusions and findings when integrated backward should show some connection with the research objectives, which were framed in line with the problem situation. This unified structure assists the reader to understand how the research problem was probed into and how the project was accomplished. As the final organized report is written after the research is over, the researcher can relate the facts and present the findings in a manner that would appeal to the reader.

17.4.2 Outline

The best way to organize a report is to develop an outline of the main sections. The outlining stage gives a natural progression to the various stages of report writing. The outlining stage concentrates on how it should be presented to make an impact on the readers. In trying to establish the relation among the various parts, the outline should introduce the complete scope of the report. As said earlier, the outline should contain the main headings of the various sections along with their subheadings and sub subheadings. This task is now made easy with the help of special software that helps in drawing a proper outline for a project report.

Two styles of outlining can be generally identified, that is, the topic outline and the sentence outline. The topic outline includes a keyword or phrase that reminds the writer of the nature of the argument represented by the keyword. The sentence outline on the other hand gives a description of the ideas associated with the specific topic. A traditional outline structure for a technical report is shown below:

1. Major Topic Heading
    - A) Main Subtopic Heading
        - 1) Sub Subtopic Heading
            - a) Further Details

A newer form of outlining is the decimal form. Decimal form of outline is shown below:

1) Major Topic Heading
    - 1.1 Main Subtopic Heading
        - 1.1.1 Sub Subtopic Heading
    - 1.2 Main Subtopic Heading
        - 1.2.1 Sub Subtopic Heading

17.4.3 Writing the Draft

Different authors have different styles of presenting their work. Some prefer to write the report themselves doing the additions or deletions while others depend on a good editor to transcribe their reports into the required format. The quality of a report depends upon the Readability and Comprehensibility, Tone, Final Proof.

17.4.3.1 Readability and Comprehensibility

A report has to be properly understood by the readers to achieve high readership. Therefore, a researcher should take into account the needs of the reader before preparing the report. The basic requirements of a report are readability and comprehensibility. The following points should be noted in this context:

1. It is necessary to avoid ambiguous statements.
2. The report should be checked for grammar.
3. As far as possible, simple words that convey the meaning clearly should be used.
4. Sentences should be reviewed and edited to ensure a flow from one statement to another.
5. Larger units of text should be broken down into smaller ones without altering the original meaning.
6. Visual aids should be provided wherever required for better understanding.
7. Visuals should not be inserted at the end of a section. They should be placed within the section for better comprehensibility.
8. Each paragraph should contain only one idea.
9. Underlining and capitalization should be used to differentiate and emphasize the important ideas from the secondary and subordinate ideas.
10. Technical terms and jargons should be avoided, wherever possible. Wherever unavoidable, they should find a reference in the footnotes.

17.4.3.2 Tone

Proper use of tone is essential for better reading effects. This highlights the attitude of the writer and reflects his or her understanding of the reader. The report should make tactful use of details and generalizations. It should focus on facts and not the opinions of the writer. The report should make use of passive voice as far as possible and should avoid the use of first person. Recommendations should not undergo any sort of alterations to give them a positive image.

17.4.3.3 Final Proof

Final editing of the draft should be taken up after a gap of at least a day. This helps in identifying mistakes, if any, better and correcting the mistakes. Final editing requires various questions to be answered pertaining to the organization, contextual, and layout of the final report. This can be done a couple of times and looking at the report with a different focus each time. The executive summary follows the final stage of editing.

17.5 Presentation of the Research Report

A researcher can present the findings of the research either in an electronic format or as a printout. Irrespective of the medium the researcher chooses to present his or her report, he or she should ensure that the findings are presented in a professional manner to the end-user. Some of the important aspects that should be considered for presenting a report are listed below.

1. Reports should be typed or printed using an ink-jet, laser, or color printer.
2. The report should have a uniform font.
3. The findings of the research study should be placed under appropriate headings and subheadings.
4. Leave ample space between the lines and on all sides for better reading. Overcrowding creates problems and is stressful for the eyes.
5. Split larger text paragraphs into smaller paragraphs.
6. Use bullet points to list specific points.
7. Ensure that appropriate labels are assigned to every table, figure, and graph that appears in the report.

17.5.1 Oral Presentations

The findings of the research may be presented orally. Such presentations are made to a small group of people or decision-makers or managers, who are more interested in the critical findings of the report. Therefore, unlike written reports that are elaborate, oral presentations are only briefings. Oral presentations are known to continue for 20–30 minutes, but presentations extending beyond an hour are not uncommon. Such sessions are interactive, where the audience clarifies their doubts at the end of the presentation. Some distinctive features of oral presentation are explored in the subsequent sections.

17.5.2 Visual Aids: Tables, Charts, and Graphs

17.5.2.1 Audio-Visual Aids

Since audio-visual (AV) aids help in recreating reality in a miniature form through visuals and sound, greater CREDIBILITY and CLARITY can be achieved in presentation. Since both sound and sight senses are activated at the same time along with the body language, CONCENTRATION, RETENTION, and RECALL, can be obtained in presentation. AV aids can also help in collapsing DISTANCE and TIME. They help us to present to the audience materials and experiences from far-off places and from different times in the past to make the message concrete and clear.

17.5.2.1.1 Slides

1. Can be made in black and white or in color. Best for highlighting colored and halftone pictures.
2. Can be synchronized with running commentary on tape recorder.
3. Presentation sequence can be changed.
4. Useful for showing on and off during the talk.
5. Audience attention can be held for about 30 minutes.

17.5.2.1.2 Film Strips

1. Film strips give a feeling of continuity.
2. Can be synchronized with running commentary on the tape recorder.
3. Can be made in black and white and in color.
4. Once the strip is made, the sequence of the material cannot be changed.
5. Audience attention cannot be held for more than 10–15 minutes.

17.5.2.2 Tables

A research report more often than not contains quantitative data to substantiate the various findings. These quantitative findings, if presented in a narrative form, would go unnoticed by the reader. Therefore, a better way of representing them is to make use of tables to present the statistics. Tables save the writer from being caught in details, which can be boring. Data in the form of tables form a vital part of the report and make the comparisons of quantitative data easier.

Tables are of the following types:

1. General and Summary,
2. Based on their nature.

General tables are large, complex, and exhaustive. As they are very comprehensive they are usually reserved for the appendix. Summary tables, on the other hand, are concise and contain data that is closely associated to an explicit finding. This form of table can be customized to make it appealing. This can be done by retaining such important details only that will aid the reader in understanding the contents of the table. Tables should be used when graphs or figures cannot make the point.

17.5.2.3 Charts and Graphs

Charts and graphs in a research report tend to translate numerical information into visual form for better understanding of the subject matter. Like tables, charts and graphs also have a number and a title, labels for parts of the figure, and sources and footnotes. Graphs and charts, which depict a general trend are accompanied by a statement as "not to scale" to avoid any confusion. Charts can be of the following types: Line, Pie Charts, Bar Charts, etc.

Summary

The essence of a research report is the way it is presented, be it in the written format or orally. This makes it imperative for the report to the inclusive of all the necessary details. Short reports are concise and are made for those researches that have a well-defined problem, limited scope, and employ a clear-cut methodology. Long reports are detailed and usually comprise technical and management reports. The various components of these reports are: prefatory information, introduction, methodology, findings, conclusion and recommendations, appendices, and bibliography. We have discussed the various parts of a report. They are divided into cover and Title Page, Introductory Pages, Text, and Reference Section. Cover and Title Page have these components: (a) title of the subject or project; (b) presented to whom; (c) on what date; (d) for what purpose; (e) written by whom. The Introductory pages contain (a) Foreword; (b) Preface; (c) Acknowledgment; (d) Table of contents; (e) Listing of Tables and Illustrations and (f) Summary. Reference section follows the Text. It contains (a) Appendices; (b) Bibliography and (c) Glossary. Each of these heads and subheads are explained with the help of examples.

An oral presentation generally concentrates on the summary of the project with emphasis on the findings, conclusions, and recommendations.

Review Questions

1. Describe the concept and meaning of evaluating, interpreting, and reporting the data in research?
2. Explain the steps of evaluating or analyzing of data as well as preparing a report in research?
3. What are the basic requirements that are needed to be included as well as avoided while processing the data?
4. Define a Report.
5. Explain the need for reporting.
6. Discuss the subject matter of various types of reports.
7. Identify the stages in preparation of a report.
8. Explain the characteristics of a good report.
9. Explain different parts of a report.
10. Distinguish between a good and bad report.

18

Ethics in Research

18.1 Introduction

In a competitive market place, every company aspires to reach new customers as soon as possible with the least investment in research. Though many companies invest in research to gain better insights into the market and customers, who constitute the market, the way the research is conducted and the way findings are used raise many ethical concerns.

As a result, companies question the authenticity of research findings and procedures used for conducting research, most ethical questions that come up with regard to research are concerned with the following:

1. Whether the findings are biased and
2. Whether respondents' information is being used for unauthorized purposes.

18.2 Ethical Decisions

A typical research project involves the researcher, the client, and the respondents. The research objective cannot be achieved unless all three participate in the project whole heartedly. The researcher has to ensure that no respondents or client is put to any uncomfortable or compromising situation because of the research. Each participant has some responsibilities and rights.

For example, while a respondent is expected to provide true answers, the researcher should ensure privacy for the respondent.

Every profession has some ethical dimensions and research is no exception. The extent to which researchers follow ethics is debatable. The question is: Q: Where is the line to be drawn?

This important question has no clear-cut reply. Here, we shall try to understand ethical issues involved in market research by looking at various rights of the respondents, the client, and the researcher.

18.3 Ethical Treatment of Respondents

In any primary research, respondents play a very important role. However, sometimes respondents get little or nothing in return for their valuable contribution. Still, they are entitled to certain rights. The researcher has to ensure that these rights are not infringed or invaded.

The respondent's rights are analyzed below:

Benefits, Deceptions, Informed Consent, Debriefing Respondents, Right to Privacy, Online Data Collection, etc.

18.3.1 Benefits

A respondent might be induced to give false answers to questions because of various reasons. These false answers make the research findings unreliable. Hence, it is necessary to limit the prevalence or response bias. This can be done in various ways, one of which is demystifying the purpose of research to the

DOI: 10.1201/9781315167138-18

respondents. An interviewer should ensure that he or she introduced himself or herself to the respondent before proceeding to the questions. The details provided by the interviewer should not leave any doubt in the respondent's mind with regard to the genuine aims of the survey.

If a respondent knows the person he or she is speaking to and the survey's purpose, it is more likely that the respondent would stick to facts wile answering. While conducting some research surveys, respondents may be offered small inducements in cash or kind, but it should be ensured that the inducements are not disproportionate or intimidating.

At times, the survey's real purpose is concealed from the respondent to avoid any bias. This leads to the next factor known as deception.

18.3.2 Deception

Deception is the intentional misleading of subjects in a survey by withholding full information about the nature of the experiment or misrepresentation of facts.

Intentional misleading might include the following:

1. Withholding information about, the purpose of the research,
2. The role of the investigator,
3. Omitting procedures that are actually experimental.

Deception is sometimes necessary in survey research to:

1. Prevent respondent bias,
2. Prevent performance alteration, i.e., Hawthorne effect,
3. Protect confidentiality of the third party.

Deception is acceptable if there is limited or no risk to the respondent. It should be ensured that participants are never deliberately misled to enhance response rates.

The researcher should take the responsibility to:

1. Redesign the interview, so that respondents are not required to give false information, if possible and
2. Ensure that participant's rights are protected.

18.3.3 Informed Consent

Providing the respondents with information, to enable decision-making as to whether they want to participate in the research study is known as informed consent. As such, it entitles participant to have the right to:

1. A description of the nature and purpose of the research and its sponsorship, types of questions, subject areas covered, procedures, likely benefits to the participant and or society, the use of final findings and the probability risks and discomforts.
2. The written document of informed consent, disclosing compensation, and medical treatment available in the case of research-related injury, and anonymity and confidentiality of participant details and information provided.
3. Pertaining to research details, question investigators on any clarification.
4. Without any expectation of benefits, voluntarily authorize his or her participation in the research study, preferably in writing.

Complete informed consent might not be possible at time, when the aim is to gather factual and accurate data.

For example, a hidden camera in a Big Bazaar surveying the effects of merchandising on customers is less likely to give accurate data, if customers were told of the set up. Informed consent is most required in research dealing in medical or psychological implications.

18.3.4 Debriefing Respondents

Pertaining to the research, it might become necessary to derive certain factual and honest information from the interviewee by using either a disguised questionnaire or withholding some specific information.

In such situation, once the interview is over, it is important to debrief the respondent of the actual facts. Debriefing does not give justification for the "Unprincipled" aspects of any investigation but performs the following essential functions:

1. De-hoaxing, where the respondents are notified of the survey's deceptive aspects.
2. De-sensitizing, which helps respondents to get rid of any stress or uncomfortable feelings that the survey might have induced.

Debriefing can help reduce respondent stress, but where reactions are harsh follow-ups are necessary to ensure that no harm comes to the respondent. Debriefing can help to develop a long-term relationship with the respondents, who can then be contacted for future surveys. Debriefing brings to the surface the various lines of thought prevalent in the respondents-mind during the interview. This helps researchers to gain useful information and may serve as a guide to modifications in future research designs.

18.3.5 Right to Privacy

Recently, privacy in research has become a topic of major concern. Participants in research expect that the information they provide will be treated confidentially and, if published, we should not be traced back to them. Many times participants are reluctant to divulge any information unless they are absolutely sure of the confidentiality and privacy of the research.

Research findings, whether published or not, should always maintain the participant's anonymity.

Again, the question of where to draw the line comes into the picture. Where will an interviewer draw the line and decide whether a question is private to an interviewee or not?

This is debatable, as there are no established rules that can be simply adopted. Questions about sexual inclinations, criminal actions, and other private matters can be asked in certain circumstances, making efforts, in all cases, to keep the self-esteem of the individual intact.

18.3.6 Online Data Collection

Researchers are slowly shifting their research from offline methods to online methods, with better chances of privacy and anonymity for apprehensive participants.

Online research data are found to contain more factual and honest information as without any invasion of their privacy, respondents feel more comfortable sitting in front of their computer and answering questions. As respondents know that their identity is confidential, it is easier to obtain information even on sensitive issues.

18.4 Rights and Obligations of the Client

A client is an important part of research, as he or she is the one who sponsors the research. The client is thus entitled to some rights. In turn, the client also has certain obligations toward other parties involved in the research. These are explained by the following points. They are Right to confidentiality, Right to quality research, Clint-ethics, Open relation with research supplier and interested parties, Privacy, etc.

18.4.1 Right to Confidentiality

Keeping information given by or about a client during a professional relationship secret from others is known as client confidentiality. Clients want anonymity for various reasons, which is why they outsource research to outside agencies. They may also want to avoid the influence of the corporate image on the interviewee's responses during the test marketing of a new product or they might not want competitors to come to know of their strategies. Even the data gathered and the conclusions and interpretations made are exclusive to the client. It is the obligation of a researcher to preserve the Client's anonymity and the information gathered. This type of confidentiality is known as "Sponsor Nondisclosure" and "Findings Nondisclosure," respectively.

18.4.2 Right to Quality Research

This right entities the clients to:

1. Protection against abuse of position,
2. Protection against unnecessary research,
3. Protection against unqualified researchers, and
4. Protection against misleading presentations of data.

Most researchers have high expertise in their respective fields. It might happen that a client wishes the researcher to make use of a sophisticated research technique or conduct specific research that is not related to the research problem at hand. Here, it is the researcher's duty to clearly state the futility of engaging in say, highly technical primary research, when the right option would be less expensive research techniques or secondary research.

The researcher should try to enhance goodwill by providing ethical recommendations rather than resorting to unethical means at the sponsor's expense. Hence, the researcher should not make unethical use of his or her expertise to outwit the client or follow unethical and expensive means to conduct the survey.

Researchers usually have expertise in specialized fields. They might be requested to take up a survey that is not in line with their specialized knowledge. This might result in higher costs, time delays, and decreased accuracy. A researcher should make it clear to the client when he or she does not have the standard of expertise required for the survey. The researcher can even refer the client to other researchers, who could do the job better, earning goodwill from the client.

The researcher prepares a research report containing statistics and findings. This information can be distorted or fabricated to create the impression of greater accuracy, whereas the actual data do not warrant it.

Such "Decorated Information" in reports can be due to unethical means adopted by the researcher such as:

1. Overly technical jargon,
2. Unnecessary use of complex analytical procedures, and
3. Incomplete reporting.

An ethical researcher should refrain from presenting the data in a misleading way.

18.4.3 Client Ethics

A client should abstain from the following:

18.4.3.1 Issuing Bids, when a Research Supplier has been Predetermined

It cannot be termed unethical to have a preference for a particular researcher or supplier, which may be due to friendship or proven efficiency. However, it is certainly unethical to continue to request for proposals from various other suppliers when the contract has already been granted to the preferential supplier.

18.4.3.2 Obtaining Free Advice and Methodology via Bid Requests

Clients are known to request detailed bids from various suppliers or researchers. These bids consist of complete methodology and a sample questionnaire from each supplier in response to the research problem. The clients then assemble questionnaires and either give the contract to the supplier with the lowest bid known as low-ball pricing or may directly employ field services to gather data.

18.4.3.3 Making False Promises

Clients should refrain from using "Pseudo-Pilot Studies" and making false promises to suppliers.

In Pseudo-Pilot Studies, a client makes a false promise for granting a forthcoming comprehensive project to a particular researcher in return for the latter giving low-ball pricing for the current research. Such promises rarely materialize and result in the researcher making losses.

18.4.3.4 Unauthorized Request for Proposals

These are cases where client representatives seek proposals from various suppliers for a research project without prior authorized consent for grant of funds.

The following are some instances of unauthorized requests for proposals, which leave suppliers at a loss:

1. A client representative asking for proposals first and then trying to convince the management to approve the project.
2. The management although not too enthusiastic about the research proposal of a representative directs him or her to collect proposals from various suppliers. The management might do this to avoid discouraging the representative.
3. Though a supplier's application meets the representative's requirements, it might be rejected right away because of differences between the management and the representative.

18.4.4 Open Relation with Research Supplier and Interested Parties

To encourage the researcher to do the job and frame appropriate objectives for the survey, the client should explain the following:

1. The nature of the problem in its broadest terms.
2. State the time,
3. Financial constraints, and
4. Other related information.

This will help the researcher to design the survey project keeping constraints in view. Prior detailing of requisites and constraints also helps reduce researcher bias, if any.

The sponsor should avoid distorting the researcher's original findings to present itself in a superior position in order to influence a transaction or decision.

For example, a researcher conducting a survey for an automobile company to gauge the mileage efficiency of its bike may find that it scores over competing bikes only under a certain conditions.

The automobile company might quote the findings and promote the bike as the best performing bike in the market, without stating the particular conditions for optimum efficiency.

It is indeed unethical, if the advertisement does not mention under which conditions the research was conducted.

18.4.5 Privacy

This highlights the client's obligation toward subjects. A client should follow ethical means and avoid buying mailing list data from data providers who have compiled it, using unethical means. It is unethical in any way on the client's part to use these data.

18.5 Right to Protection against Client Tactics

The credit for the successful completion of a project invariably goes to the supplier, as he or she is the one who designs the entire research according to the client's requirements.

As the researcher plays a very important role in the entire research, he or she is entitled to some rights. He or she also has to fulfill obligations toward the procession.

Following are the rights and obligations of the supplier. They are listed here:

1. Right to protection against client tactics,
2. Right to safety,
3. Right to ethical behavior of assistants,
4. Misrepresentation of research,
5. Protecting the right to confidentiality of both client and subject,
6. Abuse of respondents.

18.5.1 Right to Protection against Client Tactics

A supplier is vulnerable to a client's unethical practices. As such, the supplier needs protection against the following:

1. Improper solicitation of proposals,
2. Disclosure of proprietary information and techniques,
3. Misrepresentation of data.

Unless informed of the specific criteria for the selection of research proposals, it is considered unethical on the client's part to discard proposals on unspecified grounds.

A client should specify the grounds, on which proposal will be selected in advance.

A proposal should not be disqualified on grounds that were not specified by the client, after the researcher has spent valuable time preparing the proposal.

Suppliers have the right to confidentiality of their proprietary information and techniques. Clients are sometimes known to assemble the best ideas from competing research proposals and later disclose it to the preferred supplier for use in the research study. This is unethical.

Research suppliers have the right to protection against deliberate misrepresentation of findings by clients for a better image or as tactic to improve sales.

Clients to their advantage may sometimes distort actual findings. As it can misled parties using the information as an inducement to get into a transaction. This too is unethical. This can also be easily verified by competitors and lead to the research organization losing its reputation.

18.5.2 Right to Safety

A research assistant has to venture out to different areas during field work. This may even involve a visit to unsafe areas. In such a case, two research assistants might be sent together to the place rather than alone individual. In case an interviewer feels threatened in a particular area, he or she should be replaced by another interviewer, who is more comfortable.

18.5.3 Right to Ethical Behavior of Assistants

Research data collected can be distorted or fabricated by research assistants and field workers. This can happen when the assistants do not act according to the plans and techniques outlined for sample selection and data collection. This leads to bias and finally leads to incorrect findings. Research data are also known to contain wrong information due to cheating, where the interviewer, to save time or avoid asking sensitive questions, might deliberately skip questions and fill in the answers on his or her own.

Assistants should also refrain from disclosing any client or respondent related details to any third party without prior approval. As the assistants are under the researcher's or supervisor's direct control, the researcher can demand ethical behavior from the assistants.

18.5.4 Misrepresentation of Research

Researchers have an obligation toward the clients and the respondents to use honest means in data analysis and reporting. In any way, they should not try to overemphasize the significance of the findings, if the data do not support such a stance. Noncompliance with right methods or any significant errors discovered during the research should be bought to the client's notice.

18.5.5 Protecting the Right to Confidentiality of both Client and Respondent

Researchers, field assistants, and supervisors are obliged to maintain anonymity of both the client and the respondent. This is extended to confidentiality of the research findings and any other information specific to both the client and the respondent. Any breach of faith in this regard is unethical. This means the supplier has to refrain from disclosing:

1. Any specific information with regard to the client's general business affairs and the research findings to any third party.
2. Names of respondents or tracing the information provided to any specific respondent.

Hence, details about the client and the respondent should, in no cue, be disclosed to unauthorized sources, except with prior permission from the parties involved. This obligation is fulfilled by signing a confidentiality and nondisclosure statement.

18.5.6 Abuse of Respondents

Respondents can feel abused in many ways. Confusing questions, poorly trained interviewers, and difficult-to-read questionnaires are some aspects that will upset respondents. Others aspects that also upset respondents are Lengthy interviews, Over-interviewing, Continued pressing for interviews, and Lack of privacy.

Respondents may get bogged down by interviews, which extend beyond reasonable limits. At times, some special questions not related to the current survey may be requested by the researcher and stretch the interview beyond its time limit and appear to be irrelevant to the respondent. Asking personal questions that are in no way related to the problem also adds to the respondent's discomfort.

All this may result in the respondent getting irritated and giving wrong information in a hurry to finish with the interview.

It might happen that a particular stratum of a locality or a geographical area presents the best sample for conducting a survey. This can be due to a well-represented population fulfilling the essential survey criteria. This leads to the same population being interviewed by different researchers for diverse surveys. This may lead to what is called as over-interviewing of the people of that area and may result in low response rates.

Telephonic interviews have also become a source of "Abuse" for respondents. Researcher's resort to this form of interviewing because of the low cost involved. They fail to realize that the constant ringing of the telephone or mobile can be irritating, and that many people just disconnect once they realize, it is a call for a survey or sale.

Sometimes respondents find themselves duped when they realize that their right to privacy has been invaded. This is a result of unethical behavior on the part of the supplier, who discloses the details of the respondent to a third party, who uses it as a sales call lead. The respondents then realize that the specific information given to the interviewer has been unethically passed on to a third party without their consent. This results in respondents refusing to take part in further surveys.

A couple of corollaries of the above situation exist called "sugging" and "frugging." "Sugging" means "Selling under the guise of research." Companies try to get product related specific information under the guise of a survey, using disguised questions.

Upon positive responses, they re-contact the respondent for actual sales. It is not until the respondent is approached again that he realizes he has been "sugged."

"Frugging" relates to "Fund-Raising under the Guise of Research." Here, the interviewer approaches the respondents with a tactful set of disguised questions, which try to probe the respondents' opinions toward charity donations. The respondents are then approached for charity donations, and they realize they have been "frugged."

Summary

The fundamental principles that define values and determine moral duties and obligations are known as ethics. The goal of research ethics protection of all those involved in research. The very conceptualization of research has to incorporate several ethical issues involving the researcher, the client, and the respondent. Each party has to fulfill a set of obligations toward the others in return for its share of rights.

Uphold in these rights and obligations is ethical behavior. A respondents' rights include knowledge of benefits, protection against deception, and right to informed consent, privacy, and debriefing. The client's rights and obligations consist of right to confidentiality, right to quality research, open relationship with the research supplier and interested parties, privacy, and advocacy research. The rights and obligations of the researcher or supplier include the right to protection against client tactics, the right to safety, and the right to ethical behavior of assistants, misrepresentation of research, protecting the right to confidentiality of both client and subject.

Review Questions

1. Explain the various ethical concerns of respondents, research client, and researcher himself or herself involved in the research.
2. What are the needs to make ethical decisions?
3. What are the ethical concerns for respondents?
4. What are the ethical rights and obligations of the client and the rights and obligations of a researcher?

Index